Stephen Smale

Stephen Smale

Reaching Higher Dimensions

Interviews and text by
Steve Batterson and Kate Springer

香港城市大學出版社
City University of Hong Kong Press

AMS American Mathematical Society

ISBN: 978-962-937-690-1

Published by

City University of Hong Kong Press

Tat Chee Avenue

Kowloon, Hong Kong

Website: www.cityu.edu.hk/upress

E-mail: upress@cityu.edu.hk

Printed in Hong Kong

Stephen Smale

CONTENTS

Series Preface ix

Preface | Lenore Blum xiii

Preface | Felipe Cucker xv

Preface | Ding-Xuan Zhou xix

Preface | Steve Batterson xxi

Acknowledgments xxv

Credits xxix

1. One Room Schoolhouse 1
2. Marxism and Mathematics at Ann Arbor 19
3. Early Mathematical Audacity 61
4. On the Beaches of Rio 83
5. Berkeley to Columbia and Back to Berkeley 105
6. The Lone Ranger of the Antiwar Movement 139
7. The Summer of 1966 189
8. Smale Versus the National Science Foundation 227
9. The Aesthetic Side: Minerals and Photography 263

10. Adventure and Physical Risks 293
11. Other People 321
12. Smale the Mathematician 345
13. Six Years at City University 367
14. New Challenges in Chicago 397
15. Back to Hong Kong 423
16. Biology, Berkeley, and a Lasting Legacy 453

Mathematical Appendix A: Smale's Thesis 489
Mathematical Appendix B: Everting the Sphere 498
Mathematical Appendix C: Chaos and the Horseshoe 504
Mathematical Appendix D: The Higher Dimensional Poincaré Conjecture 514

Notes 519

Series Preface by Kenneth Lo

The mission of education has always been the preservation and dissemination of knowledge, a profound responsibility that our society holds dearly. Universities, often referred to as "ivory towers," are seen as the pinnacle of this mission. Yet, the scholars within these towers, engrossed in their research and exploration of knowledge, can sometimes seem distant or unapproachable to those outside academic circles. Even internationally renowned scholars are often known only as names printed on academic works, their true identities obscured by the weight of their scholarly authority.

City University of Hong Kong (CityUHK, formerly City Polytechnic of Hong Kong, which officially became a university in 1994) moved to its permanent campus on Tat Chee Avenue, Kowloon Tong in 1989. It stands in the heart of the community, fostering connections with the world. For our professors and scholars at CityUHK, imparting knowledge and inspiring the imagination and creativity of young minds is not just an academic pursuit, but a practical mission.

Scholars, despite their titles, are human beings with their own unique experiences and stories. In 2021, CityU Press initiated the Legacy Series, featuring a collection of interviews with outstanding senior CityUHK scholars and distinguished academics who have made significant contributions in their fields. This series aims to humanize these scholars, presenting their personal journeys, experiences, and educational lives to the world.

Each scholar featured in this series has a unique personality and experiences, and may have encountered hardships or faced challenges. They share a common thread—a passion for academic research and a spirit of perseverance. Though they each have their own distinct characteristics and insights, they are all committed to promoting academics. Their shared goal is to cultivate the next generation of students, embodying the spirit of "dedication and community" that is central to CityUHK's motto.

It is worth mentioning that, from the beginning, the Legacy Series has been a testament to CityUHK's commitment to teaching and learning. Supported by teaching and learning funds from the Talent and Education Development Office (TED, formerly known as the Education Development and Gateway Education Office), this series has involved CityUHK students in nearly every step of the publishing process. This includes assisting in inviting scholars, observing writers during data collection, preparing for interviews, visiting scholars, transcribing interview recordings into text, and final proofreading. This hands-on experience has added a deeper layer of meaning to the series, aligning with TED's mission to support teaching and learning, cultivate outstanding students, and inspire innovative abilities.

CityUHK fosters an open learning and research environment, encouraging scholars to develop their academic expertise and share their knowledge with students. Beyond providing the public with an understanding of academic life, I have another aspiration for the Legacy Series. I hope that young students, whether they are interested in academic research or entering a specific field, can learn from these scholars' stories. My aim is for students to feel inspired and encouraged, not intimidated or doubtful. These scholars with outstanding academic achievements are not out of reach. They have paved academic pathways, showing that with hard work and perseverance, every student can embark on a journey towards higher and further academic research.

I extend my deepest gratitude to my fellow CityUHK scholars who have shared their experiences in this series. Their outstanding research and academic careers have not only resulted in individual educational achievements, but have also significantly contributed to society. Their work leaves an indelible influence on young researchers, both today and in the future.

Whether you are a teacher or student at CityUHK, or a member of the general public, I sincerely invite you to read these stories of CityUHK scholars. As you trace their personal paths of learning and thinking, you will uncover the historical periods and places they have traversed, and the people and events they have encountered. This series is not just a record of individual life courses, but a chronicle of an entire generation.

Kenneth Kam-Wing Lo
Director, Talent and Education Development Office
Chair Professor, Department of Chemistry
City University of Hong Kong

Preface by Lenore Blum

A fascinating and inspiring story of how Steve Smale, a bright yet seemingly unexceptional country boy, educated in a one room school house in rural Michigan, became one of the most brilliant and influential mathematicians on the planet. His years at the University of Michigan and his desire to see the world, traveling to Eastern Europe, hitchhiking across North and Central America, no doubt played a role in building illimitable self-confidence and shaping his persona: at once audacious, ambitious, a calculated risk taker, yet unpretentious and approachable.

The original chapters, published when Steve was 70, take us through his then achievements in mathematics (astonishing in breadth and depth: from differential topology and "turning the sphere inside out," to dynamical systems and finding the "horseshoe" on the beaches of Rio, to solving the Poincaré Conjecture for dimensions >4 leading to the Fields Medal, the mathematician's Nobel prize, to economics and complexity theory), his political activism, his audacious mountain climbing and sailing (from San Francisco to the Marquesas and back), his world renowned mineral collecting and photography.

But no stopping there. This expanded edition continues into Steve's 94th year with a dazzling array of travels and adventures (his years in Hong Kong, hiking Tiger Leaping Gorge in Yunnan Province and Death Hollow Canyon in Utah, sailing the Caribbean), his collecting Chinese minerals and more influential mathematics (his 18 problems for the 21st century, learning theory, modeling language evolution, flocking, analyzing high-dimensional data, mathematical biology, and his current obsession to understand the Kuramoto model for studying the emergence of synchronization).

Lenore Blum
Distinguished Career Professor of Computer Science, Emerita
Carnegie Mellon University

Preface by Felipe Cucker

Towards the end of 1989, during a visit to Montevideo, a colleague passed me a copy of the Blum–Shub–Smale paper on real computation. Like wings flapping in the butterfly effect phenomenon—a metaphor in chaos theory, one of the many branches of mathematics deeply impacted by Steve—this seemingly inconsequential event would, with time, have an enormous impact on my life. The fact that I am now writing this preface in Hong Kong speaks to that impact.

Were it not for Steve, I would not be living in Hong Kong now. Steve came to Hong Kong in the fall of 1995 after retiring from Berkeley. Shortly after, in February 1996, Lenore Blum, Mike Shub, and I visited him to continue working on a book we were writing. By the time the book was published in early 1998, I had been living in Hong Kong for more than a year, during which my collaboration with Steve had expanded in various directions. And I decided to stay.

While our book was being printed, another book of a different kind was being written. Steve Batterson, an academic grandson of Steve, was penning a biography about him. There were good reasons for such an endeavor. Certainly, Steve's work in the almost four preceding decades had had a profound impact on various branches of mathematics. But Steve's life had other aspects that merited mention: he had been an active member in the movement against the Vietnam War and had troubles with the National Science Foundation; he was an adventurous sailor and a no less adventurous hiker and mountain climber; he had gathered a world-class collection of minerals and was a fine photographer of his collection. Batterson succeeded in narrating a story where all these facets of Steve's life formed a single, resplendent crystal.

Batterson's book was printed by the American Mathematical Society in the year 2000. Steve was then 70 years old, and one could be excused for assuming that there would be little to add to his biography. Excused for sure, nonetheless mistaken. Since the publication of Batterson's book, Steve wrote two of his three most-cited papers, his collection of minerals focused and developed to become what is probably the finest collection of Chinese crystals in the world, his photos joined hands with those of Jeff Scovil to form two coffee table books on that collection, and he continued hiking and sailing. It was, therefore, a natural initiative of the City University of Hong Kong to update Batterson's book. The generosity of the American Mathematical Society—which granted CityU Press a reprint license to update the book—paired with the fine professional qualities of Kate Springer—who wrote the four chapters in this updated edition—turned this initiative into the book you have in your hands.

I hope to be pardoned if I take advantage of this writing opportunity to include some personal memories and comments. I first met Steve during a visit to Berkeley in May 1991. I had gone there just to discuss some work I had done based on his, Lenore Blum, and Michael Shub's theory of computation over the real numbers. I did not foresee that the visit would mark the beginning of a long-lasting relationship including, but going well beyond, mathematical collaboration.

The nature and extent of this collaboration shaped my research for decades. Steve has always been a source of mathematical questions (and most of the time, answers to these questions as well). It was common to have him knocking on my door to propose we work on one of them. His ideas towards a solution would spread over a page from his legal pad which Steve would pass to me. These pages (Steve called them "scribbles") contained more than one diamond in the rough and the time we spent polishing these diamonds was filled with the joys of mathematics.

We eventually wrote several joint papers, and some of them received a substantial attention from the mathematics community (the two highly cited papers mentioned above are also my two most-cited papers). But the influence of Steve's ideas in my research goes beyond our joint papers. It was Steve who first proposed that I work on condition-based algorithms (a path that would lead to my book *Condition,* written with Peter Büergisser), and it was a paper by Steve (with Partha Niyogi and Shmuel Weinberger) that led me, in a series of papers with various co-authors, to develop condition-based algorithms for the computation of the homology of semialgebraic sets.

Steve's ideas influenced mathematics in more ways than just the writing of papers. The four-week conference he organized in Park City in 1995 led to the creation of the Foundations of Computational Mathematics society, which now attracts and disseminates the best research on the subject through a triennial conference and a journal. This conference, with the various hikes organized by Steve in Park City's stunning natural surrounds, brings other aspects of Steve's character to these pages.

Steve loves good food and good wine. And he loves enjoying them with friends. I treasure the memories of dinner with Steve in restaurants such as Chez Panisse in Berkeley or Le Procope in Paris as much as I cherish the numerous occasions my wife and I had dinner with Steve and Clara in their apartment in Hong Kong. I remember them arriving at my house in the French countryside, where we spent a week walking in the neighboring vineyards and talking mathematics, with a case of Puligny-Montrachet. Steve disguised this show of good guest manners by pretending that he was unfamiliar with the local wines.

Seizing the opportunity for good times with friends has not been limited to the pleasures around a table. In 1994 Steve invited me to spend two weeks in the apartment he and Clara were renting in Rio de Janeiro so that I could enjoy the atmosphere of Carnival and the beaches of Rio, which were witness to our mathematical dialogues. In 2012 he chartered a 50-foot sailboat to island-hop in the Caribbean with his family. He also invited my wife and me. She could not go but, despite that and my incapacity to help with the sailing, I was welcome on a tour of St. Martin, Anguilla, and St. Barts, with Steve at the helm. In 2016 Steve repeated (and so did I) the experience, this time in Grenada and the Grenadines.

Looking back over the more than 30 years since I met Steve, I have manifold reasons to be grateful to him. These lines can barely serve as a way of thanking him, but I hope they serve to express my gratitude.

Felipe Cucker
Emeritus Professor, Department of Mathematics
City University of Hong Kong

Preface by
Ding-Xuan Zhou

When Shiing-Shen Chern heard from Steve Smale about his move to City University of Hong Kong (CityUHK), mentioning that CityUHK was a relatively unknown institution, Steve responded, "I don't need a famous university to work in." Instead, Steve's presence elevated CityUHK's reputation, particularly in the field of mathematics. His time at CityUHK was marked by groundbreaking research and influential collaborations.

This book provides a glimpse into Steve Smale's life and work at CityUHK from 1995–2001 and 2009–2016, and his influence on the academic community and beyond. The book does not delve too deeply into Steve's scientific contributions. It is written for a general audience, offering insights into his professional journey, collaborations, and the lasting impact of his work.

Steve's influence on young scholars and CityUHK goes far beyond his two appointment periods. My own experience serves as a testament to this. I began collaborating with him on the approximation error of learning theory in 2000. Our paper was subsequently invited by Roderick Wong for publication in the inaugural issue of the journal *Analysis and Applications* in 2003. Although Steve left CityUHK in 2001, he continued to welcome me to UC Berkeley or TTIC every year, including three times during my sabbatical in the spring of 2005.

In addition to these visits, we attended numerous conferences in Minneapolis, Tübingen, Barcelona, Santander, and College Station and even co-organized a workshop in Oberwolfach with Kurt Jetter. Our collaboration was further bolstered by intensive email, fax, and phone communications, resulting in three joint papers on Shannon sampling and learning theory, as well as two articles on diffusion kernels, Markov sampling, and one on language evolution with Felipe Cucker.

As Steve's research collaborator, colleague, and friend, I have benefited immensely from working with him. He set an exemplary model for a focused researcher and demonstrated how to choose research areas, topics, and problems by exploring interesting, beautiful, and valuable connections among different fields. He also taught me how to write research articles "economically" (Steve's word).

Working with Steve has been one of the most rewarding experiences of my career. His dedication, curiosity, and adventurous spirit have not only inspired me but also many others in the field. His approach to research and life continues to influence my work and decisions. In the spring of 2021, I spent my sabbatical at the Statistical and Applied Mathematical Sciences Institute and Duke University. Due to the pandemic, I could not follow my original schedule to visit Steve or other collaborators. However, Steve and I had several long phone conversations. He asked me how many doses of the vaccine I had received and encouraged me to explore opportunities outside Hong Kong. His encouragement, along with support from others, influenced my decision to accept an invitation from a headhunter and, subsequently, an offer of school headship from the University of Sydney in 2021.

Ding-Xuan Zhou
Professor and Head, School of Mathematics and Statistics
University of Sydney

Preface by Steve Batterson

Stephen Smale* startled the mathematical world in 1957 by proving that, in a theoretical sense, it was possible to turn a sphere inside out. Until Smale's work there was a prevailing intuition that such an act required that the sphere be torn or crimped. Even more sensational were his theorems a few years later, persuading mathematicians that a 6-dimensional world was simpler than a 3-dimensional world. For this work Smale received international acclaim and was awarded the most prestigious prize available to a mathematician. Mathematics, however, is a subject that has stood outside the mainstream of intellectual discourse. The discoveries of its giants have generally been known and appreciated only by those within the field.

Whatever barriers exist to deter entrance into higher mathematics did not prevent Smale's occasional exit. His vision and influence extended beyond mathematics into two vastly different realms. With Jerry Rubin in 1965 he initiated a program of civil disobedience directed at ending the Vietnam War. As a mineral collector Smale accumulated a museum quality collection that ranks among the finest in the world. Despite Smale's unique resumé of diverse accomplishments, the name of this contemporary genius is virtually unknown outside the narrow worlds of mathematicians and mineral collectors. One objective of this book is to bring his life and work to the attention of a larger community.

* pronounced as snail with an m.

I first became aware of Stephen Smale when I was a mathematics graduate student in the early seventies. As I began to study the theory of dynamical systems, Smale's name was everywhere. He had set the agenda for development, made the important conjectures, and proved the big theorems. John Franks, a former student of Smale, was my thesis advisor, making Smale my academic grandfather. Reading papers on dynamical systems and working on my thesis problem, I was continually struck by Smale's insight into the subject. Not only had he identified the structures that I was struggling to assimilate, but he had seen how they might fit together into a theory. By that time Smale himself had moved on to mathematical economics, leaving behind a rich collection of problems for his progeny.

Smale's mathematics is distinguished by its breadth as well as its depth. It was his work on the theory of computation that inspired me, in the mid-eighties, to shift my own research in that direction. At Smale's invitation I visited Berkeley for a semester in 1990. During that period we became better acquainted. I learned that he had been a communist during the McCarthy era. More surprising was that, in addition to his ambitious research program, the sixty-year-old mathematician had just recently begun devoting considerable energy to making high quality photographic prints of his minerals. One night at dinner he confided to me that the biggest driving force in his life at that instant was the hope that he might someday be known as a great photographer. I immediately thought "Why isn't it enough to be acknowledged as a great mathematician?"

I wanted to know more about Smale's life and what he had accomplished with his remarkable mind. Ten years earlier Smale had begun to write an autobiography, focusing on his protest activity. Unfortunately he only completed a portion of the project. Reading the chapters whetted my appetite for more. Since my childhood I have enjoyed the biography genre, but there are few biographies of mathematicians. Thinking of Smale as a possible subject, it became clear why the life stories of mathematicians are rarely recorded. To place their lives in perspective requires some appreciation of their theorems. Biographers are not trained in mathematics, and mathematicians do not learn to write biographies. It was unlikely that a suitable author would emerge to write Smale's biography.

Although I am foremost a mathematician, I became intrigued by the notion of working on a Smale biography project. Smale is a significant figure in intellectual history and his story should be recorded and known. Few people have any idea as to what research mathematicians do, or even the intellectual skills and personal qualities that contribute to success in this mysterious subject. The life of Steve Smale offers a case study in the development of a great mathematician. This book examines what made him that rare individual who succeeded in proving profound theorems. Additional perspective is provided by comparing Smale's approaches to politics, minerals, and other endeavors.

Several factors made me hesitate to write this book. My admiration of Smale's mathematics and antiwar work raised a concern that I was not suffciently objective. There were doubts about my qualifications as well as a reluctance to delve into his personal life. In the final analysis these issues were outweighed by my genuine interest in Smale's life and the realization that biographers, while struggling for objectivity, are never indifferent to their subjects. When informed of my intentions, Steve was gracious, making himself and his papers available to me. I believe, however, that he may have doubted that this project would be completed.

My goal has been to produce a book that makes Smale accessible to nonmathematicians. Not to provide the details of his proofs, but rather to convey an appreciation for the problems he attacked. One of the challenges is the vertical nature of mathematical knowledge—where each concept depends on several others which themselves depend on still others and so on in long cascades. In the twelve chapters of narrative there is little technical discussion, but there are a few selected localities with intuitive descriptions of abstract mathematical concepts. To illustrate more of the flavor of Smale's work, appendices are included at the back of the book. These sections explore, at varying levels of sophistication, the nature of four mathematics problems to which Smale made significant contributions. The presentation is directed at nonmathematicians and mathematicians whose expertise is outside the topic under discussion. The appendices are not essential to the text nor do they encompass the full range of Smale's mathematics. Readers may wish to skip them entirely or read a portion before moving on to the next section.

Steve Batterson

Author, *Stephen Smale: The Mathematician Who Broke the Dimension Barrier*

Acknowledgements

Chapters 1–12

No biography can be more interesting than the life it describes. Steve Smale has served as the subject of this book as well as contributing a considerable amount of time for interviews. To him, I am deeply indebted.

A pleasant surprise in researching this book was the gracious manner in which so many of the sources received me for interviews. I thank the following people for their conversations with me about Steve Smale:

Eric Asselborn
Frank Bardacke
Joel Bartsch
Lenore Blum
Raoul Bott
Merle Bowman
Bill Carpenter
Chandler Davis
Norman Doorenbos
Sanford Elberg
Si Frazier
Matthew Gaffney
Ron Gillies
Vincent Giuliano
Morris Hirsch
Barbara Gullahorn
Jane Jackson
Joe Jewett
Rob Kirby
Bill Larson
Saunders MacLane
Gene Meieron
Marilyn Milligan
James Munkres
Charles Pugh
Frank Raymond
Art Rose
Marty Roysher
Jerry Rubin
Hans Samelson
Stephen Scheer
Jack Scruggs
Holecek John Searle
Ed Shaffer
Myron Sharpe
Michael Shub
Clara Smale
Helen Smale
Judy Smale
Laura Smale
Nat Smale
Steve Smale
Steve Smith
Edwin Spanier
John Stallings
Wayne Thompson
Robert Thrall
Jack Weinberg
Steve Weissman
Dave Wilbur
Robert Williams

Several archival sites were extremely helpful in offering access to their materials and aiding my efforts to find them. In particular I would like to acknowledge the following individuals and their institutions: Pam Wasmer of the Indiana State Library, Bill Roberts of the University Archives of the University of California at Berkeley, George Mazuzan of the National Science Foundation, Nancy Bartlett and Ann Frantilla of the Bentley Historical Library, and Ralph Elder of the Center for American History at the University of Texas at Austin.

This research was supported in part by an award from the University Research Committee of Emory University.

Several individuals have assisted by reading chapters, offering pointers, doing translations, and other invaluable services. This includes, but is not limited to:

Juan del Aguila
Irene Browne
Dan Carter
Bob Devaney
Dwight Duffus
John Franks
Fred Gehring
David Goines
Phil Holmes
John Juricek
Mark Kitchell
Harvey Klehr
Steven Lighthill
Linda Mitchell
Steve Mitchell
Narasimhan
Lynn Narasimhan
Wilma Palmer
Karen Stolley

I would like to thank Sergei Gelfand, Chris Thivierge, Jennifer Sharp, and Connie Pass of the AMS for their skilled assistance in transforming my manuscript into a book.

Midway through the process of writing this book Ellen Neidle became my wife, as well as a vital sounding board and de facto editor. If I were not so defensive about making all the changes that she suggested, the exposition would be better.

When I thought that I had finished the book, Edwin Beschler took up the task of editing the manuscript. The book and I have benefited from his insights and suggestions.

Steve Batterson
Atlanta, Georgia
May 11, 1999

Chapters 13–16

It has been a joy and an honor to chronicle the past quarter-century of Professor Stephen Smale's life. As this project was far from a solo venture, I would like to thank the many people who contributed their time, insights, and expertise.

First, I must extend my heartfelt thanks to Prof. Smale, who entertained questions of all manner and magnitude over the past six months. He contributed a substantial amount of his time and often pointed me in the right direction when I needed more information.

I am also grateful to my teammate, Carla Thomas, who managed the early stages of the project while I was on maternity leave. She conducted a few interviews and provided valuable feedback after I drafted the new chapters. Thank you for supporting me while I found my footing as a mother of two and for your excellent suggestions.

It is a testament to Steve's character that many of his family members, friends, and colleagues welcomed my calls and emails. In particular, I would like to thank three mathematicians—Felipe Cucker, Ding-Xuan Zhou, and Indika Rajapakse—who participated in multiple interviews, reviewed the new chapters, and shared pointers on everything from timelines to mathematical descriptions. Without their extensive knowledge of Steve and his many pursuits, I could not have reconstructed the past 25 years of his life.

In addition, I am grateful to everyone else who made time to answer my questions and share their memories, including (in alphabetical order):

Lenore Blum
Robert Devaney
Way Kuo
David McAllester
Charles Pugh
Dylan Smale
Judy Smale
Laura Smale
Nat Smale
Shmuel Weinberger

Several individuals have supported this project behind the scenes by reading the new chapters, sharing feedback, and offering copyediting and fact-checking support. I would like to acknowledge the following contributors:

Michelle Rahardja Chandra
Chris Gladd
Anna O'Connor
Craig Sauers
Margaret Springer
Carla Thomas

Finally, I would like to thank Sydney Stonner, my editor at CityU Press, who helped refine the chapters with a keen eye for detail. She has been an excellent sounding board throughout the process, and I am grateful to have had the opportunity to work with her.

Kate Springer
Philadelphia, Pennsylvania
June 14, 2024

Credits

Chapters 1–12

Chapters 1–12 are reprinted from *Stephen Smale: The Mathematician Who Broke the Dimension Barrier* by Steve Batterson. City University of Hong Kong Press is most grateful for the kind permission of the author and the American Mathematical Society (AMS).

The AMS gratefully acknowledges the kindness of these individuals, institutions, and publishers in granting the following permissions.

Stephen Smale

Photograph of Blue Cap Tourmaline by Steve Smale on p. 283.

Photograph of Nevada Sulphur by Steve Smale on p. 289.

Photograph of Steve and friend on train to Campeche on p. 51. Photograph was taken by Haskell Rothstein.

Photograph of Steve Smale and Gerard Debreu on p. 351.

Photograph of Steve Smale and Charles Pugh on p. 301. Photographer was probably Welington de Melo.

Photograph of Steve and Clara Smale on p. 274. Photographer was Bob Sullivan.

Photograph of Clara and Steve Smale and President Clinton on p. 361. Photograph was sent to Steve by the White House.

Photograph of Clara, Steve, and Laura Smale on p. 361. Photograph was sent to Steve by the White House.

Quotes from *Chaos: Finding a Horseshoe on the Beaches of Rio*, by Steve Smale (preprint), courtesy of Steve Smale.

Quotes from various papers and unpublished manuscripts written by Steve Smale, courtesy of Steve Smale.

Judy Smale

Photograph of Steve and Judy Smale on pony on p. 9. Photograph was likely taken by their father, Lawrence Smale.

Photograph of Steve and Judy Smale sitting in grass on p. 10. Photograph was likely taken by their father, Lawrence Smale.

Wilma Palmer

Photograph of Horton Country School students on p. 12. Photograph is from the estate of Art Leech.

Springer-Verlag New York

Quotes from *On the steps of Moscow University* by Steve Smale in The Mathematical Intelligencer, Vol. 6, No. 2 (1984), pp. 21–27.

Quotes from "Some autobiographical notes" by Steve Smale in *From Topology to Computation: Proceedings of the Smalefest* (M. W. Hirsch et al., editors) Springer-Verlag (1993) pp. 3–21.

Quotes from "Personal reminiscences" by Mike Shub in *From Topology to Computation: Proceedings of the Smalefest* (M. W. Hirsch et al., editors) Springer-Verlag (1993) pp. 296–299.

Cambridge University Press

Quotes from *The Beginnings of the Nobel Institution* by Elisabeth Crawford, Cambridge University Press (1984), pp. 51–53, 217, 251.

Bentley Historical Library of the University of Michigan

Excerpts from March 18, 1952, Investigation: Testimony of Smale (p. 35), Box 58, Folder 9 of the Ruthven Papers.

Excerpts from April 15, 1952, Joint Judiciary: Case of Smale (pp. 50– 54), Box 58, Folder 9 of the Ruthven Papers.

Excerpts from May 7, 1952, Sub-Committee on Discipline: Smale appeal (pp. 1–4), Box 58, Folder 9 of the Ruthven Papers.

Excerpts from December 13, 1954, letter from Gardner Ackley to Ralph Sawyer, Box 4, Folder S of the Niehuss Papers.

University Archives, University of California, Berkeley

Excerpts from May 11, 1965, letter from Steve Smale and Jerry Rubin to Chancellor Martin Meyerson, CU 149, Box 65, Folder 133.

Excerpts from October 1, 1965, letter from District Attorney Coakley to Governor Edmund Brown, CU 149, Box 65, Folder 134.

Excerpts from October 3, 1965, letter from University of California President Clark Kerr to Chancellor Roger Heyns, CU 149, Box 65, Folder 134.

Excerpts from August 31, 1966, note by Vice Chancellor Robert Connick on phone call from Leland Haworth of the NSF, Box UARC, Folder 955-63-5.

Excerpts from January 16, 1967, memorandum from Vice Chancellor Robert Connick to University of California Vice President Angus Taylor, Box UARC, Folder 955-63-5.

The Center for American History of The University of Texas at Austin

Quotes from the March 14, 1956, letter from R. L. Wilder to A. W. Tucker at Princeton, Box 49 of the Raymond Lewis Wilder Papers,

Archives of American Mathematics, The Center for American History, The University of Texas at Austin.

Quotes from the February 1956 letter from R. L. Wilder to the California Institute of Technology, Box 49 of the Raymond Lewis Wilder Papers, Archives of American Mathematics, The Center for American History, The University of Texas at Austin.

Mineralogical Record

Excerpt from article on the 1976 Tuscon Mineral Show competition for the McDole Trophy by Wendell E. Wilson, *Mineralogical Record*, May–June 1976, pp. 136–137.

Tata Institute of Fundamental Research

Photograph of banquet during a 1964 conference at the Tata Institute in Bombay on p. 65.

Ten Speed Press

Quotes from *The Free Speech Movement* by David Lance Goines, Ten Speed Press (1993). Copyright © 1993 David Lance Goines.

A K Peters, Ltd.

Reproduction of Transparent Peach Eversion from the "Outside In" video project (1990–1995) generated at the Geometry Center, University of Minnesota, and distributed by A K Peters, Ltd., Natick, Massachusetts. Reprinted by permission.

Chapters 13–16

City University of Hong Kong Press gratefully acknowledges the kindness of these individuals and institutions in granting the following permissions.

Laura Smale

Photograph of Clara Smale on p. 371.

Photograph of the Smale family sailing on p. 408.

Photograph of Clara Smale on p. 433.

Photograph of Steve and Clara Smale on p. 447.

Indika Rajapakse

Photograph of Steve on p. v. The photograph, part of Smale's collection, was digitized by Indika and Dylan Smale.

Photograph of Steve and Clara Smale on p. 393. This photograph of Smale's was also digitized by Indika and Dylan Smale.

Photograph of Steve hiking on p. 465.

Photograph of Steve and Indika on p. 472.

City University of Hong Kong

Cover photograph.

Getty Images

Photograph of Tai Long Wan Bay on p. 384.

Hong Kong Institute for Advanced Study

Photograph of Steve on p. 472.

One Room Schoolhouse

01

A. Attending the 45th reunion

On August 16, 1966, the 36-year-old mathematician Stephen Smale arrived in Moscow to receive the Fields Medal at the International Congress of Mathematicians. Smale had earned this award, often described as the "Nobel Prize of Mathematics," by bringing a profound new understanding to the subject of higher dimensional topology. Normally there is little interest in the Fields Medal outside the upper echelon of the mathematics community. In 1966, however, Smale's trip to Moscow frustrated attempts to serve him with a Congressional subpoena. On the same day as the Fields Medal ceremony, the House Committee on Un-American Activities began a hearing in Washington to investigate radical antiwar protests by Smale and others. The unusual combination of mathematical achievement and political activity raised the profile of a diminutive mathematician with a distinctive, high-pitched voice. Ten days later, Smale held an ad hoc Moscow press conference in which he condemned the United States involvement in the Vietnam War and compared it to the Soviet invasion of Hungary.

Soviet attempts to control him led to a front page story in the New York Times. Amidst the Cold War tensions, many Americans felt that Smale's speech, in the enemy capital, was an act of treason. Members of Congress were outraged, and Smale's federal research grant was suspended. When the furor subsided, Smale resumed his mathematical work. That research continues today (in 2000), as he approaches the age of 70. In the intervening years he rose to the top rank of collectors of exotic minerals, became a first-class photographer of these specimens, and balanced all his interests with a love for outdoor adventures. Somewhere at the core of Smale's life lies the genesis of a brilliant mathematical mind, a passionate defender of political rights and freedoms, and an aesthetic consciousness that perceives the beauty in the patterns of both intellectual constructs and natural objects.

At what point in Smale's life did his extraordinary intellect become apparent? Other notable scientists were precocious mathematical achievers. A striking example was Nobel physicist Richard Feynman who was the star of the New York high school math team circuit.[1] John von Neumann, the great mathematician and computer pioneer, was a child prodigy in his native Hungary. At the age of 17 von Neumann completed his first mathematical publication, and the following year he won a prestigious prize, awarded on the basis of a competitive mathematics exam among the nation's high school graduates.[2] Was Steve Smale the *mathematical genius* of the Grand Blanc (Michigan) High School Class of 1948 and, if not, were there any indications of future greatness?

An opportunity to explore these questions arose in 1993 as I was beginning to research this biography. Following up on some sketchy information from Steve, I spoke to his class valedictorian, Jack Scruggs, who had not seen Smale since their college days together at the University of Michigan. Shortly thereafter Scruggs received a notice for his 45th high school reunion. Accompanying the announcement was a plea for information on the whereabouts of Smale and others who had not been in contact with the reunion committee.

After Scruggs informed the organizers of my project, I was invited to attend the reunion and speak at the banquet, even though a prior commitment precluded Steve's participation. I had recently returned from my own 25th reunion and was intimately acquainted with the ritual as well as the mixture of emotions that accompanies seeing old friends after so many years apart. With Steve's reunion, the 45 years of separation from his peers provided an intriguing ingredient. The recollections of his classmates were largely untainted by knowledge of his subsequent accomplishments. I began to view my Michigan trip as an archaeological dig.

Thus on a Friday evening in October, I arrived in Grand Blanc for the initial function. It was an informal get-together at Little Joe's, a pizza and beer place on the main street. Approximately 30 members of the 104 person class were present. This was my opportunity to receive the recollections of a significant portion of Smale's class. I quickly learned that his classmates knew very little about Steve, either from high school or the intervening years. Some were vaguely aware that Steve had previously engaged in some sort of radical activity, but all seemed surprised that his life was now the subject of a book.

The high school composite of Smale, obtained from several interviews, was of a bright guy who was viewed as somewhat of a loner. Nobody even suggested that he was a child prodigy, but the comments were consistent with his third-place rank in the class. The most lasting impression of Steve was that of an avid chess player, engaged in competition before school and at lunch. In most cases the opponent was Fred Burr, Steve's closest friend and a drowning victim in the fifties. None of the women recalled any contact with Steve.

I returned to Grand Blanc the following morning for the Liar's Club breakfast. The Liar's Club consisted of friends who met for breakfast each week at the local Halo Burger. The group's name was derived from the tradition of exaggerating progress reports during the annual pumpkin growing contest. Among the regulars were Joe Jewett, Steve's high school biology teacher, and Art Leech who had been a few grades ahead of Steve at their small elementary school. Shortly after I arrived at the Halo Burger, a man in his seventies approached and introduced himself. It was Joe Jewett, who brought the news that Art Leech had died the previous night. Despite the saddened mood among the Liars, Jewett did his best to convey a picture of Grand Blanc High School in the forties.

As I left Halo Burger to explore the Grand Blanc area landmarks of Steve's childhood, I was bewildered by the seeming incongruity of his early life. During his first eighteen years, while immersed in a middle American culture dominated by agriculture and the automobile industry, there was no hint of mathematical genius. Yet in his late twenties, Steve stunned the mathematical world by successively everting the sphere, creating a mathematical function known as the "horseshoe," and solving the higher dimensional Poincaré Conjecture. While the trip to Grand Blanc had given me a sense of the atmosphere in which Steve had grown up, it had left unresolved several questions concerning his intellectual development. The next step was to examine the Smale family life.

B. Ancestry

Sidney Smale, Steve's paternal grandfather, was born in England in 1870. One year later he accompanied his parents to the United States, where the family eventually settled in Flint, Michigan. Denied a formal education by the exigencies of the day, the 11-year-old Sidney obtained employment in the Flint store Smith, Bridgman, and Company. During the next 12 years, he rose from the entry position of cash courier to become head of the bookkeeping staff. When the Union Trust and Savings Bank was organized in 1893, Sidney accepted a position as teller.

The following year the young banker married Alice Maud Hughes. Maud belonged to a prominent Flint family and was a direct descendent of a Continental Army officer in the Revolutionary War. The marriage produced four children in its first six years. In 1899 Steve's father, Lawrence Albert, became the third child.

The appealing American story of a self-made and educated young immigrant ended in 1901 when Sidney Smale died of diabetes. His obituary, on the front page of the Flint newspaper, described a man of deep religious commitment, already launched on a trajectory toward social and business success. Maud, who survived her husband by 15 years, attempted to impart the Methodist tradition to her children. She experienced mixed success. One of her sons became a devout Methodist, while Lawrence rebelled after reading the essays of Emerson. The reaction was so profound that his own son, Steve, would later remark. "I was 20 years old when I first set foot in a church. That was Notre Dame in Paris."[3]

Lawrence was a man who never quite found his niche. One year after his mother's death Lawrence graduated from high school and then enlisted in the Navy as the United States entered World War I. After the War he enrolled in the Michigan College of Mines, aspiring to become a mining engineer. In one of several abrupt life changes, Lawrence dropped out of college, intent on seeing the world. The next two years were spent in a sequence of hopping freight trains, working boat passages, and taking odd jobs. During this period his itinerary included San Francisco, Alaska, Hawaii, Yokohama, Shanghai, and Hong Kong. Eventually he reached Manila, penniless and unable to find work. At that point the former sailor successfully appealed to his government to provide passage home.

Following his return to Michigan, Lawrence enrolled in Albion College. Shortly after matriculation he published the first issue of an intended periodical, *The Maelstrom.* Despite printing less than ten pages of text, Smale managed to impudently defy college regulations, present paradoxes on the existence of God, and trash the presidents of the college and the United States (Calvin Coolidge). Perhaps the most provocative piece was a short note suggesting homosexual prostitution as a work study option. The latter led to the charge of distribution of obscene literature as Lawrence was quickly arrested and expelled from Albion.

Smale returned to Flint where he obtained employment at AC Spark Plug. Working as an assistant in an experimental ceramics lab, his job involved testing spark plug play. Meanwhile Helen Morrow moved from Canada to Flint and became a secretary to the advertising manager at AC Spark Plug. The circumstances surrounding the future Mrs. Smale's departure from Canada were the product of her own unfortunate family life.

Helen's Canadian parents, Archibald and Pauline Diesfeld Morrow, were married in 1889. They settled in Gault, Ontario, where Archibald was a high school classics teacher. Helen was born in 1905, the last of four children. She remembers her father as a "horrible man" who expelled both of her brothers from high school.[4] When Helen was 11, her parents separated with Archibald moving to a boarding house.

Remaining with her mother, Helen completed high school and one year of business college. Helen's brother-in-law was an engineer who worked in a Gault boiler factory. Expecting to lose his job after the War, he and Helen's sister moved to Flint seeking new opportunities. AC Spark Plug provided a job. In 1925 Helen and her mother followed. The two women moved into a Flint apartment supported by the income from Helen's job. A few years later Lawrence and Helen were brought together by mutual friends. After a short courtship the couple married. The union endured, except for a brief separation, until Lawrence's death in 1991.

Family descriptions of Lawrence portray a kind man with highly unconventional views who had difficulty finding contentment. Despite Helen's protests, there were several house moves, ostensibly for economic reasons. Lawrence possessed formidable practical skills and devoted a substantial portion of his energy toward renovation of each new residence. Steve's father was both an atheist and a Marxist. He

Steve and sister Judy in 1934

belonged to a political organization known as the Proletarian Party, a Michigan group that broke off from the United States Communist Party shortly after its formation in 1919. While Lawrence's beliefs were sincerely held, his involvement was substantially more theoretical than action oriented. As Helen recently reflected, "the most we did in the Pro Party was have picnics."[5]

As Helen and Lawrence began their family, it is significant to observe that the environment for their children differed sharply from that of their peers. Although neither parent graduated from college,

Steve and Judy in 1936

both were well read intellectuals of sorts. Lawrence possessed an interest in science and placed a value on independence which he sought to impart to his offspring. Economically, the country was entering the Depression, and it was a time of considerable deprivation and unemployment. Lawrence had a stable job which was adequate to support a family. Conflicting with these mundane needs were his hatred of the job and a burning desire to become a successful writer. Lawrence elected to sublimate the creative for the practical until his children entered college.

C. Youth

Lawrence and Helen's first child was born on July 15, 1930. The son was named Stephen, with no middle name. Lawrence was contemptuous of tradition and determined that his son would not be named in honor of another person, particularly a relative. Many years later, as Lawrence explored his own genealogy, he was disappointed to learn that the brother of his great grandmother was named Stephen Crocker.

Two years later, with the family still living in Flint, daughter Judy was born. When Steve was five, the Smales moved to a small ten acre farm in a rural region south of Flint. The farmhouse was located at the intersection of two dirt roads in a sparsely populated area. Even today the roads remain unpaved. The house required a lot of work, as it lacked plumbing and electricity. Lawrence devoted his spare time to the renovation, and the house became more habitable after a year.[6] Since the amount of land was inadequate to sustain a farm, the Smales leased a few acres to the owner of a neighboring farm. Lawrence set up a chicken and egg business that would eventually fail. This was in addition to his job at AC Spark Plug, where Helen had worked until Steve's birth. Helen recalls that the young couple worked extremely hard during those years.

For their elementary education, Steve and Judy attended the Horton School. This was a one room country schoolhouse, located on a dirt road one mile to the east of the farmhouse. There were a few students enrolled in each of the eight grades and kindergarten. Remarkably, the entire school program was administered by a single person, typically a young woman with some college training. The teacher was responsible for the custodial work and lunch production as well as conducting lessons in a variety of subjects to all the grades of students. Not surprisingly, there was a high turnover rate among the teachers. During years one through eight, Steve had seven different instructors. In the classroom, seats were assigned to group the grades together. Usually the teacher conducted a lesson for two grades at a time, while the other students were occupied with assignments or study.

Each student received regular letter grade evaluations in areas such as mathematics, English, history, penmanship, spelling, geography, science, and citizenship. Descriptive assessments were made of attitude, recitations, and conduct. Inspection of Steve's report cards reveals that

Steve (left end of 2nd row) in the 3rd grade with the rest of his elementary school. Judy is 3rd from left in front row. In the back row Bill Carpenter is on left and Art Leech is 3rd from right.

he received mostly *A*s and *B*s along with an occasional *C*. The only *C* in mathematics occurred in February of the second grade. Deportment ratings were favorable with the exception of the first term of the fifth grade when Steve was cited for "whispering too much." Most of the *C*s were for penmanship.

The overall picture is that of a very good, but not outstanding, student in a less than challenging situation. Steve viewed the education as good, but lacking in various technical respects. The encyclopedia was an important reference source, from which Steve recalls learning a method for solving three linear equations with three unknowns.

For recreation, Steve and Judy played in the woods near their house. The children wrestled until Judy became aware of her brother's strength, which was somewhat surprising with his small physique. During his younger days Steve was a bit of a prankster. In one instance he removed the pointed end from a dart, prior to throwing it at his terrified sister.

Bill Carpenter was Steve's classmate and friend. The Carpenter family owned a farm near the school. There was a creek on the Carpenter farm where Steve, Judy, and Bill swam in the summer. Softball was the most popular sport. Although the sparsely populated rural area did not contain enough children to support two full teams, they improvised a game that included a pitcher, batter, and catcher, with the remaining players in the field. Batting was the most desirable position, leading to occasional disputes over the completion of a turn. One day Steve lost a portion of a tooth when he attempted to grab the bat from Bill's hands. It was a genuine accident, without malice, leading immediately to remorse by both boys.

It is interesting to contrast the attitudes toward education in the Carpenter and Smale households. During harvest season, Bill remained home to work in the fields. College was completely out of the picture. His father "thought it was foolish for a farm boy to go to school past 16. He needed them working."[7] Lawrence cultivated scientific interests in Steve and Judy. Both children received chemistry sets. Although Judy showed little curiosity, Steve enjoyed conducting experiments and converted a portion of the chicken house into a laboratory. To promote an interest in astronomy, Lawrence constructed a telescope.

Under Lawrence's influence, Steve began other intellectual activities that he would pursue for many years. Father and son played chess, and Steve rapidly demonstrated strong potential. Unlike some competitive family situations, Lawrence was pleased when his son surpassed him. Late in elementary school, Steve began to practice at the Flint Chess Club. When there was an opportunity to participate in a chess tournament in Chicago, Steve revealed his independence by traveling alone to the city. In high school and college, he would continue to compete seriously in the sport, achieving success at the national level.

Dominating Lawrence's own intellectual interests were Marxist political theory and the craft of writing. Until his death at the age of 91, he yearned to produce a great novel. Steve often accompanied his father on trips to the Flint Library where the young man selected novels and scientific texts.

Upon completion of the eighth grade in 1944, 605 Genesee County students took a competitive achievement test. Steve obtained the highest score. Given his lack of formal education, the result is somewhat noteworthy. However there are many counties in the United States, and the credential certainly did not ordain genius.

When Steve reached the ninth grade, he enrolled in Grand Blanc High School. Grand Blanc was a small town a few miles southwest of the Smale house. With a student body of about 250, it was quite a change from Horton School. The combination of Steve's shyness and his rural background made for a difficult social transition. He felt awkward, and it is not surprising that his classmates viewed him as a loner. Two significant friendships did, however, develop.

Norm Doorenbos was two years older than Steve. Despite the age difference, their common interests in chemistry and chess brought the two together. Doorenbos would eventually earn a Ph.D. in chemistry at the University of Michigan,[8] but in the midforties they were two intellectually inquisitive teenagers, performing experiments in the loft of the Smale chicken house. An almost inevitable by-product of such adolescent curiosity is some sort of explosion. One particular experiment produced nearly disastrous consequences. Doorenbos and Steve placed some sodium peroxide powder in a glass container.[9] A subsequent explosion shattered the glass, just missing Judy who had entered the chicken house.

Doorenbos was a member of the American Chemical Society and received the journal, *Chemical and Engineering News*. Along with the technical articles were notices by companies soliciting various chemicals.

Norm and Steve established a partnership, intent on meeting the demand. Although they responded to several advertisements, no job ever materialized.

While the attraction to Doorenbos was intellectual, Steve did develop another more personal friendship. Fred Burr lived with his aunt and uncle in a poor section of Grand Blanc known as Wigville. Burr was a shy, offbeat teenager in whom Steve found a kindred spirit. Burr played chess and possessed a remarkable knowledge of music. Without much of a home life of his own, Burr enjoyed spending time in the Smale household where music and card games were popular pastimes. Burr was Steve's only close friend in high school. Under Burr's influence, Steve developed a love of classical music which he would retain and pass on to others.

Steve's ninth grade biology teacher was Joe Jewett who also taught a farm course and supervised the 4-H Club. A conflict with Jewett led to Steve's first political action.[10] Steve was familiar with the highly publicized Scopes Monkey Trial which had taken place 20 years earlier. In that proceeding a Tennessee high school biology teacher, John Scopes, was convicted of violating a state law by teaching the theory of evolution. The Grand Blanc biology text included a chapter on evolution, but Joe Jewett omitted the material. Steve was disappointed and reacted by circulating a petition seeking reinstatement of the chapter. The effort received virtually no support from his classmates.

In 1993 I asked Jewett about the conflict.[11] Although he had no recollection of the incident, Jewett acknowledged that he had skipped the evolution as well as the human reproduction chapter. He recalled that he deleted the material from the syllabus out of devotion to his Fundamentalist Baptist mother who was vehemently opposed to the teaching of evolution. Jewett, uninfluenced by Steve, eventually incorporated both topics into his course. Although teaching human sexuality in biology was a big step for him, Jewett always included the subject of animal breeding in his (exclusively male) farm course.

While Steve was in high school, Lawrence moved the family to a house in downtown Grand Blanc. Another *fixer-upper*, it was a large house with several rooms that would be rented as apartments. The house attracted quite a bit of attention when Lawrence selected red paint for the exterior. The color was so striking, compared to the neighboring houses, that some of Steve's classmates mentioned it at the reunion 45 years later.

In the twenties Lawrence had toured the country via freight trains and hitchhiking. Later he enjoyed taking his family on automobile trips. When he was temporarily laid off from work by a strike in 1946, father and son took the opportunity to travel to Mexico, with Judy and Helen remaining at home. Steve acquired his father's taste for the road, often hitchhiking on vacation excursions.

During Smale's high school experience, chess provided the only hint of future success. Here Steve displayed both talent and drive, honing his skills and competing in national tournaments. In another endeavor he was less successful. Smale's friend, Fred Burr, was an accomplished pianist. When Steve expressed an interest in learning to play, Lawrence purchased a cheap piano. Steve took lessons and practiced for a few months, but quit after achieving merely a minimal level of competence.

The editor of the 1948 Grand Blanc High School yearbook, the *Echo*, was Jack Scruggs. Scruggs was also the class valedictorian. Although Scruggs and Smale are positioned next to each other in several yearbook pictures, they recall little interaction in high school. For the pictures Steve was dressed in a coat and tie, unlike many of his peers. Next to each senior picture is a quote, selected by the yearbook staff, to characterize the subject. The quote linked to Steve seems prescient and suggests that, while he left few fingerprints at his high school, there were early indications of rebellion and intellectual independence. "I agree with no man's opinions—I have some of my own."[12]

Steve graduated third in his class. Both Smale and Scruggs received Regents-Alumni Scholarships that provided full tuition for four years at the University of Michigan at Ann Arbor. Steve's physics teacher was dubious about his prospects at the competitive University, and advised that Michigan State might provide a more appropriate opportunity.[13] This lack of support did not deter the intellectually self-assured Smale. He was already making his own decisions and elected to matriculate at Ann Arbor. His former chemistry partner, Norm Doorenbos, had just completed his sophomore year. Doorenbos was also acquainted with Jack Scruggs through participation in the band. When Doorenbos learned that Scruggs and Smale were coming to the University, he invited them to share his room for the following year.

In retrospect, neither Smale nor anyone else was the *mathematical genius* of the Grand Blanc Class of 1948. Certainly the top scholars experienced success in their careers. Scruggs obtained a Ph.D. in chemistry at the University of Michigan and rose to the position of Vice President of Research for Phillips Fibers. The salutatorian was an athlete who became a college athletic administrator. Many members of the class were employed in the automobile industry, where Michigan area jobs were plentiful in the fifties.

High school mathematics problems are largely computational. It is not uncommon for a high school student to be anointed as a *mathematical genius* when it is accepted that he or she can solve any problem and, most importantly, obtain the solution in a lightning-fast manner. Although computational facility may correlate in some way with success at higher level mathematics, quickness is neither necessary nor sufficient. Many exceptional high school mathematics students do not succeed at careers in mathematical research. Steve was not quick, but later it became apparent that he possessed a powerful independent insight and other qualities that do not register when evaluating mathematical talent in high school. At the age of 18, Steve was just a bright guy who was somewhat of a loner and played a lot of chess. Unlike Feynman, von Neumann, and a multitude of other high school students who impressed their peers, there was no indication of genius.

Marxism and Mathematics at Ann Arbor

02

A. Freshman year

When Smale and Scruggs arrived from Grand Blanc to attend the University of Michigan, they entered a strikingly different cultural and intellectual environment. Grand Blanc was a Waspish world of autoworkers and farmers in which Smale and Scruggs were distinguished by their scholarly accomplishments. At Michigan, it seemed to Jack that everyone was a high school valedictorian. There he first met a Jewish person.[1] Later, both Steve and Jack would share rooms with African-Americans.

It was 1948 and the Michigan student population included a large influx of veterans from World War II. To address the increased housing demand, double rooms were converted to triples. The young men from Grand Blanc shared such a room in Lloyd House, a dormitory housing approximately 100 students. By accepting Doorenbos' offer, Steve avoided the frequent freshman trauma that results when universities randomly match strangers as roommates.

Although the Grand Blanc students shared a serious interest in science, Steve differed from his roommates in other respects. Jack and Norm were religious individuals, whereas Steve was an atheist who had never been inside a church. Politically, Steve would move far to the left of his conservative roommates. Despite the differences, none of the three can recall any serious conflicts. It appears that, pairwise, each relationship was friendly. For recreation, Jack and Steve often played three cushion billiards at the student center.

In the triple room, Smale slept in the lower berth of a bunk bed, with Norm in the upper. Still a prankster, many of Steve's schemes were directed at his more serious bunkmate. In one instance Steve and Jack removed several of the hooks attaching Norm's bed to the frame, leaving just enough support for the mattress. Norm recalls arriving home late and being encouraged to retire immediately. As Doorenbos jumped into bed, Steve scurried out of the lower, and Norm's bed gave way.[2]

Freshmen arrive with varying degrees of preparation for college. To aid students in course selection, Michigan administered placement tests in certain subjects. Jack was a capable student who would obtain a Ph.D. in chemistry eight years later, but he felt insecure in the new environment. Scruggs skipped the chemistry test, conservatively electing to begin in the first course. As an entering freshman Smale took the placement tests, advancing in both chemistry and mathematics, subjects that he had studied independently while in high school.

Steve's ambitious approach toward course selection reveals two aspects of his character that would serve him well as a mathematician. He possessed a quiet confidence in his ability and a willingness to accept carefully chosen risks. Most people learn mathematics by revelation, in the sense that understanding tends to come suddenly rather than gradually. To reach that point one must persevere through ignorance to the moment when the concept becomes *clear.* Some individuals with math anxiety or insecurity fail to push to the instant of revelation. Confidence should be distinguished from arrogance, which involves an outward display. In interviews, Steve's friends from this period were quick to volunteer that he was not arrogant, but he was intellectually self-assured. Risk taking is a thread that will run throughout this book. For a mathematician, there is a risk in undertaking a problem that has already resisted the attack of prominent scholars. To Steve such challenges would be an attraction. In 1948, however, the small risk, associated with enrolling in an advanced class, proved to be well chosen.

At large universities such as Michigan, freshmen rarely encounter professors, except in large lecture sections with hundreds of students. Much of a student's direct teaching interaction occurs with graduate teaching assistants. By virtue of his accomplishment on the placement test, Steve enrolled in a special calculus section for advanced freshmen. The class of approximately 20 students was taught by an actual professor. Bob Thrall was 34 years old, an exceptional teacher with the ability to inspire students in mathematics, and a rapport that extended beyond the classroom. Steve described him as his "first memorable teacher."[3]

Two of the calculus students, Art Rose and Vince Giuliano, had graduated from the same high school in Detroit. Both would later room and travel with Steve. Rose's experience illustrates the fragility of the notion of mathematical ability. As one of the best students in a good high school, Rose entered college with confidence that he would excel in mathematics. Rose quickly discovered that he was in a new world. First, the special section was designed to combine two years of course material into a single year, and Thrall's pace was a quantum leap from high school. In Rose's previous experience, his effort and ability had always been sufficient to master the material and place him among the best. Now he identified "five mathematical geniuses," including Smale and Giuliano, whom he perceived as vastly superior.

Although Giuliano and Rose attended the same high school, they had reached Ann Arbor by different routes. Giuliano was a talented but erratic student. After performing poorly his first two years of high school, Giuliano dropped out for a year, returning to achieve at a high level. The University of Michigan was not impressed, and Giuliano's application was rejected. The decision was reversed following the intervention of an uncle who served in the state legislature. Despite the questionable circumstances of his admission, Giuliano thrived in the challenging environment of Thrall's calculus class. Forty-five years later Giuliano reflected on the class:

> *I was, at the time, an addict of pistachio nuts. Where I lived, to go to class you had to go through the Michigan Union. The Michigan Union had, in the basement, a pistachio nut machine which at the time would give you 14 nuts for a penny, which was a pretty good deal. I would generally go, on my way to class, and buy 1, 2, or 3 cents worth of pistachio nuts and put them in my pocket. During class I would crack them open and eat them one at a time. There was this guy who was in the class who I didn't know, but I generally sat next to him, and it seemed sort of the right thing to do to pass him a pistachio nut or two. He would eat them. This went on silently for several months. The guy was Steve and somehow, after several months of this, we started talking and we met outside of class.*
>
> *Bob Thrall's class was a wonderful thing to me. Thrall was incredible. He was positive, dynamic, the absolute opposite of the stereotype mathematician—outgoing, enthusiastic, believed in what he did, and was willing to lead students to go as far as they were willing to go. It was because of Bob's class that I decided to stay in math. I had originally gone there intending to major in physics.*

It was also an influential class for Steve who worked hard, was challenged, and received an *A* each semester. While Rose classified both Smale and Giuliano as mathematical geniuses, Thrall was more cautious in his assessment. Thrall recalled that Smale and Giuliano demonstrated keen minds that placed them at the top of a strong class, but the level of material was still inadequate to determine whether they possessed what he termed the "flair" for mathematical accomplishment. To a scholar of Thrall's experience, the ability to solve calculus problems quickly did not necessarily portend success as a research scholar.

Thrall's influence extended beyond mathematics. For example, he taught the game of *go* to the class. Eventually Steve would shift his attention from chess to *go*. During his undergraduate years, Smale continued to play chess quite seriously. When Steve organized a university chess club, Bob Thrall agreed to serve as faculty sponsor.

Steve completed his freshman year with a 3.5 grade point average despite a *C* in English composition. Both he and Scruggs were recognized by the freshman honorary society, Phi Eta Sigma. A frequent visitor in their Lloyd House room was a precocious classmate, Don Brown, the only child of two New York lawyers. Brown was a remarkably talented young man who was well aware of his ability. It is interesting to compare Jack Scruggs' impressions of Smale and Brown. He viewed Steve as a bright guy, probably more capable than himself, but never would he have predicted Smale's mathematical accomplishments. In contrast, Scruggs was in complete awe of Brown. Don had "the most brilliant mind I've ever known" and "would have been one of the world's great achievers at something."[4]

While Smale and Brown became close friends, Doorenbos was put off by Brown's brash attitude. Doorenbos was a junior chemistry major, and Brown was a 16-year-old freshman. On one occasion, Brown challenged the upperclassman to quiz him in chemistry. Doorenbos was to ask three questions from his current knowledge of chemistry, and Brown claimed that he would succeed on two out of three. Brown won in an impressive display of intellectual daring and skill.

Throughout Smale's college days, his choice of courses was overwhelmingly in the direction of mathematics and science, with one exception. Steve selected a substantial number of philosophy courses, one of Don Brown's many interests. They engaged in a friendly competition to determine who would receive the best grade in philosophy, with the least effort.[5]

Occasionally Brown took off for a long weekend, visiting his home in New York. He usually hitchhiked, a commonly accepted means of transportation during this period. On one of these trips, tragedy struck in Massalan, Ohio. The car in which Brown had received a ride collided with a train at a railroad crossing, killing both occupants.

B. On the left

Steve remained a student in Ann Arbor for eight years, receiving a mathematics B.S. in February 1952 and completing work on his Ph.D. in 1956. In examining Steve's life during these years, the traditional breakdown between undergraduate and graduate work fails to capture adequately the more significant divisions in his educational and personal life. This is not especially surprising since Steve began graduate school in the middle of his fourth year, continuing work in the same department.

Smale's college development falls into three periods. The first, his freshman year, was characterized by serious study, academic success, and a few male friends. In the second period, during the following four years, Smale's grade-point average dipped as his energy shifted from scholarship to left wing political activity. The third and final phase was distinguished by a departure from politics, engagement in mathematics, and marriage.

Consider the historical context for Steve's Marxist evolution. The years, 1949 and 1953, marked the Soviet development of the atomic and hydrogen bombs. This timing corresponds to Smale's first and fifth years at Ann Arbor, respectively. Cold War tensions increased as Americans worried about communist expansion. These concerns were exacerbated when Mao Tse-Tung gained control of China in 1949, and the Korean War commenced the following year. On the national scene, in 1950, Joe McCarthy began his demagogic ascendance with allegations of communists in government. As Americans obsessed over scenarios of disloyalty, the growing anticommunist hysteria enveloped university campuses.

To gain some understanding of this development it is instructive to examine the United States political dynamic. Harry Truman had become president in 1945 and faced an anticommunist coalition of Republicans and southern Democrats.[6] In the 1946 House and Senate elections Richard Nixon and Joe McCarthy effectively employed communist smear tactics against their opponents. When the Republicans made significant congressional gains, Truman became vulnerable to attacks that he was soft on communism. On March 22, 1947, he upped the political ante by issuing an executive order barring communists and "sympathetic associates" from the federal payroll. Moreover, the order authorized the attorney general to compile a list of subversive organizations.

Meanwhile the president faced two groups to his left, the Progressive Party and the Communist Party (CP). In the 1948 election Henry Wallace headed the Progressive Party ticket. This former vice president deplored the Cold War and advocated conciliation with the Soviet Union. When he was endorsed by the Communist Party, the alliance made a nice target for Truman. Thomas Dewey was the Republican candidate. As the election approached, the Truman Justice Department indicted 12 leaders of the American Communist Party under the Smith Act. (The trial produced a conviction that was upheld by the Supreme Court in 1951 as the CP went underground.) Truman won the election, and Wallace received just one million votes.[7]

The communist threat to domestic security was an issue that resonated among the American public. As national leaders made political capital with allegations concerning Reds in the federal bureaucracy, state legislators seized analogous opportunities. The hunt for communists on the state payroll required a target. What could be better than the image of intellectual professors in public universities, corrupting the impressionable youth of America?

In 1948 a state Senate committee chaired by Albert Canwell conducted hearings into subversive activity at the University of Washington. The two stage government-university proceedings set the agenda for faculty and administrators in the following years. The case is covered in detail in a book by Ellen Schrecker and is summarized below, beginning with the Canwell Committee.[8]

> *Eleven professors appeared. Two … denied that they had ever been in the Party. One … talked about his years in the Party and named names. None of the others did. One man admitted that he had been in the CP for a year, but claimed he couldn't remember any of his comrades. Four others … revealed that they, too, had been communists, but refused to name names. Three professors, Joseph Butterworth, Herbert Phillips, and Ralph Gundlach, were completely uncooperative. They refused to answer any questions about their political beliefs or associates. As a result, Canwell cited them for contempt. The prosecutor refused to charge Butterworth, but the other two went through years of litigation. Phillips was tried and finally acquitted; Gundlach, tried and convicted, actually spent thirty days in jail.*[9]

Each of the professors had tenure, which meant that their employment was permanent unless they demonstrated incompetence, moral turpitude, or some such egregious offense. The University was immediately confronted with determining whether any of the testimony (or lack of it) justified dismissal or discipline of a tenured professor. This required a procedure and decision. At that time, membership in the CP and refusal to answer questions of a government committee were not the sort of offenses detailed in institutional regulations. A special faculty committee acted as a grand jury and recommended that six of their colleagues face charges before the University Committee on Tenure and Academic Freedom. At the hearing, Phillips and Butterworth acknowledged CP membership. A defiant Gundlach continued his intransigence as evidence against him was presented. The Committee was divided over whether CP membership provided grounds for dismissal. In 1949, following the recommendation of the president, the regents terminated Phillips, Butterworth, and Gundlach. The other three faculty were acquitted.

University administrators observed the Washington case, fully cognizant that they could face similar decisions. Before the year was over, the presidents of Cornell, Stanford, Harvard, and Yale each had declared that communists were unfit for faculty positions at their institutions.[10] In 1950, the Congressional Democratic majority eroded as Nixon and McCarthy continued their attacks on Reds. The country was then at war with the communists of North Korea.

In 1952, the Senate Internal Security Subcommittee questioned left wing faculty from New York area colleges. Several took the Fifth Amendment and were subsequently fired by their institutions.[11] When the Republicans gained control of Congress the following year, it became an open season on academic communists. Showing a special interest in university faculty was Harold Velde, the new chair of the House Committee on Un-American Activities (HUAC).* The HUAC had originated in 1938 as a special committee with a small budget. Following an adroit parliamentary maneuver by John Wood of Georgia in 1945, the HUAC achieved permanent status. Two years later, a subcommittee conducted the high profile hearings leading to contempt citations of the Hollywood Ten.[12]

Over 100 college faculty members were questioned by investigators in 1953.[13] *Are you now or have you ever been a member of the Communist*

* The more common abbreviation of HUAC is used rather than HCUA.

Party? Many were former communists, confronted by the dilemma of whether or not to cooperate. To satisfy the Committee required confession, contrition, and naming names. This was the course followed by mathematicians Norman Levinson and W. Ted Martin.[14] Those refusing to cooperate, fully or in part, risked contempt citations and job termination. Legal remedies offered the uncertainty of years of appeals while the individual was blacklisted from academic positions. During this period CP association could destroy an academic career, regardless of scholarly accomplishment. Membership rosters were rapidly declining.

These factors exerted their influence over the Ann Arbor left wing scene in the early fifties, forcing much of the planning underground. Just as the Democratic and Republican Parties promote their own youth groups, the Progressive and Communist Parties had corresponding campus organizations, known as the Young Progressives (YP) and Labor Youth League (LYL). The LYL was a small, secret organization where most of the University of Michigan radical left wing activity was organized. This included support for civil rights as well as opposition to the Korean War and McCarthyism. Members were encouraged to obtain positions of influence in open organizations, enabling them to promote the agenda set in the LYL. A select few were invited into a student communist cell, deliberately covert and separate from the faculty group. There was no card, just token annual dues. The infrequent meetings generally involved the study of Marxist theory rather than sinister plotting. Membership in the CP was a quiet, personal status symbol rewarding especially diligent left wing activists.

Of course joining the LYL and CP was more complex than becoming a member of the YP or a Spanish club. There were no meeting notices posted on bulletin boards nor could one simply sign up at an activities fair. To gain admission to the LYL required establishing credibility with its current members. This could be accomplished by attending political functions and espousing appropriate views in interactions. During Steve's first years at Michigan his political activity was inconspicuous, limited to occasional events such as a meeting in support of Henry Wallace. When the Korean War began, Smale became more engaged in seeking out forums to demonstrate his opposition. The sincerity of Steve's beliefs became apparent, and he was assimilated into the LYL community.

Smale became regarded as a reliable participant, willing to undertake occasional initiatives and accept the risk of exposure. While Steve was not among the most prominent two or three Marxists on campus, he rose to the next group, "outing" himself as an LYL member in letters to the student newspaper. Steve enjoyed the notoriety associated with having his name in the paper. Eventually he joined the student CP cell. To reveal his membership offered the trade-off between the benefit of further recognition and the risk of the devastating repercussions associated with the McCarthy era. With his CP credential, Steve exercised discretion. He did not even tell his wife about his previous membership until sometime after their marriage. For many years he was cautious in this regard. Despite numerous subsequent investigations by the Michigan State Police, FBI, and HUAC, Smale's CP membership remained a secret until he revealed it in the eighties.

Climbing the LYL social ladder required obedience to the Stalinist version of Marxism and tended to select for the doctrinaire. There was little slack for experimentation or independent thought. When Smale returned from a trip to New York and naively informed his comrades of a visit to the Socialist Workers Party (Trotskyist flavored Marxism), he risked censure.[15] That Steve was able to, more or less blindly, accept the CP line was largely out of character with his subsequent intellectual style of developing his own point of view. In retrospect Steve came to understand his actions by identifying parallels between religious faith and his own indoctrination into the CP:[16]

> *There are strong tendencies for just about everyone to want something basic to believe in without questioning. Most often this takes the form of traditional religion. But even atheists*

and revolutionaries must rest their beliefs on some foundation. They are not immune to the problem of reasoning in circles. But still to accept the Communist Party?

Consider my frame of reference at that time. I was sufficiently skeptical of the country's institutions to the point that I couldn't accept the negative reports about the Soviet Union. I so believed in the goal of a utopian society that brutal means to achieve it could be justified. I was unsure of myself on social ground, and the developing social network of leftists around me gave me security. Then, these were the times of McCarthyism, the Rosenberg executions, the Korean War hysteria; the CP was the main group giving unqualified resistance to these forces.

In most families, religion exerts an early influence on a child's life. This was totally absent in the Smale household, while Marxist dogma was present. Given the strong correlation between religious practices of parents and offspring, the emergence of "red diaper babies" is not surprising.

For a young man with just a few male friends, Marxist activity opened up opportunities to meet men and women in both structured and unstructured settings. Steve recruited some of his earlier acquaintances, merging them into his new leftist network. For the first time in his life, Smale was part of a large circle of friends, all with compatible ideology. Finally, in understanding the attraction of communism to Smale, antimilitarism was a recurring theme in his political activity. He demonstrated against the Korean War as it began in the summer of 1950. Later, the Cuban Missile Crisis and Vietnam War prompted more striking responses.

As Steve's involvement in politics increased, his commitment to academics decreased. Although he took overloads each semester as a sophomore, grades of *C* in both Calculus II and Physics (Electricity and

Light) indicate a diminution of effort from the freshman year. The trend continued as a junior with extra courses and a *B*– average. Unimpressive grades in physics, including a failure in nuclear physics, led Steve to major in mathematics where he was earning *A*s and *B*s. The overloads and summer course work enabled him to complete his B.S. a semester early. At that point Steve was deeply immersed in political activity, but his undergraduate mathematics work was sufficiently strong to obtain admission to graduate school. Continuation as a student included the fringe benefit of a deferment from the Korean War. During the first three semesters of graduate school, politics continued to occupy a higher priority than academics.

An unusual feature of Smale's college years was his creative use of the summer for travel. These plans were enabled, in part, by a $2500 inheritance from his grandfather in 1950. That summer Steve spent two months in New York, taking philosophy courses at New York University and playing chess. Determined to stretch the money as far as possible, he lived frugally in a tiny, windowless Greenwich Village apartment.

In the fall of that year Judy entered the University of Michigan. Both Smales could rely on their Regent's Scholarships and inheritance to finance the remainder of their education. Meanwhile, with his children provided for and away at college, Lawrence quit his job at AC Spark Plug. Leaving a job he despised, Steve's father intended to earn a living as a writer, an aspiration to which he devoted the remainder of his life. As Lawrence unsuccessfully attempted to sell his stories, the couple's income was reduced to the meager return from four rental units in their house. Frustrated after a period of failure, Lawrence left Helen and obtained a job delivering trailers. The family crisis summoned Steve and Judy from college to reconcile their parents. Lawrence returned home, continuing the delivery job and his writing.

The following summer Steve devised a more elaborate itinerary. In August of 1951, a communist-organized World Festival of Youth and

Students for Peace was planned for East Berlin. The stated purpose was to unite 2,000,000 young people in a rally for peace. For a 20-year-old American communist with antiwar beliefs, the festival would provide a perfect ending to a summer of European travel. Given the political climate of the time, Smale's participation also carried a substantial risk of jeopardizing his future career opportunities.

Steve's roommate Vince Giuliano had already been invited to accompany his grandparents on a vacation to Europe. When Giuliano learned of Steve's plans, they agreed to meet in Europe midway through the summer. Smale, first by himself and later with Giuliano, toured the continent. One of the highlights for Giuliano was a memorable climb of Mount Vesuvius. To circumvent American travel restrictions, they reached East Berlin via a route through Vienna and Prague. Most of the participants at the Peace Festival represented communist countries. Smale and Giuliano were part of a very small United States delegation.

The beginning of the *New York Times* description of the Festival parade is interesting for both its substance and tone: "More than a million people tramped through East Berlin today shrieking defiance of the United States and devotion to communism. It was a frightening demonstration."[17] Other aspects of the festival included calls for world peace and tributes to Stalin. It was a long way from Ann Arbor and made a tremendous impression on Smale and Giuliano. The close scrutiny of Eastern European society, however, did raise doubts in each man, that, in the leftist culture of the day, neither confided in the other.[18]

The Berlin Festival inspired Steve to found a campus group, the Society For Peaceful Alternatives. Such activity was consistent with the CP policy of encouraging its members to assume leadership roles in complementary open clubs. Smale also managed to combine the peace agenda with his predilection for pranks. He organized a clandestine, guerilla operation aimed at the ROTC office. At 3 am, as lookouts were deployed along each approach, the word *Peace* was painted onto a cannon. When the mission was complete, an anonymous phone call informed the campus newspaper. However, no publicity was forthcoming. After participants in the early morning regimen of the ROTC discovered the prank, the cannon was quickly restored and the story suppressed.

In his first semester as a graduate student, Steve was involved in a contentious civil liberties case at Michigan.[19] He was an active member of the Young Progressives which attempted to bring two controversial speakers to campus, Arthur McPhaul and Abner Green. McPhaul and Green held positions in the Civil Rights Congress and the American Committee for Protection of the Foreign Born. Both organizations had been listed as communist fronts by the attorney general. A university regulation required that advance approval of speakers be obtained from a committee consisting of five professors. In the past only three petitions had been declined, in each case the proposed speaker was an avowed communist. Neither McPhaul nor Green had declared their connection to the CP, but their activities and organizations were under a great deal of suspicion and scrutiny.

The University Lecture Committee considered the request in March 1952. Green had just completed a six month jail term for contempt of court. He, as well as author Dashiell Hammett and two others, had refused to cooperate with an investigation of the Civil Rights Congress Bail Fund.[20] During the previous week McPhaul had appeared in Detroit at a HUAC subcommittee investigation of alleged communist involvement. McPhaul took the Fifth Amendment, asserting his constitutional protection against self-incrimination. This was a common response which the HUAC and others typically interpreted as a confession of guilt.

Rejecting the petitions for speeches by Green and McPhaul, the University Lecture Committee stated that they required evidence that the talks would not be subversive, placing the burden on the applicants. The decision did not close the issue. After an unsuccessful attempt to obtain an off-campus venue for his speech, Green dined at a co-op and spent the night at the campus center building known as the Union. On March 6, the *Michigan Daily* contained an article reporting that McPhaul would be on campus for the day, visiting residences and attending a private dinner.

Steve was among the 30 dinner guests who heard McPhaul speak at the Union. Did the evening with the banned speaker involve violations of university regulations? It was common for private groups to have dinners and speakers at the Union, without prior approval of the University Lecture Committee. On the following day, the Dean of Students authorized an investigation by the Student Affairs Committee. When they found that a member of this committee was also among the dinner guests, University President Hatcher appointed a special student-faculty group to conduct the inquiry.

The obvious scapegoat was the organizer, but investigation revealed that the dining room had been reserved under the pseudonym of Henry Gerard. Of the 18 students who attended the dinner, 14 were identified. These included Steve and a reporter for the *Michigan Daily.* The participants, as well as other students, were examined by the Investigation Committee, and a stenographer recorded their testimony. Steve was among the least cooperative.[21]

> ***Prof. Blume:*** *(usual opening remarks)*
>
> *Now, Mr. Smale, did you know about this dinner before it was given?*
>
> ***Mr. Smale:*** *I don't know that I want to answer that question under the present set-up here.*
>
> ***Question:*** *If you do know any of the facts concerning the dinner, we would like very much to have them. For instance, who arranged for the dinner—if it was Mr. Girard* [sic], *or somebody representing themself as Girard.*

Answer: *I don't agree with some of the things that have been done. The Lecture Committee picked the McPhaul case to be tough about. Most groups on campus can have speakers without even going through the Committee.*

Question: *I am not saying that we do or do not know who the sponsor of the McPhaul speech was, but if someone or some people rented a room at the Union under a false name in order to circumvent the ruling of the University Lecture Committee then there is a probable violation. We do not know whether there has been conduct unbecoming a student or a group, but in either event it is subject to university discipline, if a violation did occur. We are not trying to accuse anyone falsely or unjustly.*

Dean Walter: *Were you at the dinner, Mr. Smale?*

Answer: *On the same grounds I am not satisfied with this investigation, I don't think I can answer that question.*

Prof. Blume: *When you came in here, I tried to give you a clear understanding of the nature of these hearings. I hoped that you would see that it is strictly a problem of student discipline and nothing further. We expect you as students to cooperate with us. We would like to have you tell us whether or not you were at the dinner.*

Answer: *I cannot say whether or not I was.*

Prof. Blume: *Then if you are not willing to give us any information, we may as well drop it right there. This is all we can do.*

The Committee submitted its recommendation to President Hatcher finding that the 14 students had violated the following rule:

"No permission for the use of university property for meetings or lectures shall be granted by any student organization not recognized by university authorities, nor shall such permission be granted to any individual student."

After a month delay Hatcher referred the case to the student Joint Judiciary Committee for a determination of punishment. While the defendants noted that the rule bore on the organizer rather than the participants, the Joint Judiciary Committee began a new set of hearings. The two-week inquiry revisited the witnesses and continued the interrogation, receiving varying levels of cooperation. When Steve appeared, he declined to answer each of the following questions:[22]

1. *Did you attend the dinner?*
2. *Were you invited to the dinner?*
3. *Did you hear discussion in a co-op house about the fact that McPhaul was going to speak at a private dinner?*
4. *Who are your roommates?*

While Steve was convinced that attending the dinner was within his rights, he felt that the questions of the Joint Judiciary Committee were aimed at incrimination of himself and his friends. His refusal to cooperate put him in jeopardy of an uncertain discipline. There was even a possibility of expulsion which would not only end Smale's academic career, but also expose him to the draft. Each student was under a great deal of pressure, in some cases exacerbated by family and career considerations. Steve had the support of his parents, but, for the young risk taker, the principle was the overriding factor.

Finally on May 3, verdicts and punishment were announced. All students were cleared of attending the dinner two months earlier. However, Steve and four others were placed on probation for "failure to give the Judiciary the cooperation students should reasonably be expected to give a student disciplinary body." Appeals were unsuccessful and Smale was obliged to resign offices as secretary-treasurer of the Chess Club and treasurer of the Society For Peaceful Alternatives. At the appeal hearing, excerpted below, Steve disclosed that he did attend the dinner, separating the issue of self-incrimination from that of testifying about the participation of others.[23]

> ***Answer:*** *At the time of the hearing with Joint Judiciary, I took the stand that I would not incriminate myself to the extent of saying whether or not I attended the dinner. Now I will say that I was at the dinner.*
>
> ***Mr. Blume:*** *Did you honestly give to the Joint Judiciary Council all of the information that was called for by their questions except the naming of other students?*
>
> ***Answer:*** *No. There were four questions that I did not answer. For example, I was asked the names of my roommates. My refusal to answer that would be outside of what you have just limited it to. I was asked whether or not I was invited to the dinner. Now I will say that I was invited, and that I attended. I did refuse to give the names of my roommates, however.*
>
> ***Question:*** *What is your general attitude in the way of cooperation as far as Joint Judiciary is concerned?*
>
> ***Answer:*** *When it involved incriminating others or even myself, I didn't go along with that.*
>
> ***Question:*** *That is the heart of the thing right there, and that is what we are interested in here today.*
>
> ***Mr. Robertson:*** *If you had been informed that you were expected to give full, frank testimony concerning your own*

> *participation, do you think your conduct before the Council would have been any different than it was?*
>
> ***Answer:*** *If it had been explicitly said to me that it was necessary for me to answer questions regarding myself and not those involving others, then perhaps I would have gone along with the questions. I can't say for sure. I felt differently then about it than I do now.*

While the two months of university investigations seemed a long ordeal for the students, it paled in comparison to McPhaul's struggle with the HUAC. The Committee contended that the Fifth Amendment did not shield McPhaul from producing certain documents that they were seeking. After Congress approved a contempt citation, McPhaul was convicted and received a nine month prison sentence. In 1960 the case finally reached the Supreme Court which affirmed the verdict in a 5-4 decision with Douglas, Black, Brennan, and Warren dissenting.[24]

Having spent his previous summers in New York and Europe, Steve made plans to head west in 1952, recruiting Art Rose to accompany him. The two had met in Bob Thrall's class, where Rose was overwhelmed by Steve's facility with the material. They became friends and, eventually, shared a room. In their relationship Smale exerted a profound influence on Rose, introducing him both to leftist politics and to chamber music.[25]

Rose was enrolled in a program that combined undergraduate and medical school training. He had traveled little, and San Francisco sounded appealing. Rose also knew that the medical program would dominate his life in subsequent summers. To reduce expenses, they obtained a driveaway car to Los Angeles and then hitchhiked to San Francisco. There Steve obtained jobs in a can factory and as a hotel elevator operator while Rose worked in a restaurant and then at the Stanford Medical School. For most of the summer they lived in North Beach. Especially memorable for Rose was a trip to the Grand Canyon where they walked to the bottom, despite their inadequate preparation for the trek.

Returning to Ann Arbor in the fall, Smale began his second semester of graduate work. Graduate students are expected to earn *A*s and *B*s in all their course work. After three semesters, Smale had dropped two courses and received a *C* in another. This erratic beginning raised questions about his ability to complete the program. In June 1953 Steve received an ultimatum from the department chair, T. H. Hildebrandt. To remain a graduate student, it was necessary for Smale to improve his grades.[26]

That summer Steve worked as a welder at an automobile plant in Flint and made a trip to Grand Teton National Park. He also prepared for the German language exam. One of the requirements for a mathematics Ph.D. was to demonstrate a reading knowledge in two foreign languages. He had already completed the French exam. To practice for the second test, Steve read a German topology book. In addition to acquiring new language skills, Smale solidified his background in an area where he had only made a token effort in the previous year's course work.

In following Steve's career it is useful to have a basic understanding of the organization of mathematics and its study. The subject is divided into various areas. While there is no universally accepted breakdown of the fields, one reasonable division is into the areas of algebra, analysis, and topology. Undergraduate mathematics programs provide glimpses into these subjects, but students are rarely in a position to focus their interests until sometime during their graduate career. In their early training, mathematics Ph.D. students typically take courses in each of these areas. Such broad exposure permits informed specialization, and prepares the student for the interdisciplinary overlaps that they are likely to encounter. At the graduate level the material tends to be quite abstract, involving the development of structures and proofs of theorems. It is a significant departure from the computational mathematics taken in high school and early in college.

For example consider the subject of algebra. In high school one learns how to solve equations such as $7x = 5$. In slow motion:

$$7x = 5$$

$$\frac{1}{7}(7x) = \frac{1}{7}5$$

$$\left(\frac{1}{7}7\right)x = \frac{5}{7}$$

$$1x = \frac{5}{7}$$

$$x = \frac{5}{7}$$

Notice that our goal is to isolate x by canceling the 7. Solving the equation required bringing in the number $\frac{1}{7}$. This has the feature that it is an *inverse* of 7 in the sense that their product is 1. The number 1 is called an *identity* because it yields the result a whenever it is multiplied by another number a. Finally the *associative property* of multiplication permitted us to shift the parentheses from the second line to the third line. Thus to solve the equation we employed the properties of identity, inverse, and associativity on the set of numbers under the operation of multiplication. In (abstract) algebra one departs from a specific set of numbers and the traditional operations such as multiplication or addition. In their place is an abstract set and operation(s) that are endowed with properties such as identities, inverses, and associativity. The set could be some type of number or it could be other objects such as matrices or functions. Rather than working with a specific set and operation, one relies on the properties to obtain further results.

The field of analysis is a generalization of calculus. Two central aspects of calculus involve employing limits to compute the area under a curve and the slope of the tangent line to a curve. In analysis, as in algebra, numbers are replaced with abstract objects and structures that share the essence of numbers. The difference in the fields occurs in the nature of the structures. While algebra is directed toward the solution of equations, analysis infrastructure is motivated by the computation of limits.

Topology is an area that is not normally studied until advanced undergraduate or early graduate work. Here, roughly, two objects are equivalent, provided one can be stretched, shrunk, or transformed into the other without tearing or attaching. For example a circle can be changed into a square by gradual stretching in certain spots. On the other hand a circle cannot be deformed into a figure eight unless one does some attaching. The topological notion of equivalence is called *homeomorphism*,* and it can be defined rigorously from other structures in topology. Once the definitions are established, it can be proved that a circle is homeomorphic to a square, but not to a figure eight. These examples (the circle, square, and figure eight) are known as *spaces*. A fundamental goal of topology is to classify all spaces up to homeomorphism. This means to decide which spaces are equivalent and which are not.

C. Mathematics, marriage, and the HUAC

When Steve returned to Ann Arbor in 1953, he was a marginal, underachieving graduate student. He had been coasting. Ahead of him lay more course work and then independent research on a thesis. Could he respond constructively to Hildebrandt's warning and focus his

* Homeomorphism is one of the few technical terms that will recur in this book.

attention on mathematics? At that point Raoul Bott fortuitously entered the picture.

Bott was a young mathematician, beginning his third year on the Michigan faculty. Raised in Slovakia, he had immigrated to Canada at the start of World War II. After receiving a Ph.D. in applied mathematics at Carnegie Institute of Technology, Bott obtained a postdoctoral position at the Institute for Advanced Study in Princeton. There he became interested in topology. Hildebrandt was an autocratic chair who oversaw a strong group in topology at Michigan. When he offered Bott a low level Instructorship, Raoul turned down tenured offers from Sarah Lawrence and Kenyon where the research environment was less stimulating.[27] He remained at Michigan for nearly a decade, becoming an eminent mathematician and then moving to Harvard. A tall, jovial man with considerable charm, he was described by a former colleague as someone who "could look at you and see what is in your heart."[28]

In 1953 Bott was given his first opportunity to teach an advanced graduate course. The topic was algebraic topology. Note the word *algebraic* is an adjective. While mathematics is divided into algebra, topology, and analysis there are overlaps, and it is often productive to apply the techniques of one area to another. To prove that two spaces are not homeomorphic, one can associate algebraic structures to each space and show that these are not equivalent. Algebraic topology itself has different subareas, based on the nature of the associated algebraic structure.

Bott planned to develop the material from a recent seminal paper by the French mathematician Jean-Pierre Serre.[29] To reach that frontier, he first needed to provide background in singular homology, a branch of algebraic topology. The class attracted three enrolled graduate students along with a large number of auditing faculty. Remarkably, two of the students, Steve and James Munkres, became notable topologists. Bott likes to refer to the third student, James Berry, as "the really smart one."

This is a bit of a joke since Berry did not complete the degree, but Bott felt that he was talented in other directions.[30]

Munkres grew up in a small town in western Nebraska. After graduating in a high school class of 10 students, he attended a liberal arts college, Nebraska Wesleyan. While Ann Arbor provided the transition from small town life to big time topology for both Smale and Munkres, they were otherwise dissimilar. Munkres voted for Eisenhower in the 1952 election. Insecure about his background and lacking in confidence, Munkres worked tirelessly to mitigate his perceived inferiority.[31] He became viewed by the Michigan mathematics faculty as their best student in topology.

Bott had a daunting pedagogical task: construct lectures at a level and pace that maintained interest among his colleagues, but did not discourage the heterogeneous trio of students. Furthermore, he needed to provide an environment in which the students could ask questions without being intimidated by the faculty. Bott realized that this was unrealistic within the classroom, and organized weekly luncheon meetings with the three. Munkres and Smale shared the predicament of being competent topology students in a course directed at a more mathematically sophisticated audience. Such experiences often draw students to pool their resources and work together. This did not occur. The two students worked independently and remained cordial, but never close.

Steve responded to the challenge of Bott's course, engaging himself in algebraic topology. With Steve's quiet nature, the change was not apparent. As Bott recalled, "He sat in the back and it wasn't clear whether he was paying attention, but he always looked benign."[32] Munkres' impact was more tangible. He took copious notes, carefully polishing and rewriting them outside of class. The final product was mimeographed and distributed to everyone.

Although Smale was overshadowed by Munkres, for the first time since his freshman year, he was working diligently. This was the initial point of a spectacular path toward mathematical achievement. One can only speculate on how to weight the factors contributing to the turnaround. Steve believes that he had finally reached a point in his life when he had to think seriously about the future. He also specifically cites Bott's teaching and Hildebrandt's admonition, but discounts the significance of any particular magic in the subject of algebraic topology.[33] As he became more immersed in mathematics, the leftist activity declined. Two events soon illustrated that political involvement could impinge on the teaching career of a mathematician. They would not be the last.

In May 1954 the HUAC tour returned to Michigan, ostensibly to investigate communist activities in the state. Steve attended the hearings on May 10 in Lansing, as three Michigan faculty and two graduate economics students testified under subpoena.[34] The graduate students were Mike Sharpe and Ed Shaffer who, along with Steve, had been disciplined in the McPhaul incident. Among the faculty was mathematician Chandler Davis.

Davis, like Steve, was a "red diaper baby." During the previous year Davis' father, Horace, invoked the Fifth Amendment at a Senate Internal Security Subcommittee hearing. The elder Davis was then fired from his position as a tenured economics professor at the University of Kansas City. Chandler was just four years older than Steve. After receiving his Ph.D. in mathematics from Harvard, Davis accepted a tenure track offer from Michigan. Having the degree and faculty position, Chandler was situated on the academic ladder precisely where Steve saw himself a few years in the future. Davis and Smale shared a significant feature in their backgrounds that would have been of great interest to the HUAC. Both had recently belonged to the Communist Party in Ann Arbor, albeit in different cells. Neither was aware of the other's membership until decades later.

While the other four witnesses were accompanied by attorneys and pleaded the Fifth Amendment, Davis appeared alone and followed his own principled course. He answered the initial questions about his academic resumé, but when the focus quickly shifted to Chandler's political beliefs and his associations, he invoked the First Amendment in refusing to respond. Davis claimed that he was protected by the guarantees of freedom of speech and assembly and declined to exercise his Fifth Amendment rights. While opinions varied as to whether his testimony was courageous, naive, or illegal, the transcript of the hearing demonstrates remarkable poise under considerable pressure.

It appears that the HUAC was largely unaware of Chandler's communist background. Davis was questioned extensively about his role in the production of an anti-HUAC pamphlet and his knowledge of communist activity at Harvard. Another topic was the confiscation of his passport by the State Department. Throughout the proceeding, Davis was careful in distinguishing the questions that he deemed proper. His determination not to reveal the names of any associates conflicted with the HUAC approach of obtaining names. After several forays in this direction, the Committee chair Kit Clardy baited Davis by asking whether he was acquainted with President Hatcher. Chandler cited the First Amendment and Clardy reprimanded the audience for its subsequent demonstration.[35]

Following the hearing, all three Michigan faculty members (Davis, Mark Nickerson, Clement Markert) were suspended with pay. Although the mathematics department was supportive, Chandler's fate was to be decided over the summer by a chain of two university senate committees, President Hatcher, and the regents.[36] While Nickerson and Markert had invoked the Fifth Amendment with the HUAC, they cooperated with the university committees. Each admitted that they were former CP members. After Davis continued the unresponsive posture that he had adopted with the HUAC, the committees recommended his dismissal and the reinstatement of his colleagues. Hatcher concurred with the decisions on Davis and Markert, but concluded that Nickerson had not sufficiently recanted his communist views. In August the regents fired both Davis and Nickerson.

On a parallel front, Davis faced the legal consequences of his testimony. He was cited for contempt by the House of Representatives and then indicted on 26 counts by a federal grand jury. After a trial, Davis was found guilty and sentenced to six months in prison and a $250 fine. In 1959 the Supreme Court declined to review the case, having just settled the similar Barenblatt case. Davis went to Danbury Federal Penitentiary the following year.[37] Throughout the six years between the hearing and prison, he was blacklisted and only managed to find various temporary positions. In 1962 he emigrated to Canada and accepted a position at the University of Toronto that he held until retirement. Nickerson had earlier obtained a position at the University of Manitoba.

Lloyd Barenblatt was a former Michigan graduate student and Vassar instructor of psychology. His HUAC contempt conviction was reviewed twice each by the Court of Appeals and the Supreme Court. When it was finally resolved in 1959, the 5-4 alignment of the justices was identical to that of McPhaul the following year. The opinions were authored by John Harlan and Hugo Black, close friends with different perspectives on the primacy of the First Amendment. Writing for the majority, Harlan stated his view of the constitutional considerations: "Where First Amendment rights are asserted to bar governmental interrogation, resolution of the issue always involves a balancing by the courts of the competing private and public interests at stake in the particular circumstances shown." In the final analysis he found a compelling national security threat from the CP: "this court in its constitutional adjudications has consistently refused to view the Communist Party as an ordinary political party, and has upheld federal legislation aimed at the communist problem which in a different context would have raised constitutional issues of the gravest character." In his dissent Hugo Black argued "I do not agree laws abridging First

Amendment freedoms can be justified by a congressional or judicial balancing process."

If Steve were unable to identify with Davis' plight, he could not have escaped comparing himself to the two subpoenaed students. After all, each of the three was a McPhaul probationer and a graduate student who had "outed" himself as a left wing advocate. Moreover, Sharpe was the leader of Steve's CP cell. At the McPhaul dinner, he played the role of the infamous Henry Gerard. The dinner was actually the idea of Shaffer, who had a long history of communist activity. While Shaffer and Sharpe were the most prominent Marxist student leaders on campus, Steve was among the next few. If the HUAC had selected two or three more students for the hearing, it is likely that Steve would have been called.

Both Shaffer and Sharpe invoked the Fifth Amendment as the Committee expressed interest in their LYL involvement and other activities. If either had elected to cooperate with the HUAC, Steve's name would undoubtedly have arisen. Considering the examination of Shaffer and Sharpe, it is reasonably clear how the Committee would have dealt with Smale as a witness. They would have focused on several aspects of his leftist activity:

1. Had he ever belonged to the Communist Party?
2. The 1952 East Berlin youth festival.
3. LYL involvement.
4. His associates.

Steve was fortunate that his role in the HUAC hearings was limited to that of an observer. Under slightly different circumstances he could have been subpoenaed or fingered. The impact on Shaffer and Sharpe was more tangible as the University considered academic sanctions. The chair of the economics department, Gardner Ackley, responded on behalf of his colleagues by suggesting that blacklisting was more appropriate than expulsion.[38]

> *We freely confess that individually we could not, and will not recommend either of these men for teaching or research positions. It has been urged that if we cannot recommend them, we should not grant the degree. But there is a great difference between a private letter of recommendation, giving personal evaluation, and the formal, legal act of recording course grades and granting degrees.*

Sharpe became discouraged at his prospects, withdrew from the University, and eventually started a publishing company.[39] Shaffer was among the Michigan contingent who found academic homes in Canada. If Smale had received the public scrutiny of the Committee, one can only speculate on the subsequent impact it might have had on Canadian mathematics.

In the fall of 1954, Steve moved into Ed Shaffer's apartment. He quickly learned that his own leftist activities had not escaped the notice of the authorities. Following the previous year of serious mathematics study, Smale had been awarded a teaching assistantship. The Michigan State Police learned of the employment when they conducted a formal "investigation of Smale as a suspected communist."[40] A meeting between the police and a dean produced a commitment that the University would take action against Smale's employment. After five class meetings, Hildebrandt terminated Steve's teaching assistantship, citing administrative pressure.[41] In a good news–bad news session, the chair shifted Smale's support to a research contract. Steve had benefited from the advocacy of topologist Ray Wilder who had intervened on his behalf with Hildebrandt.[42] It is interesting that the conservative Hildebrandt made the accommodation and the previously doctrinaire Smale accepted the settlement. Mathematics and pragmatism took precedence with both individuals.

Steve with friend on train to Campeche during summer 1954 trip to Central America.

Steve's mathematical awakening, although profound, did not completely dominate his life. For the summer of 1954, he planned an excursion to Argentina. First Steve hitchhiked to Mexico City where he was joined by a male and a female friend from Ann Arbor. A civil war in Guatemala obstructed their overland effort to proceed south. At the waterfront in Veracruz they sought another solution, asking boat owners: "*A donde va este de barco*?" (Where is this boat going?) When this failed to produce a way around Guatemala, the three attempted an end run along the Yucatan. There Steve temporarily left his discouraged friends, making his way as far as British Honduras before rejoining them in Merida. After an extended stay on the Yucatan, they obtained passage to Louisiana on a shrimp boat. Arriving in New Orleans, Steve discovered that the National Open Chess Tournament was scheduled to begin the following day. He finished fiftieth in a field exceeding 100.[43]

Up until this point, all of Steve's close friends were male: Fred Burr, Don Brown, Vince Giuliano, Art Rose. When Judy arrived in Ann Arbor, brother and sister were able to increase each other's social contacts. Through Judy, Steve became acquainted with a woman whom he described as a "quasi-girlfriend." Otherwise his experience with women was limited. The trip to Mexico provided the opportunity for some romantic interaction with his female companion. He also connected with an American student in the Yucatan, leading to a subsequent correspondence.

The summer experiences had increased Steve's confidence and comfort with women. Shortly after returning to Ann Arbor he met Clara Davis (unrelated to Chandler Davis), a friend of Ed Shaffer. Smale was attracted to the library science graduate student and acted on his feelings. Clara responded, and they were married a few months later.

Clara's parents, Harry and Marion Hill Davis, were both born in Ontario. Her father's ancestors were American Tories who emigrated to Canada during the Revolution. Harry and Marion moved to Alberta and began a wheat farm as homesteaders. In the midtwenties they lost the farm and moved their family to the Detroit suburb of Dearborn, where they had a relative. There Harry made a career change into the insurance business. On February 2, 1930, Clara Evelyn was born, the last of eight children.[44]

Allergies and illness interfered with Clara's early schooling, but she was otherwise a good student. When she was awarded a Regent's Scholarship, Clara enrolled at the University of Michigan in 1947, one year prior to Steve. At first she lived in Ann Arbor with an older sister. Clara joined a sorority and worked in the office of the *Michigan Daily*. Majoring in economics, she lacked any particular career aspiration. After graduating with a *B* average, Clara remained in Ann Arbor for a year, performing secretarial and clerical work.

She returned home to Dearborn when both of her roommates married. Harry Davis, a heavy smoker, was diagnosed with cancer and died the following year. Still without a vocation, Clara first worked for a lawyer. Later she obtained more satisfying employment in a library, prompting a return to the University to pursue a master's degree in library science. Back at Michigan in 1954, Clara became a housemother in a cooperative, a dormitory situation in which students worked part-time to defray a portion of their expenses. The cooperative scene at Michigan attracted the more radical students. Clara followed politics, but did not engage in activity. She was a liberal who voted for Eisenhower in 1952.

When Clara met Steve, she was aware of his reputation as a "Stalinist." He appealed to her as an "interesting guy who made me laugh." Recalling her initial concerns about Steve's politics, Clara "thought he was fun to be with, but I would never get serious." She was mistaken. The relationship proceeded quickly as did Steve's brief phone calls asking for dates.

Although Smale sometimes gives the impression of caprice, he is an individual who carefully considers important decisions and then moves decisively. Marriage was certainly an important question, and this was his first committed relationship. Steve wanted to marry soon and Clara agreed. On January 31, 1955, the ceremony was performed in an Indiana marriage mill with his parents in attendance.

During the courtship Steve prepared for his preliminary exams. Graduate mathematics programs normally require students to take one or two sets of examinations outside of their course work. The tests might be oral and/or written and constitute a rite of passage which indicates that the student is prepared to pursue the independent work that lies ahead. The exact nature and timing of these exams vary among departments. In December Steve successfully completed a battery of oral tests, covering different mathematical areas. This cleared him to proceed to the final Ph.D. stage, the thesis.

To obtain a Ph.D. in mathematics, a student must complete a substantial piece of original research. The work is conducted under the supervision of an advisor. The student writes up the results into a thesis which usually must be approved by a committee chaired by the advisor. There is no set length for the document, and typically they are in the 50–100 page range.

For nonmathematicians, the notion of mathematical research is often difficult to fathom. After all, what is unknown? One major target of twentieth century research was the famous problem known as Fermat's Last Theorem. To describe it, we will assume that *a, b,* and *c* are positive integers (whole numbers). The familiar Pythagorean relation $a^2 + b^2 = c^2$ is satisfied by $a = 3$, $b = 4$, and $c = 5$ as well as many other triples of positive integers. Now suppose the exponent is changed to an integer greater than 2. Do there exist any solutions to the equation? Fermat's Last Theorem asserts that there is never a solution.

Fermat's Last Theorem. *If n is an integer greater than 2, then there do not exist any positive integer solutions to the equation:*

$$a^n + b^n = c^n.$$

The problem has a 350-year history, during which professional and amateur mathematicians sought a proof or an example to disprove the statement (a counterexample). Despite a variety of attacks, Fermat's Last "Theorem" maintained its status as a conjecture. In 1993 Princeton Professor Andrew Wiles announced that he had proved the theorem. The story aroused unprecedented excitement among mathematicians and nonmathematicians, even appearing on the front page of the *New York Times.*[45] Wiles' complex argument was then submitted to the scrutiny of the experts in the field. A gap in one of the proofs was subsequently identified.[46] Several months later, Wiles succeeded in rehabilitating his argument. This time the proof withstood critical analysis. Fermat's Last Theorem, a conjecture for 350 years, is at last a theorem.[47]

While there are a multitude of unsolved problems in mathematics, the nature of the discipline often places them beyond the reach of the nonexpert. The subject tends to rely on abstruse terminology and a vertical sequence of prerequisites. For example, problem I might rely on concepts A, B, and C that themselves rely on other results, and so on. Mathematicians in one subspecialty are typically unable to read papers of those in another.

When a graduate student reaches the thesis stage, formulation of the problem becomes crucial. Important considerations are that it should be deep enough to qualify as a thesis, but not so difficult as to be intractable. Moreover, the solution must not have appeared in the literature, even in an obscure journal. Although a few graduate students obtain their own thesis problem, customarily it is provided by the advisor. Often the topic evolves somewhat as the research proceeds, and the final product may not fully address the original issues.

Clearly, the advisor has a significant role in the thesis stage. The protocol in obtaining a mentor involves the student approaching the faculty member, who may accept (usually) or decline (rarely). The choice of advisor is an important decision, involving a number of considerations for the student. Among the foremost factors are the subspecialty of the professor and the level of rapport that might be expected from the relationship.

On these grounds, Bott was the obvious choice for Smale, who was interested in topology. Michigan, however, had several topologists, and Bott was the most junior among them. Having an established senior scholar as an advisor has several advantages. Often they have a stock of good problems, but also important can be their influence in obtaining a job for their student. Ray Wilder was the most prominent topologist at Michigan, yet Smale selected Bott rather than Wilder.

Although Bott was not yet famous, he did provide Steve with an excellent problem: *Classify, up to regular homotopy, regular closed curves in an arbitrary manifold.* This concise statement is, of course, incomprehensible to a nonmathematician. Some of the terms are even unknown to mathematicians in other specialties. In Appendix A the general reader will be provided with an intuitive explanation of the jargon and problem.

One of the purposes of graduate course work is to provide the student with the background and tools to approach the frontier of research. In order to complete a thesis, a student must make the transition from *learning* to *doing* mathematics. For many, it is a difficult process. Previously the student has read developments of others and worked on problems with known results. Now he or she is faced with a situation in which the nature of the answer must be determined, as well as supported by the creation of a rigorous argument. The primary weapon is the brain. Frequently things proceed in the following order: deep thought, idea of attack, working through the idea, failure of the idea, starting again. Most ideas do not work, and the process can be incredibly frustrating. Sometimes the advisor offers guidance, ideas, or encouragement.

Bott had already given some thought to the problem. He had in mind an approach involving techniques from analysis. However, Steve moved in a different direction, creating his own tools. In completing the research, Smale demonstrated three important qualities that would be central to his future success. First was the desire to look at problems in his own way. Second was the "tenacity" of his attack which, at this stage, struck Bott "more than brilliance."[48] Finally, there was Steve's deft application of the fiber space concept he learned in Bott's course.

Although Bott was his thesis advisor, Smale had considerable contact with Ray Wilder. Solving problems posed by Wilder, Steve wrote two papers in addition to his thesis. As Smale approached the job market, his ability to do mathematics had been validated.

To apply for a job, candidates send out letters and resumés and obtain recommendations from professors familiar with their work. Steve was confident that he would receive a position in a strong department. He was disappointed and surprised by the outcome. Nothing materialized for quite a while. Finally, he received an instructorship at the University of Chicago. While Chicago had one of the finest mathematics departments in the country, Steve's offer originated from the *College*. The College was a separate part of the University specializing in general education. There the emphasis was on teaching, rather than research. After Steve accepted the position at Chicago, he received a temporary offer from Princeton.

Steve's application to Chicago had been submitted directly to the mathematics department. Saunders MacLane, then a member of the Chicago faculty, remembers receiving Bott's letter and considering the candidate.

After concluding that the case was not sufficiently strong for the mathematics department, MacLane forwarded the application to the College. It was unusual for him to take such an action, and it indicates that Smale looked reasonably strong, though not exceptional.[49]

Several factors contributed to Smale's lack of success on the job market. Bott's imprint carried less clout than that of more established scholars. While the substance of Smale's thesis was a strong piece of work, the actual theorems did not possess enough intrinsic interest to draw many readers past the headlines. As a result, Steve's research accomplishments were underrated. The level of enthusiasm in the letters of recommendation was further moderated by Steve's inauspicious earlier period at the University of Michigan.

Certainly Michigan mathematics chair Hildebrandt's letter was a frank second hand appraisal of Smale as a student. He described Steve as a bright young man who was performing well after a disappointing start. Although Hildebrandt expressed confidence that Smale had completely overcome his earlier difficulties, the chair did feel obliged to disclose the possible causes, specifically mentioning "liberal" student activity and a disciplinary incident. While the word "liberal" certainly understates the degree of Steve's left wing involvement, it was also a code word for the black list. Depending upon the reader, the characterization could have disqualified Steve from consideration. Smale was fortunate that Ray Wilder learned of the political allusion and suggested discontinuing Hildebrandt as a reference. The other matter mentioned in Hildebrandt's letter involved a fight that took place during a party at Smale's apartment and led to his being placed on probation. Transmitting information from the administration, Hildebrandt described a triangle in which Smale and another man fought over a woman. In Steve's account, he was a bystander who was disciplined because he hosted an unchaperoned party, violating university regulations.

The body of Wilder's letter to Princeton is reproduced below. Unlike Hildebrandt, Wilder had had substantive mathematical contact with Smale and was in a position to provide a more informed opinion. Still the Wilder assessment was guarded, particularly in comparison to some of his other evaluations. In a letter to the California Institute of Technology one month earlier, Wilder described one of his own thesis students as having "conclusively demonstrated unusual creative ability" and predicted a "brilliant future." Caltech offered the position to Wilder's student over Smale and other candidates.[50]

> *Stephen Smale, one of our graduate students, asked me to drop you a note regarding his mathematical promise.*
>
> *I believe he shows lots of promise. I did not think so, however, until this year. Maybe his getting married was the turning point; for he has been developing very rapidly this year. He has been participating in my seminar, and shows himself very quick to pick up suggestions for investigation, as well as*

quick to bring in results. Possibly you heard his paper at the recent meeting in New York; this was a result of the seminar, and embodied a homotopy analogue of the Vietoris mapping theorem. He formulated the conditions himself (as well as the proof, of course).

I haven't the slightest idea of Bott's opinion of him; as you probably know, he is writing his dissertation under Bott and Bott is at the Institute this year. I assume you will get his independent opinion of Smale. Perhaps Bott does not know Smale's wife; she is a very charming and capable young lady—a professional librarian. She is working in Dearborn—commutes every day—to help out with the family expenses.

It has not been possible to obtain the Smale reference letters authored by Bott and another professor, Hans Samelson. Since Smale was Bott's first student, it may be that Raoul's inexperience in producing recommendations was a factor. Furthermore, Bott was at the Institute for Advanced Study during Smale's last year of graduate school, limiting their contact during a period of growth. In light of Smale's subsequent accomplishments, one can read his thesis and identify a great deal of talent. However his referees in 1956 did not have the benefit of future vision, and they were fully aware of his uneven past. From this perspective Steve had demonstrated an ability to solve problems, and he definitely possessed promise as a mathematical scholar, but no one could have predicted the development that would occur during the next few years.

Early Mathematical Audacity

03

A. The conference in Mexico City

During Smale's eight years in Ann Arbor, he acquired an unusually cosmopolitan perspective of the world. Through his own initiative Steve saw much of Europe and Mexico. He made numerous excursions to New York and traveled throughout the western United States. Politically, Smale embraced communism and rejected his own country's actions in Korea.

On the mathematical side, Smale's experiences were considerably more provincial. The subject of topology was undergoing intense development by mathematicians in Europe and the United States. Although Michigan had a strong group in topology, staying abreast of the advances was a major challenge, especially with the long lag between mathematical discovery and publication. Interaction with scholars at other universities was vital. As a graduate student, however, Smale's mathematical travel was limited to one Princeton trip to visit his advisor at the Institute for Advanced Study. A further opportunity arose in the summer of 1956, following the completion of Smale's graduate career at Michigan and prior to the start of his position at Chicago. A major conference on algebraic topology was scheduled to take place in Mexico City. The lecturers were the people whose work Smale had seen in journals.

Clara joined Steve for the trip, his third to Mexico. The young couple managed an economical itinerary, first obtaining a driveaway to Dallas in a Cadillac, then proceeding by bus. It was Steve's first mathematical conference, but it made a lasting impression.[1] All the luminaries were there, and the rising stars as well. Among them was 25-year-old John Milnor who spoke about his new groundbreaking result on spheres. Milnor's work, later recognized with the 1962 Fields Medal, was opening a new branch of topology, known as differential topology. Smale was not yet sufficiently prominent to speak in Mexico City, but his impact on differential topology was on the horizon.

Attending a mathematics conference can be a stimulating experience. It is a rare opportunity to participate in a subculture of shared interest. Contrast this with the typical university situation. There the mathematician's work might be understood by a few colleagues, but, given the vertical nature of the discipline, his or her research is likely outside the expertise of the others. Beyond the department or university, the subject is regarded as incomprehensible or even arcane. At a conference, a mathematical subspecialty unites a tribe. There is often a plethora of lectures on the subject and there are opportunities for networking. Not only did Smale learn of recent developments and meet the leading practitioners, but the exchange of ideas moved in both directions. Steve discussed his thesis with René Thom who had already heard of the work from Bott. Thom, destined to receive the 1958 Fields Medal, was probably the first person to appreciate fully the significance of Smale's thesis. It was a timely connection as Thom was on his way to the University of Chicago as a year-long visiting professor. Other Chicago mathematicians were at Mexico City, permitting Steve an introduction to his future colleagues.

After the conference, the Smales joined the Botts and several other participants for an excursion to Acapulco. The group rented a limousine for the trip and booked cheap hotel rooms. The hot, windowless accommodations reminded Bott of a cave. After one night, everyone upgraded to a nicer hotel, except the Smales. Steve and Clara had little money, but Steve knew his way around Mexico and was skilled at traveling on a budget. Bott recalled:[2]

Steve (3rd from right) at a banquet during a 1964 conference at the Tata Institute in Bombay. René Thom is on the left, Raoul Bott in the center with Lars Gårding and Michael Atiyah on the right. (Courtesy of the Tata Institute of Fundamental Research. Reprinted by permission.)

> *He lived in a quite different style in Mexico than the rest of us ... They lived for a fraction of the price we did because they lived with the people. Steve could eat the stuff from the street ... Steve was always tough. He could take it without complaining.*

B. The University of Chicago

Following the Mexico City conference, the Smales moved to Chicago. The University of Chicago is located in Hyde Park, a section south of the downtown area. The Smales rented an apartment a few blocks further south. Clara liked the campus, which reminded her of Ann

Arbor, but she considered their neighborhood too grim.[3] Having sold their car, she was dependent on the trains to reach her job at the Chicago Public Library downtown. It was a difficult commute, especially since Clara was then pregnant with their first child.

Steve joined the faculty of a prestigious institution. To understand Steve's status, or lack of it, requires a bit of history. The University of Chicago was established in the late nineteenth century. They recruited superb mathematicians for their faculty and built an outstanding graduate program. With the second generation of faculty, in the 1930s and early 1940s, the department lost its luster. The decline of mathematical scholarship happened to coincide with the emergence of an innovative new liberal arts undergraduate program under President Robert M. Hutchins.[4]

The Hutchins College was built around a curriculum of interdisciplinary general education courses, sometimes called great books courses. Students read primary sources and were given substantial freedom during the academic year. The program permitted, and encouraged, talented students to enter the University prior to high school graduation. Progress toward a degree depended entirely on successful completion of comprehensive examinations, regardless of the student's course performance. For well prepared students, there was an option to skip specific courses by demonstrating competence on the corresponding exams.

To teach the College courses, Hutchins recruited a faculty dedicated to its mission of undergraduate liberal arts education. This faculty was separate from the traditional departments, such as mathematics, where the priorities were research and graduate education. Having placed his imprint on undergraduate education, Hutchins turned his attention to the disciplinary departments. Chicago's role in the Manhattan Project positioned the president to bolster his physics and technical departments by retaining Enrico Fermi and some of the other brilliant scientists.

To restore mathematics to its former glory, Hutchins pursued a different approach. In 1946 he installed Marshall Stone of Harvard as mathematics chair, despite the opposition of the department.[5] Stone was an astute judge of mathematical talent who followed a single-minded determination to bring the world's best scholars to Chicago. Overcoming colleagues and administrators, Stone made a stunning sequence of appointments. André Weil, Antoni Zygmund, Saunders MacLane, and S. S. Chern quickly elevated Chicago mathematics to a position of leadership. The standing of a graduate program generally follows the scholarship of its faculty. The Stone Age[6] revitalized the Ph.D. program and offered new opportunities to undergraduates beyond the Hutchins College courses.

When Hutchins departed in 1951, the College lost its champion and entered a transitional phase. Meanwhile the department of mathematics was at the top of its game. Smale arrived in 1956 with the rank of instructor in the College. Appointment to the College offered no portent for a mathematician. It was a place where dedicated teachers taught mathematics to bright students who held other interests. Scientifically inclined students skipped the College mathematics course which emphasized set theory rather than calculus or great books. As an instructor, Smale was at the bottom of the faculty caste system, in a department with negligible research status. The junior faculty in the department of mathematics were the young Ph.D.s who had demonstrated exceptional scholarly promise.

The Hutchins College faculty environment was unlikely to incubate a great mathematician. Nevertheless, a position in the College provided an entrée into the first class Chicago mathematics department. Smale took full advantage of this opportunity, attending seminars and interacting with topologists and analysts. He was fortunate that his first year at Chicago coincided with the visit of René Thom from France. By attending Thom's lectures, Steve added an important concept called transversality* and other cutting edge mathematical tools to the collection begun by Bott in graduate school. That Smale was able to glean this material was especially impressive since Thom was a profoundly original thinker but a barely comprehensible expositor.

At Mexico City, Thom had taken an interest in Smale's work. They continued the dialogue in Chicago where Thom increased Steve's status by visiting him at his College office. Other acquaintanceships from Mexico City expanded in Chicago. Smale interacted with algebraic topologist Edwin Spanier and his graduate students Morris Hirsch and Elon Lima, all of whom had roles in enabling Smale's mathematical future. Spanier preceded Steve as a topology graduate student at Michigan, receiving his Ph.D. in 1947, and also as a faculty member at Chicago, where he was rapidly ascending the academic ladder.

Steve was now embarked on the faculty stage of the ladder. One of the most critical junctures for a mathematician is the first year as a faculty member. Completion of the doctoral thesis establishes the ability to do research, but conditions are somewhat controlled in graduate school. An advisor often selects the problem and provides some degree of help and support. With the transition to faculty status the mathematician is expected to formulate his or her own problems and proceed with little support. Moreover, there are increased teaching duties and new service responsibilities. In the fifties, the publish or perish attitudes of universities were much less pervasive than they are today. For most mathematicians, the thesis was the climax, rather than the beginning, of the research enterprise.

Several years earlier Steve had faced a similar test of career commitment. Following the admonition from Hildebrandt, he shifted his energy from Marxism to mathematics. In the fall of 1956, there were

* This is one of many, though not all, mathematical terms whose name conveys little meaning to the nonmathematical reader.

new questions concerning his ability and drive to begin an independent research program. Certainly a return to politics was unlikely as Steve had been inactive for years. The Hungarian revolution occurred in October, followed by the brutal Soviet crackdown. The Russian tanks delivered the *coup de grace* to Steve's courtship of communism. Still there were avenues other than research for the energy of a young mathematician beginning a career and family.

Steve quickly dispelled any doubts about his research potential. In a matter of months he announced a startling discovery. Smale showed that, in a mathematical sense, the sphere could be turned inside out. To mathematicians the result, later known as everting the sphere, was profound, counterintuitive, and fascinating. For the first time in his life, Steve's work was regarded as world class.

What does it mean to turn a sphere inside out, and why is it remarkable? More details are available in Appendix B, but here is a rough idea. Consider a function* whose domain is a hollow unit sphere and with range in 3-space.

One example is the function which maps each point of the sphere to itself and another example sends a point of the sphere to its reflection across the plane of the equator. To turn a sphere inside out requires an intermediate collection of functions which *nicely* deform the first example into the second. In the intermediate functions, different points may be mapped to the same target. Thus portions of the sphere are allowed to pass through itself. The key is the word "nicely." There is a straightforward method of performing the deformation, but it lacks the niceness property. The niceness involves aspects of smoothness rather than complexity. Thus it was plausible that there existed a complicated, but nice, method of completing the deformation. However, most

* Recall that a function is simply a rule that takes points of one set, called the domain, and assigns to them points from another set, called the range.

mathematicians considered this unlikely, particularly in view of its proven impossibility for the analogous problem with the circle.

Using sophisticated mathematical techniques, Smale proved the existence of a nice deformation of the sphere. The proof, however, was so abstract that it yielded little insight into a comprehensible prescription for constructing the actual intermediate functions. Spanier was "tremendously impressed."[7] The result was so surprising that Bott initially was convinced that there was an error. When he finally understood that Smale was correct, Bott was also impressed. "That's when I thought he was quite remarkable. I had never dreamt that you could go further and do immersion theory on that scale."[8]

Immersion is a technical term for the type of function occurring in Smale's thesis and his subsequent work on spheres. For the thesis problem, Bott had suggested that Smale generalize earlier work of Hassler Whitney. When Smale reached Chicago, he sought to push the ideas further, changing the function domain from a circle to a sphere. In many respects a sphere is mathematically more complicated than a circle, and Bott was dubious that the apparatus could be extended to that degree. Not only was Steve successful in adapting the techniques, but he established a totally unexpected solution.

One of the intellectual obstructions to mathematical research is the conjectural nature of the work. If you believe that the sphere cannot be turned inside out, then your efforts are likely to involve different approaches than if you believe that an eversion exists. To reach one of these beliefs for a geometric problem, mathematicians typically seek to devise a rough, intuitive version of the deformation. Upon finding a good candidate, the task is to prove that it meets the mathematical criteria. With the sphere, neither Smale nor anyone else had such a candidate. Since Whitney had proved its nonexistence for the circle, it was reasonable to believe that no such deformation of the sphere existed. Conventional mathematical wisdom was to pursue the problem in that direction. Smale had the audacity to invest his concentration in a direction in which others saw little likelihood for a payoff. It is impressive that he possessed the intellectual power to abstractly establish the existence of a geometric process without explicitly seeing the geometry. Once Smale showed that an eversion did indeed exist, others sought to find it. Arnold Shapiro succeeded in obtaining a complicated realization. Shapiro's concept was described in a *Scientific American* article,[9] along with this historical note about its dissemination.

> *In fact, Smale's paper contains no pictures at all. The intricacy of the pictures, which were in a sense implicit in Smale's abstract and analytical mathematics, is amazing. Perhaps even more amazing is the ability of mathematicians to convey these ideas to one another without relying on pictures. This ability is strikingly brought out by the history of Shapiro's description of how to turn a sphere inside out. I learned of its construction from the French topologist René Thom, who learned of it from his colleague Bernard Morin, who learned of it from Arnold Shapiro himself. Bernard Morin is blind.*

During his first year at Chicago, Smale emerged as both a mathematician and a mathematical mentor. For an instructor in the College, the latter was at least as unlikely as the former. The student was Morris (Moe) Hirsch whose *de jure* advisor was Edwin Spanier. Smale and Hirsch had met at the topology conference in Mexico City.

Hirsch was a few years younger than Steve. Like Smale, he was raised in a left wing environment, but the particulars were different. The Hirsches were a New York Jewish family, active in the Zionist movement. After just two years of undergraduate work at Saint Lawrence University, Hirsch entered the Chicago graduate program in 1952. Hirsch's early graduate school career was from the Smale script, an inauspicious beginning followed by an inspiring topology course. Rather than politics, the distraction for Hirsch was music. He quickly assimilated into the Chicago popular music scene, playing in jazz groups. It was Spanier's topology class that shifted Hirsch's attention toward mathematics.[10] The next step was to advance from course work to research. Hirsch traveled to Mexico City for the conference, but was unprepared for the level of abstraction and sophistication in the talks. When he learned of Smale's thesis, the topology seemed more accessible. "Back at Chicago I would go to his office every day to explain my much simpler proof of his result; and every day he would explain why I didn't have a proof. Eventually he taught me to understand his work, which dealt with the then unnamed field of differential topology."[11]

Smale's success with immersions of spheres inspired the Hirsch thesis problem. The idea was to push a further generalization where the domain for the immersion was more arbitrary than a sphere. Formally, Spanier was the thesis advisor, as Smale had no standing for that designation. In practice both Spanier and Smale played roles in guiding Hirsch, but Smale was the person that Hirsch "bugged."[12]

Another milestone occurred in Steve's personal life. Nathan (Nat) Smale was born on June 26, 1957. Steve was less of an iconoclast than his own father, but he did inherit some elements. There was no middle name for Nat nor was his birth an opportunity to perpetuate the name of a family member. Steve and Clara were regular listeners of a jazz program by Nat Hentoff, who may have inspired the naming of their son.[13]

With Nat's birth, Clara ended her library career and assumed the traditional role of a supportive faculty wife. Coping with the new circumstances was a difficult adjustment for her. While Steve went off each day to the excitement of the University, she was "stuck with the baby."[14] Just getting out of their third floor apartment was problematic. They still did not have a car and she was not particularly keen on the neighborhood. The baby carriage was stored in the basement, requiring a sequence of trips, first down for the carriage, then up for the baby, and back down for the excursion.

Smale had arrived at Chicago in 1956 with little notice in the mathematical world. For the first 26 years of his life, he was regarded as a bright guy, but far short of extraordinary. By the end of his first postdoctoral year, Smale's stock had appreciated considerably. If mathematics had a Rookie of the Year award, Smale would have been a contender. As Steve began his second year at Chicago, the several papers that he had written on immersions gave him a strong enough resumé to look beyond the Hutchins College for a more suitable position. For an ambitious young topologist it was natural to consider the city of Princeton. Located in the middle of New Jersey it was a world center for mathematics, especially topology. Princeton University had an outstanding mathematics department that included Milnor and other topologists. On the outskirts of the town was the Institute for Advanced Study, home for several of the world's greatest mathematicians. The Institute, as it is called by mathematicians, featured a small permanent faculty and a larger number of yearly visitors, known as members. Admission to this exclusive group was a mark of status among mathematicians. Bott was a member two years earlier while Smale was completing his thesis.

Funding for a residency at the Institute required support from the Institute endowment or from an external grant. The National Science Foundation had recently begun awarding postdoctoral fellowships to promising young mathematicians. Successful applicants received financial support for two years, with no teaching duties. The fellowships were designed for recent Ph.D.s to visit institutions with expertise in their subspecialty. There they gained insight from the senior scholars and assimilated into a culture that valued research in their field. The release from teaching duties permitted full time concentration on scholarship. In the fall of 1957 Steve submitted applications for a National Science Foundation Postdoctoral Fellowship and a membership at the Institute for Advanced Study. Spanier wrote the following text in support of Steve's application to the Institute.[15]

> *This letter is in support of the application of S. Smale for membership in the Institute for next year. Smale has done some excellent work on immersions of spheres in Euclidean spaces and on the self-intersections of an immersed manifold. There is no question in my mind about his ability as a mathematician and I confidently expect his work in the future to be equally important.*
>
> *Smale is a quiet person and a hard worker. He is an active participant in our seminars and is always interested in talking mathematics. We are glad to have him around here and I believe he would fit in well at the Institute also. I recommend him highly for membership.*

Given the inflation that periods of tight job markets have triggered in recommendations over the past forty years, it is difficult today to measure the enthusiasm behind Spanier's November 26, 1957, letter. More significant was his successful appeal, on Steve's behalf, for a position in the Chicago Mathematics Department. Chicago had achieved its standing by identifying and recruiting mathematical talent. Although Steve had not applied for a position, they were now approaching him. By the time Smale received the offer, however, he had already been accepted for an Institute membership. With his National Science Foundation award, Smale could spend two years at the Institute. He declined the Chicago offer.

C. The Institute for Advanced Study

The Institute for Advanced Study is the ultimate ivory tower. It evolved from the munificent desires of the Bamberger family, following the 1929 sale of their New Jersey department store.[16] Originally the Bambergers intended to endow a medical school in the Newark vicinity. They sought advice from medical education authority Abraham Flexner, who took the opportunity to push his own expensive vision for higher education. He conceived a nonmedical institution devoted to graduate study and postdoctoral research, sort of a first class university without undergraduates. The consultation succeeded in merging the Flexner plan with the Bamberger millions, and Flexner was authorized to implement his dream.

Among the most critical early decisions were the choice of venue and faculty. The Bambergers were devoted to New Jersey, but Newark was an unsuitable site for an ivory tower, and there was already a successful operation in the New Jersey town of Princeton. Flexner gambled that the town could sustain two independent ivory towers and saw possibilities for a rich symbiosis.

The first faculty appointments, Oswald Veblen and Albert Einstein, were magnificent choices. Einstein, the twentieth century icon for genius, provided instantaneous standing for the Institute. However, it was Veblen who profoundly shaped the scientific direction and personnel selection for the new center.[17] The Iowa born Veblen was a product of the University of Chicago's mathematics graduate program during its early glory days, later joining the Princeton University faculty. Despite his American roots, Veblen fully appreciated the European mathematics tradition. Under his influence, Princeton recruited John von Neumann and Hermann Weyl from Europe as well as the American educated Solomon Lefshetz. With Veblen and Einstein in the fold, Flexner's dream was taking shape.

When the Institute for Advanced Study opened in 1933, in its temporary home at Princeton University, graduate education was absent from the model. The mission was to provide a setting to facilitate research at the highest level, specializing in certain nonlaboratory disciplines. In the first year, there was just the School of Mathematics that included its four permanent faculty members and other shorter term visitors. The original faculty were Veblen, Einstein, John von Neumann, and James Alexander. Kurt Gödel was among the first year visitors and would eventually become a professor.[18] Physicists and mathematicians remained together in the School of Mathematics until 1966 when the physicists split off into a School for Natural Sciences. In 1935 the Institute founded two other divisions which are currently the Schools of Historical Studies and Social Sciences.

Eventually the Institute acquired its own large wooded surroundings, and has remained separate from the University. There are approximately 200 members in residence among the four schools, with about 25 holding permanent faculty positions. The typical visiting member is a university professor on leave from his or her own institution. Contrast life at the Institute with that on a university campus. At a university, faculty energy is divided between research, teaching, and service tasks such as committee and administrative work. Often there are conflicts between the competing interests. In the Institute environment, there are seminars and colloquia, but no formal courses or duties. A visiting scholar has the opportunity for total immersion in research. The Institute provides an office, secretarial support, nice places to walk, and options between solitude and collaborative stimulation. Flexner and Veblen constructed a utopia for thought. For some, the setting is too perfect. It can be maddening when the ideas are not flowing and there are no excuses.

Then there is the dynamic between the permanent faculty and visiting members. To become an Institute professor requires seminal scholarly accomplishment. When the system works well, these legendary figures continue to obtain profound results and interact with the temporary members. The discussions provide both new ideas and inspiration for the junior party. However, it doesn't always work. There are natural intimidating barriers between the two classes. How many young Ph.D.s would think that their own work is sufficiently important to initiate a conversation with Einstein and disturb his thought? Some permanent faculty have been particularly sensitive toward fostering the work of junior scholars, while others have not.

In 1946 J. Robert Oppenheimer became the third director of the Institute, bringing his unique credentials to the position. During World War II Oppenheimer was the Director of the Manhattan Project. Although he had administered one of the most extraordinary scientific developments in the history of the world, there was a certain irony in rewarding the *father of the atom bomb* with the master key to utopia. It was a rocky 20-year tenure.[19] The midfifties was a difficult period for the Institute. First there were the public allegations of Oppenheimer's disloyalty followed by the castration of his security clearance.[20] As Oppenheimer and the Institute coped with the fallout from communist smears, a generation was passing. Albert Einstein and John von Neumann, members of the original faculty and the Institute's two biggest stars, died in 1955 and 1957.

Despite these setbacks, the Institute remained a vibrant center and magnet for mathematicians. There were so many Chicago mathematicians beginning residency at the Institute in 1958 that it might have been cost efficient to charter a bus to Princeton. André Weil joined the permanent faculty and Edwin Spanier was a one year visitor. Smale, Hirsch, and Richard Palais received National Science Foundation Postdoctoral Fellowships and each selected the Institute as his base. Unlike many junior members who arrive in Princeton with few acquaintances, Smale already had his own network. Moe Hirsch was his office mate.

For any young member at the Institute, the truly exciting prospect for interaction lies with the permanent members. Marston Morse, who joined the Institute faculty in 1935, was the most promising candidate for Steve. Smale learned of his work, known as *Morse theory,* from Bott at Michigan. Over the next several years, Morse theory would be a profound ingredient in some of Smale's most famous theorems. Given the chronology of these developments, it is tempting to envision stimulating conversations between the old master and the rising star. Unfortunately, the Institute framework is not sufficiently magical to overcome barriers of ego. While Smale had enormous respect for Morse's work, which in 1977 he rated as "the single greatest contribution of American mathematics,"[21] their encounters were a disappointment to him. Morse had little interest in what others were doing. "He was always trying to push his own work."[22]

To realize substantive mathematical interaction between two scholars requires a delicate relationship that depends on both the personalities and mathematical interests of the parties. For Smale and Morse, it did not work. However, Smale's time at the Institute brought him into contact with a number of scholars. When Steve arrived in the summer of 1958, he briefly overlapped with Mauricio Peixoto, a Brazilian visitor from the previous year. The connection between Smale and Peixoto was enabled by their mutual friend Elon Lima, also from Brazil and a graduate student of Spanier's whom Steve knew from Chicago.[23] Peixoto was working on a notion called *structural stability.* He had obtained interesting results for the specialized mathematical setting of the disk, but was unable to identify a generalization to other domains. Smale quickly saw enormous potential for the development of structural stability. With Peixoto moving to nearby Baltimore for the following year, there was ample opportunity to continue their mutually beneficial discussions.[24]

What did Smale do with the structural stability concept? First Steve took the *transversality* tools that he learned in Thom's Chicago seminar and injected them into the mix. Being the first to apply the concept of transversality to the problem of structural stability required more than mathematical heavy lifting. Smale possessed the intellectual power to see that the transversality tool was appropriate for the new setting. However, Smale's vision went much further. He grasped the other central ingredients of a new theory and suggested how they might all fit together. Smale's vision was not entirely correct, but it was a valuable first approximation which set the agenda for future refinements. Everting the sphere was no fluke. He was that rare mathematician with the independent insight to open new frontiers. Over the next decade, Smale built a mathematical subspecialty from the seed planted by Peixoto.

Having resigned his position at Chicago to accept the two year National Science Foundation Fellowship, Steve was a mathematical free agent. Meanwhile his reputation and results spread among the elite mathematics departments. During his first year at the Institute, Steve was invited to deliver a lecture at Harvard. On the visit, he was offered a faculty position, to commence the following year. What a difference a big result makes! A few years earlier, Steve's applications were unsuccessful at the top mathematics departments. Now Harvard and Chicago were soliciting him. To accept the Harvard offer would have forced Steve to give up his second year at the Institute. Others might have seized the prestige and prospect for future security, but not Smale. He declined the Harvard position to remain at the Institute. Smale had always been confident of his ability to secure a job. At Michigan he had been mistaken, but now there was no problem. There were ample opportunities for the services of a truly hot young mathematician.

The University of California at Berkeley was in the process of elevating its expertise in topology. Beginning at the top, they lured Chern and Spanier from Chicago. Spanier consummated the deal while he was on leave at the Institute. When Berkeley sought Spanier's wisdom in identifying junior talent, he put forward the names of Hirsch and Smale. Compare his May, 1959 assessment of Steve with that following the sphere eversion in 1957.[25]

> *This letter is in reply to your request for an appraisal of the work of Stephen Smale. Smale received his doctorate at Michigan in 1956, spent the two years immediately after that as an instructor at Chicago and is now in the middle of two years at the Institute for Advanced Study on a National Science Foundation Fellowship. I have been acquainted with*

him and his work since he came to Chicago almost three years ago.

Smale works in that branch of mathematics lying between topology and modern differential geometry. His most important work to date has been the classification of immersions of spheres in Euclidean spaces up to regular homotopy. This is a deep and highly original piece of research and has proved to be extremely important in many contexts. Other important work he has done includes a study of immersions of manifolds in Euclidean spaces and their self-intersections (with Lashof), a classification of some involutions of the three sphere (with Hirsch), work on the triangulation problem, a proof that differentiable structures are unique on manifolds of dimension less than five, and, most recently, some results on the global aspects of differential equations on manifolds.

All of this work is first rate and in the main stream of the subject. He is not afraid to tackle difficult problems or to learn new fields. I believe he is the most promising topologist to appear in many years. (emphasis added)

Smale has already acquired quite a reputation. This is affirmed by the fact that he received offers or expressions of interest in making him offers from: Princeton University, Harvard University, The University of Chicago, Massachusetts Institute of Technology, Yale University, Johns Hopkins University, and the University of Minnesota. Typical of these offers is the one from Chicago as an Assistant Professor at a salary of $7500 with a teaching load of six hours.

In addition to his strength in research Smale has shown himself to be a good teacher both in elementary and in advanced

> *courses. He is interested in mathematics and in talking mathematics with students. He is also an active participant in seminars and, generally, a helpful person to have around. He seems to be interested in coming to Berkeley, and I recommend his appointment in the highest possible terms.*

If Steve had accepted the earlier offers from Chicago or Harvard, it would have pre-empted his opportunity at the Institute. In 1958–60 Princeton was an ideal location for a young topologist. Milnor was lecturing on his current work at the University. Together the members at the Institute along with the faculty and graduate students at the University formed a stunning collection of topologists. The timing of the Berkeley offer was more favorable. Steve had completed the first year of his postdoctoral fellowship and was permitted to finish the second. Berkeley was an interesting place and the department was moving in the direction of the Chicago/Princeton topology crowd with the appointments of Chern, Spanier, and Hirsch. Moreover, it was good to settle the employment situation since Steve was leaving the Institute, and actually the country, for the last six months of the fellowship. Lima and Peixoto had invited him to visit a mathematics institute in Rio de Janeiro. In a move that, at the time, may have called into question Smale's priorities as a mathematician, he chose Rio over the Institute. Steve would spend the first half of 1960 in Brazil and then begin a faculty position in Berkeley.

For Clara, Princeton had been a big improvement over Chicago. The environment was more family friendly, and the Smale family was expanding. Daughter Laura was born on April 22, 1959.

On the Beaches of Rio

04

A. Getting settled

The University of Chicago and the Institute for Advanced Study produced solutions to the following problem: How to cultivate great mathematical research? Of course the obvious solution was to attract a collection of great mathematicians, but the efforts of Stone, Flexner, and Veblen were distinguished both by their taste in identifying candidates, and their success at reeling them in. With brilliant minds working in a stimulating environment, either individually or jointly, there was a statistical expectation of seminal theorems. However, it was still possible for more modest, even third world, institutional models to incubate important work. After all, the most vital infrastructure was a powerful mind engaged in creative activity. Smale possessed this asset and it was extremely portable.

In December of 1959, the Smale family, including eight-month-old Laura and two-and-a-half-year-old Nat, departed from Princeton. They were destined for Rio de Janeiro where Steve planned to complete his NSF postdoctoral fellowship at the Instituto de Mathemática Pura e Aplicada (IMPA). Steve had already made four trips to Mexico, first with his father in 1946 and most recently for a differential equations conference the previous summer. His 1954 planned excursion to Argentina had been frustrated by the Guatemalan civil war. At that time Steve was unable to proceed further south than British Honduras. The trip to Brazil finally provided Steve an opportunity to see the Central and South American sights.

The family landed in Panama where they viewed the jungle from a taxi. Next Steve fulfilled a long time desire by riding the train route through the Ecuadoran Andes from Quito to Guayaquil. After a few days of needed recuperation in Lima, the Smales flew to Rio de Janeiro.[1]

Brazil was a large country that was endowed with marvelous beaches and a vibrant culture. Accompanying these assets were daunting economic problems and political instability. Food prices had risen by 70% during the first 11 months of 1959, and the government was paying its bills by printing the vastly inflated currency.[2] In early December, a small group of air force and army officers attempted to overthrow the government. Although the coup was unsuccessful, more democratic means were about to unseat the president.

Steve managed to benefit from both the Brazilian resources and problems. The dollar was strong against the weakened cruzeiro, offering the potential for the Smale family to upgrade their standard of living. The timing and failure of the coup even turned to Steve's advantage. He obtained a plush Copacabana apartment which became available with the expatriation of its previous tenant, an air force colonel who led the coup.

The Smale household followed the traditional gender roles. Clara took care of the children and the other matters at home while her husband was away at work. However, Steve's daily routine was highly nonstandard, even for a mathematician. He began his work day in the morning, contemplating mathematics at the beach.[3]

> *My work was mostly scribbling down ideas and trying to see how arguments could be sequenced. Also I would sketch crude diagrams of geometric objects flowing through space, and try to link the pictures with formal deductions. Deeply involved in this kind of thinking and writing on a pad of paper, the distractions of the beach didn't bother me. Moreover, one could take time off from the research to swim.*

Afternoons he was at IMPA. There he had a conventional office, access to a good library, and the opportunity to discuss topology with Lima and dynamical systems with Peixoto. In the evenings he sometimes joined the samba dancers in the street. It was a pleasant regimen, but not the sort of venue for a budding mathematical superstar to select over the Institute for Advanced Study. Still, Smale had all the essentials: paper, pens, library, and stimulating colleagues. What environment was the most conducive to the generation and development of profound ideas, or did it really matter?

B. Early chaos

Early in his Rio de Janeiro residency, Smale focused on dynamical systems. As with many branches of mathematics, dynamical systems involves the study of certain functions defined on certain sets. One might think of the set M as modeling the possible states of some process. The function f determines the evolution of the system with time. This may be constructed so that time changes continuously or in discrete increments, but we shall restrict our attention to the discrete situation. For example, consider a function that maps each element of M into M, so that the domain is M and the range is a subset of M. (In mathematical notation $f(x) \in M$.) We may think of f as providing the transition of states after a discrete unit of time. If the system begins in state x_0, then it moves to state $f(x_0)$ after 1 unit of time, $f(f(x_0))$ (denoted $f^2(x_0)$) after 2 units of time, and $f^3(x_0)$ after 3 units. We are interested in the long term trend, if any, of this sequence; that is, whether there is some state or states in M that the values $f^n(x_0)$ approach (or remain at) as n gets larger. Such long term trends are known as asymptotic behavior. The dynamical system consists of the function f together with the set M. Given a dynamical system, the long term (asymptotic) behavior depends on the initial state. However, different initial states might have the same or different asymptotic targets. A fundamental goal of dynamical systems is to understand the possible asymptotic behavior of the system. In a particular dynamical system there might be some number of distinguished states which serve as the only possible targets for all initial states. Alternatively there could be no particular target to which the system evolves.

A dynamical system is defined to be *structurally stable* provided the asymptotic behavior of the system is preserved under small changes to the function. Thus if a structurally stable system with three targets is perturbed slightly, then the new system also has three targets. Some systems are structurally stable and others are not. Structural stability has a rigorous mathematical definition, but it is difficult to verify for a particular system. Out of his Princeton discussions with Peixoto, Smale proposed a class of systems which he believed to be structurally stable. Among the defining properties was the requirement that the system have just a finite number of asymptotic targets. Another ingredient involved the notion of transversality. Smale's first paper on dynamical systems, entitled "Morse Inequalities for a Dynamical System," appeared in the January 1960 issue of the *Bulletin of the American Mathematical Society.* In the article Smale established properties of these systems, later known as Morse–Smale systems, showing that the number of asymptotic targets was related to the topological properties of the underlying set.

While Steve believed that Morse–Smale systems were structurally stable, he did not yet have a proof. At the bottom of the first page he presented this speculation, going still further:

> *First we remark that* [Morse–Smale] *systems may be very important because of the following possibilities:*
>
> *(A) It seems at least plausible that* [Morse–Smale] *systems form an open dense set ...*
>
> *(B) It seems likely that* [Morse–Smale] *conditions are necessary and sufficient for* [the system] *to be structurally stable ...*

Notice the difference in the qualification preceding the two statements. He believes that (B) is true, but is less certain of (A). This distinction has been lost over the years. Both statements have been attributed as Smale conjectures.[4] If Morse–Smale systems were an open dense set, then they were mathematically ubiquitous. Structural stability is a robust quality that vaccinates the system, preserving its outcome against small changes. Smale was suggesting that the simple Morse–Smale asymptotics were the generic model, with perhaps some rare exceptions.

Conjecture B was partly correct. In the late sixties Smale's Brazilian student Jacob Palis made a crucial breakthrough, and then, in collaboration with Smale, proved that Morse–Smale was sufficient for structural stability. Negative answers to (A) and the necessity of (B) came more quickly. Shortly after his arrival in Brazil, Smale received a letter from MIT mathematician Norman Levinson. Levinson referred to one of his own papers, published in 1949. He claimed that it contained a system with an infinite number of asymptotic targets, and the property persisted under perturbation. If Levinson were correct, then Morse–Smale systems were not ubiquitous as Smale suspected. Levinson was a substantial mathematician whose observations deserved serious consideration. (An irrelevant but interesting footnote is that Levinson was a former communist who had cooperated with the HUAC in 1953.[5])

Smale's next steps were to understand Levinson's example and reconcile it with his own vision for Morse–Smale systems. The thread of Levinson's study led back to the work of the British mathematicians Mary Cartwright and J. L. Littlewood. All of these papers were from the analysis branch of differential equations, but Smale's background was in topology. Mathematical scholarship, with its vertical nature, can be difficult to access across disciplinary lines. Library study, however, was a strength of Smale.[6]

> *I worked day and night to try to resolve the challenge to my beliefs that the letter posed. It was necessary to translate Levinson's analytic arguments into my own geometric way of thinking. At least in my own case, understanding mathematics doesn't come from reading or even listening. It comes from rethinking what I see or hear. I must redo the mathematics in the context of my particular background. And that background consists of many threads, some strong, some weak. My background is stronger in geometric analysis, but following a sequence of formulae gives me trouble. The mathematical literature is useful in that it provides clues, and one can often use these clues to put together a cogent picture. When I have reorganized the mathematics in my own terms, then I feel an understanding, not before.*
>
> *In any case I eventually convinced myself that indeed Levinson was correct, and that my conjecture was wrong. Chaos was already implicit in the analyses of Cartwright and Littlewood! The paradox was resolved, I had guessed wrongly. But while learning that, I discovered the horseshoe!*

The horseshoe and chaos are examined further in Appendix C. The horseshoe is a function that Smale synthesized from his understanding of Levinson's work. The name is motivated by its defining geometric picture. Begin with a rectangle R in the plane. Let *A*, *B*, *C*, and *D* denote the vertices of the rectangle (see Figure 1). In an intermediate step, stretch the rectangle horizontally and contract it vertically. Next, bend the skinny rectangle from Figure 2 into a horseshoe, placing the old left side on top. Finally put the horseshoe over the original rectangle, denoting by *A'*, *B'*, *C'*, and *D'* the locations of the corresponding original vertices (see Figure 3). Now, think of doing this all in one step: Transforming a rectangle into a horseshoe. This defines a function *h* that maps each point of *R* to another point in the plane, lying inside the horseshoe (e.g., $h(A) = A'$ and $h(E) = E'$; it is a good exercise to check where h maps other points of R). With a few more steps we could define a homeomorphism of the sphere, but Figure 3 is the essence of Smale's horseshoe map.

By showing that the horseshoe had an infinite number of asymptotic targets, Steve proved that it was not a Morse–Smale system. A deeper argument established structural stability. Together these results showed that there were robust systems which were not Morse–Smale. Thus Morse–Smale was neither ubiquitous nor was it a necessary condition for structural stability. This provided (largely negative) answers to Smale's problems/conjectures A and B, but raised new questions. Was the horseshoe a model for some class of systems which were both structurally stable and ubiquitous? Was structural stability the appropriate criterion for robustness? These problems occupied Smale and his followers throughout the sixties.

Morse–Smale, with its finite number of asymptotic targets, is an orderly system. There is a finite collection of distinguished targets. From any initial state, the system evolves toward one of these targets or toward a cycle between some of them (perhaps bouncing back and forth between two or more in some periodic sequence). The infinity of targets is just one of the complexities of the horseshoe's dynamics. There are initial states from which the asymptotic behavior is essentially random. Moreover, such states can be found in arbitrarily close proximity to ones with the Morse–Smale type asymptotic targets. Thus the slightest error in choice of initial state may result in the difference between a deterministic or random future. If real world processes are modeled by dynamical systems, then the horseshoe illustrates that some processes might resist future prediction.

The term *chaos* was introduced in a 1975 paper by T.-Y. Li and Jim Yorke.[7] The notion and images of chaos and its application are so intriguing that its study has attracted the attention of many nonmathematicians.[8] Unlike structural stability and most other

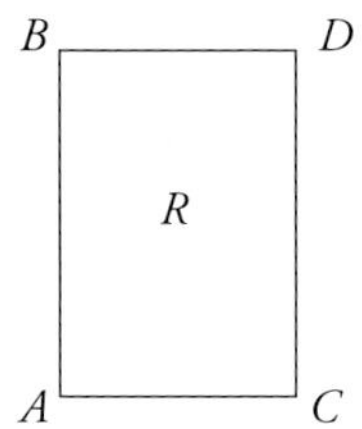

Figure 1. *The rectangle R in the plane.*

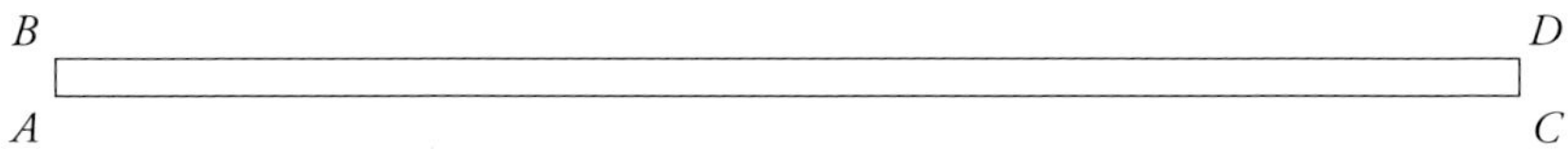

Figure 2. *Stretching the rectangle horizontally and contracting vertically.*

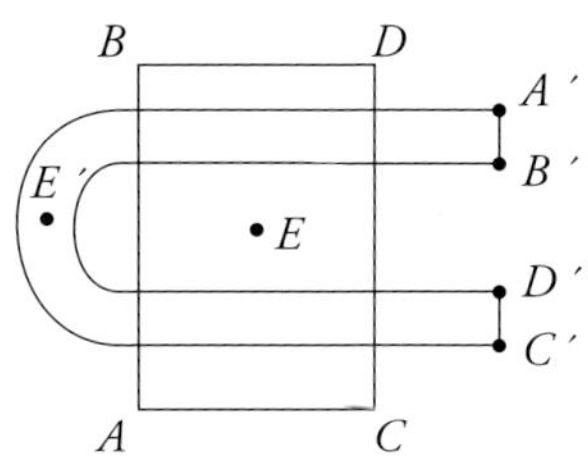

Figure 3. *The horseshoe function.*

mathematical concepts, chaos has no universally accepted precise definition. Most formulations involve aspects of randomness and sensitive dependence on initial states, both of which are present in the horseshoe. Indeed, Smale's horseshoe has become a paradigm for chaos.

The horseshoe exhibits chaotic behavior, but what triggers the chaos? One explanation is the presence of *homoclinic points.* A homoclinic point is a state which has the same asymptotic target in both future and past (negative) time. The importance of such points was appreciated in the late nineteenth century by the great French mathematician Henri Poincaré. The Harvard mathematician G. D. Birkhoff followed up in the early twentieth century, showing that there must be an infinite number of targets in the vicinity of a homoclinic point. Birkhoff received his Ph.D. from the early Chicago school, becoming the pre-eminent American mathematician of his time. Despite the important roles of Poincaré and Birkhoff in the history of mathematics, their work on homoclinic points was not an active area of interest in 1960, meaning that it was buried in classical texts that were rarely studied by mathematicians. Smale made good use of the IMPA resources. "I learned about homoclinic points and Poincaré's work from browsing in Birkhoff's collected works which I found in IMPA's library."[9] Smale took Birkhoff's ideas and pushed them much further. He showed that the relation between horseshoes and homoclinic points was intimate and quite general.

C. The great audacity

To solve a mathematics problem one needs to select a method and make it work. Smale had accumulated quite an arsenal of mathematical weapons. From Bott he learned algebraic topology, fiber spaces, and Morse theory. Chicago added Thom's transversality theory and Princeton contributed Milnor's development of differential topology. These were five powerful tools and Smale was applying them with rapidly increasing dexterity. Bott framed the thesis problem, but Steve's solution demonstrated skillful use of algebraic topology, fiber spaces, and analysis. For the sphere eversion Steve employed the same tools, replacing Bott's mentorship with his own audacity. The foray into dynamical systems was a fusion of topology, Morse theory, transversality, and analysis, all mixed together in an original recipe created by Smale.

With its abstract and vertical nature, mathematical research tends to follow traditional paths. Most mathematicians hack at the frontiers, using the standard tools from their subspecialty. While some technique from a different area might be beneficial, it is a formidable challenge to identify and adapt the technique. Suppose an algebraist attempts to apply an analysis technique to a problem in his or her field. Rarely do these efforts pay off, and such ventures might require months of intense work before their failure is apparent. To succeed involves a substantial risk and requires the intellectual power to understand the individual components and their cross-fertilization. Smale learned from the ideas of others, but, in the end, formulated problems in his own manner. He was unconstrained by conventional wisdom, and willing to take risks. He also had the insight and vision to somehow select the fertile interdisciplinary approaches.

Following his work on the horseshoe, Steve possessed a deep understanding of dynamical systems. He then saw an opportunity to combine these techniques with Morse theory in a new assault on the Poincaré Conjecture, the most famous unsolved problem in topology. The status of a mathematical problem derives from its age, intrinsic interest, and resistance to attack by prominent scholars. The Poincaré Conjecture scored exceptional marks in each category, offering a bounty of instant mathematical fame for its conquest. Smale had been powerfully drawn to the problem when he learned of it in graduate school and fashioned his first (incorrect) proof.[10] In subsequent years he made more sophisticated attempts. During his remaining time in Rio, on the beach and at IMPA, Smale focused on his new approach.

What is the Poincaré Conjecture? An explanation of the statement requires substantially more infrastructure than that of Fermat's Last Theorem. In the remainder of this section and in Appendix D, we provide some of the history and flavor of the problem along with a bit of insight into Smale's contribution. Crucial ingredients in the discussion are the mathematical notions of dimension and manifold.*

One dimensional Euclidean space is the real number line in which every point corresponds to a real number. Similarly, Euclidean 2-space, R^2, is the plane where each point is determined by a pair of real numbers, (x, y). Next is Euclidean 3-space, R^3, where points correspond to triples of numbers, (x, y, z). Notice that we may think of each Euclidean space geometrically, or simply work with the number representations. We can continue the latter analogy to Euclidean 4-space, R^4, in which points are quadruples of numbers, (x_1, x_2, x_3, x_4). Following in this manner, for any positive integer n, there is a Euclidean n-space, R^n, consisting of n-tuples of numbers. The number n designates the dimension.

In R^2 we are familiar with the unit circle, the locus of all (x, y) with $x^2 + y^2 = 1$. The (open) unit disk in R^2 consists of the points it encloses, $x^2 + y^2 < 1$. In R^3 there is a unit sphere, $x^2 + y^2 + z^2 = 1$, and a unit ball, $x^2 + y^2 + z^2 < 1$. Proceeding analogously to R^4, the unit sphere consists of the quadruples (x_1, x_2, x_3, x_4) with $x_1^2 + x_2^2 + x_3^2 + x_4^2 = 1$ and the unit ball defined by $x_1^2 + x_2^2 + x_3^2 + x_4^2 < 1$. The definitions of spheres and balls extend to R^n.

Consider the unit sphere in R^3. We say that it is locally Euclidean in the same sense that, prior to Columbus, people thought the earth was flat. If you are at a point on the surface of a sphere and possess limited vision, you would think that you are on R^2. Thus the sphere is locally Euclidean and we denote it by S^2, with the 2 signifying the dimension of the corresponding Euclidean space. In R^n the unit sphere, S^{n-1}, is $n - 1$-dimensional.

* These concepts are introduced in the following five paragraphs, which are repeated in Appendix A.

The unit ball in R^3 is also locally Euclidean, but it is 3-dimensional and denoted D^3. We will use the notation D^n for the unit ball in R^n. An *n-manifold* is a structure M which is locally like D^n. More specifically, each point $x \in M$ has a neighborhood (set of nearby points) which is homeomorphic to D^n. Thus at each point on M, it looks like R^n, but globally it might be quite different.

Thus far our inventory of 2-manifolds include S^2, D^2, and R^2. Another example is given by a hollow doughnut, known to mathematicians as a torus, T^2. To a population living on a large doughnut, locally it is indistinguishable from a sphere or plane. For each positive integer n, there are n-manifolds S^n, D^n, R^n, and T^n. One motivation for the manifold concept arises from the notion that the universe is locally Euclidean, but globally unknown.

Determining the global structure of a manifold is another matter. This is the topological classification problem. Which manifolds are homeomorphic and which are not? Let's begin with S^2. Is there a succinct description that completely characterizes the manifolds that are homeomorphic to S^2? Consider the following four specimens: a hollow cube, R^2, T^2, and S^3. Which, if any, is homeomorphic to S^2? This begs the questions of how to show that manifolds are homeomorphic and how to prove that they are not.

Certainly the production of a homeomorphism establishes that two manifolds are homeomorphic. This is the case for S^2 and the cube. The homeomorphism is the map from S^2 to the cube which deforms the sphere by taking each point of S^2 to the point on the cube that occupies the same ray from the center. We could actually write a formula for the function.

Okay, so S^2 is homeomorphic to the cube. Next, is S^2 homeomorphic to R^2? Intuitively, this seems unlikely. Deforming the sphere into R^2 appears to require tearing, an impermissible homeomorphism operation. The previous sentence, however, lacks mathematical rigor. Just because something appears to be the case does not satisfy the standard of mathematical proof. Remember, Smale proved that eversions were possible, contrary to the prevailing intuitive view. Fortunately there is a way to introduce rigor in dealing with S^2 and R^2. The sphere has a topological property, called *compactness*, essentially meaning that it is closed and bounded. The plane R^2 is not bounded and thus is not compact. Using the mathematical definitions of compactness and homeomorphism, it can be shown that homeomorphisms preserve compactness. This tells us that R^2 is not homeomorphic to S^2. Furthermore, compactness is a necessary criterion for a manifold to be homeomorphic to S^2.

The manifold S^3 is compact, but is it homeomorphic to S^2? Just thinking about S^3 is difficult, let alone trying to deform it into S^2. Fortunately there is a theorem that says that homeomorphic manifolds must have the same dimension. We conclude that S^2 and S^3 are not homeomorphic.

The 2-manifold T^2 is compact, but is it homeomorphic to S^2? Is there a way to deform S^2 into T^2 without tearing or attaching? Some effort and thought might indicate that the likelihood of accomplishing this is somewhere between impossible and improbable. The tools of algebraic topology provide the rigor. There is an algebraic measurement of both S^2 and T^2, and the results are different. Although we dodge the actual definitions, it can be shown that homeomorphisms must preserve this measurement. We now have the following list of conditions that are necessary for a manifold to be homeomorphic to S^2: compactness, 2-dimensional, same algebraic topology.

Theorem. *If a manifold is homeomorphic to S^2, then it is compact, 2-dimensional, and has the same algebraic topology.*

What about the converse? By itself, none of the three conditions, compactness, dimension, algebraic topology, guarantees that a manifold is homeomorphic to S^2. However, together they are sufficient and the following theorem was established in the nineteenth century.

Theorem. *A manifold is homeomorphic to S^2 if and only if it is compact, 2-dimensional, and has the same algebraic topology.*

To a mathematician this is an elegant theorem. In general it is difficult to determine whether or not manifolds are homeomorphic. However, when S^2 is involved, there is a laundry list of just three conditions which completely determine the answer. There are similar results for T^2 and the other compact 2-manifolds. Mathematics often develops by extension and generalization. With the classification of S^2 and compact 2-manifolds, it is natural, to a mathematician, to ask whether the result extends to 3-manifolds. For dimension 3, S^3 provides a nice starting point. The previous three conditions are again necessary for a manifold to be homeomorphic to S^3, but are they sufficient? This became known as the Poincaré Conjecture.

Poincaré Conjecture. *If a compact 3-manifold has the algebraic topology of S^3, then it is homeomorphic to S^3.*

Generalized Poincaré Conjecture. *If a compact n-manifold has the algebraic topology of S^n, then it is homeomorphic to S^n.*

Smale has reported on the history of these conjectures/problems.[11] In 1900 Henri Poincaré published his solution of the *n*-dimensional version. It was the first in a long line of erroneous solutions. Four years later Poincaré recanted and posed the 3-dimensional form as a problem. Despite this chronology, the term Poincaré Conjecture remains entrenched in the mathematical vernacular.

By 1960 the Poincaré Conjecture and its generalization had withstood the attacks of Steve Smale, Henri Poincaré and a number of other outstanding mathematicians. Most of the activity involved the 3-dimensional version which overshadowed its generalization for two reasons. In dimension three the algebraic topology condition is easier (for the record: simply connected as opposed to a homotopy sphere) and thus the problem is more accessible to mathematicians. The primary reason, however, was a prevailing conventional wisdom that the difficulty of a problem increased with dimension. Mathematicians believed that the 4-dimensional problem was harder than the 3-dimensional problem. To a 1960 topologist the logical approach was to solve the problem for S^3 then S^4 and perhaps then there would be sufficient understanding to tackle S^n for $n \geq 5$. The dimension philosophy had a profound impact which is difficult to overstate. Until the 3-dimensional version were resolved, it seemed ludicrous to go after a higher dimensional result.

Smale's new approach was directed at the Generalized Poincaré Conjecture. He was after it all. Overcoming the dimension dogma took great audacity, but conventional wisdom was not an obstruction to Steve. As he developed his proof, an extraordinary element emerged. The argument appeared to prove the Generalized Poincaré Conjecture for every dimension ≥ 5, but it failed in dimensions 3 and 4. Why did the method fail for the *easier* cases? One possible explanation for this shocking development was that the result was false in these dimensions. The possibility appeared more credible when Steve produced a 3-dimensional counterexample. This made a stunning package. Not only was he resolving the classical Poincaré Conjecture, viewed as one of the outstanding problems in mathematics, but he was establishing a generalization that was regarded as beyond reach.

Believing yourself that you have solved the Poincaré Conjecture is a long way from convincing the mathematical community. Deep mathematical arguments involve a large number of steps, each providing opportunities for mistakes or omissions. Sometimes a mathematician might see a step as intuitively evident without further detail, only to crash when unable to provide justification on demand. Smale was well aware that mathematical history is littered with such casualties. Proofs of the Poincaré Conjecture have an especially seductive effect on their creators.[12] As Smale checked his work, he discovered a serious bug in the 3-dimensional counterexample. Some mistakes admit remedies, but this one was "fatal."[13] The higher dimensional proof appeared solid. Smale wanted confirmation from another pair of eyes. After successfully rehearsing the argument on Lima, Steve was ready to go public.

With mathematics journals there is often a long lag between submission of a manuscript and its publication. The stages of refereeing, revision, typesetting, backlog, and printing typically sum to a couple of years. The *Bulletin of the American Mathematical Society* maintained a section on *research announcements* designed "to provide early announcement of significant new results." This enabled rapid dissemination (a few months) of the headlines and a summary, to be followed by a traditional detailed development in another organ. Smale submitted an announcement to the *Bulletin*, accompanied by a handwritten outline of his proof.

The mathematical world tends to greet claims of solutions to its biggest problems with a mixture of excitement and skepticism. The city of Princeton probably possessed the greatest per capita interest in the Poincaré Conjecture. When Smale's claim broke in 1960, James Munkres was at Princeton and Moe Hirsch was at the Institute. Both recall the astonishment among their colleagues. It was simply unbelievable that there was a topological proof skipping dimensions three and four, but handling all the higher ones.[14] "At the time it was the most unnatural thing you could possibly imagine," said Munkres. "It hadn't occurred to anybody." Topologists in Princeton were charged with authenticating Steve's argument, and it was quite a challenge. An outline of a deeply original mathematical development, combining previously disparate techniques, tends to raise as many questions as it resolves. With Smale in Brazil, it was extremely difficult for the readers to reach a judgement.

In the spring of 1960, John Stallings was at Oxford University. Stallings had received his Ph.D. the previous year at Princeton. When the "Smale rumor" reached Oxford, Stallings was among the skeptics.[15] After a visitor from Princeton disclosed that the proof had achieved some credibility, Stallings reacted in a somewhat perverse fashion. With a proof pending of a major theorem, some mathematicians would have tried to understand the argument while others would have moved to a different problem. Stallings went to work on the Generalized Poincaré Conjecture.[16]

> *I tried to figure it out. I tried to imagine, how would Smale do this. I thought I found a place where I couldn't quite do it like I figured he was going to do it. But then I got around it.*

Stallings had his own proof and more. He had a strong hunch that Smale had overlooked the obstacle he had managed to circumvent. Actually Stallings' proof was substantially different from that of Smale. Moreover, Stallings employed a technically weaker definition of manifold, achieving a slightly weaker conclusion. Another difference was that Stallings needed to assume that $n \geq 7$, but the bound was soon lowered to 5 by Christopher Zeeman.

At this point both Smale and Stallings had the exhilaration and uncertainty that accompanies a seminal result prior to its authentication. It remained to be seen which of the two proofs would satisfy the experts. There were showdowns shaping up for Europe in June. Smale and Stallings planned to attend conferences in Bonn and Zurich.

In retrospect Stallings candidly admits to a sense of competition. "I felt good if he felt bad or something like that."[17] Stallings listened to Smale's talk in Bonn, waiting for the juncture where he anticipated the mistake. There it was, Smale was applying an operation in too much generality, overlooking a difficult case. Stallings recalls "I thought this was a fatal error in Smale's method, and I was secretly very gleeful."[18] Stallings communicated the objection to Smale, but neither principal recalls whether it was during or after the talk. For Smale "it was traumatic." It was an unforeseen problem that he had not thought through. There was a gap in his argument.[19]

Stallings' own talk was not without trauma. His argument rested on an existing foundation of topological theory. A member of the audience grilled Stallings for further details to support his stipulations and he was unable to satisfy the demands. Still, he felt that his proof was solid and that "I was on top of the whole thing."[20]

Meanwhile Smale sought to rehabilitate his proof. With some additional work he succeeded in eliminating the gap. Even Stallings was persuaded by Smale's presentation in Zurich. The history of these events and apportionment of credit have become tangled over time. Who solved the Higher Dimensional Poincaré Conjecture?

First, it was Smale who had the audacity to challenge the dimension barrier. While the conventional wisdom of increasing difficulty with dimension remains in other areas of mathematics, it has been obliterated from the topology of manifolds. Students in this field now learn that *higher dimensions are easier because there is more room to move around.* Smale was the pioneer who broke the dimension barrier. Second, Smale solved the Higher Dimensional Poincaré Conjecture. It is true that his original proof contained a gap, but some difficulties are to be expected with works of this magnitude. It was not a fatal flaw, and the original proof stood as supplemented. Smale's result provided a vital clue to Stallings who independently devised his own proof. Stallings' contribution is an impressive piece of mathematics and quite an intellectual accomplishment. Smale, however, was there first.

Smale's work opened up higher dimensional topology for further development by himself and others. Meanwhile, the Poincaré problems remained for S^3 and S^4. In 1982 Mike Freedman of the University of California at San Diego proved the 4-dimensional version. As of this writing (in 2000) the 3-dimensional form, often called *the original Poincaré Conjecture*, remains unsolved. As it continues to rebuff its pursuers, the bounty of prestige mounts for its future conqueror. Just prove, or disprove, that every simply connected, compact 3-manifold is homeomorphic to S^3.

For Smale the acclaim was bittersweet. There were no news reports on the European conferences and email did not exist in 1960. Mathematicians learned of the Poincaré Conjecture developments through word of mouth or what they could glean from the scientific literature. Often these reports portrayed Smale and Stallings on equal footing as independent solvers, neglecting to mention Smale's precedence. In some accounts Steve was in the background or even omitted entirely.[21] The inaccuracies perpetuated over the years as Smale seethed.

Reports by the experts were part of the problem, beginning with Stallings' 1960 *Bulletin of the American Mathematical Society* research announcement that failed to mention Smale or his proof. Freedman's discussion of the 4-dimensional solution made no mention of Steve's higher dimensional theorem. When Freedman received the 1986 Fields Medal, Milnor provided the following historical perspective in his tribute:[22]

> *The cases $n = 1, 2$ were known in the nineteenth century, while the cases $n \geq 5$ were proved by Smale, and independently by Stallings and Zeeman and by Wallace, in 1960–61.*

Stallings now acknowledges that his omission of Smale was improper and of course Freedman and Milnor knew of Smale's contribution. Smale wrote to Milnor in protest, receiving an apology and corrected history. However, if these were the reports by the experts, there was little chance for the average mathematician to ascertain the facts.

Smale had had enough. At the 1989 annual meeting of the American Mathematical Society and Mathematical Association of America he gave a talk entitled *The Story of the Higher Dimensional Poincaré Conjecture* (*What Actually Happened on the Beaches of Rio*). It was an unusual topic for a program that is dominated by technical lectures. Smale proceeded to set the historical record straight, citing inaccuracies from the literature. To some in the large audience, unaware in advance of the agenda, Smale came off as egotistical and petty.

Among the mathematicians interviewed for this book, there was little sympathy for Smale's plight. Some recalled results for which they themselves had not received adequate credit. The mathematics profession seems to take an arcane pride in its record of misattributions and poor sense of history. It is the discovery, rather than the credit, that is supposed to provide the satisfaction. Certainly the discovery does provide an enormous return, perhaps known only to research mathematicians. However, when the breakthrough has the magnitude of the Higher Dimensional Poincaré Conjecture, it does seem natural to crave full credit.

Berkeley to Columbia and Back to Berkeley

05

A. Tenure at Berkeley

Faculty employment status includes the separate but related issues of rank and tenure. When a university grants tenure, the faculty member is guaranteed perpetual employment, absent some egregious act such as malfeasance or moral turpitude. Decisions on tenure are normally made in the sixth year of employment, but outstanding candidates may receive earlier consideration. Faculty without tenure may be on the *tenure track.* This deceptive terminology designates a position which has the potential for future tenure, but typically not the presumption.

Faculty ranks include the following classifications, listed in increasing order of status: instructor, assistant professor, associate professor, and (full) professor. Recent Ph.D.s normally begin as untenured assistant professors or instructors. Promotion to associate professor and tenure are often linked. Steve was an untenured instructor in the Hutchins College at Chicago. Following the sphere eversion, he declined tenure track assistant professorships in the mathematics departments at Chicago and Harvard. Prior to leaving for Brazil, Smale accepted a tenure track assistant professorship at Berkeley.

Steve arrived in Berkeley during the summer of 1960. He was about to begin his first position in a university mathematics department, but, having just proved the Higher Dimensional Poincaré Conjecture, Smale was not the typical junior faculty member. There was no doubt about his ability as a scholar. Since he had established the primary prerequisite for tenure, a probationary period was unnecessary. Steve was promoted to the rank of associate professor with tenure. In effect, he skipped the assistant professor level and the tenure track.

It had been seven years since Steve was warned that his graduate career was in serious jeopardy. During those years, politics remained on the back burner and mathematics flourished beyond expectation. As a tenured Berkeley faculty member, Smale now had a great deal of freedom to set his own course, both mathematically and politically. In 1960, Berkeley was not yet the political hotbed it would become in a few years. Steve's focus remained on mathematics, but some political issues did arise.

California state employees were required to take an oath disclaiming radical activity of certain sorts in the past (five years), present, and future.[1] Smale had been separated from the Communist Party for a sufficient period of time that he could, in good faith, sign the oath. However, acquiescence to the procedure raised certain principles as to political freedom and privacy. These issues were important to Smale during the McPhaul hearings. Several Berkeley faculty jeopardized their careers by protesting the policy as it was formulated and evolved in the early fifties. Smale signed the oath.

While Steve was politically inactive, he was not apathetic. When some students asked him to serve as faculty sponsor for the campus branch of a new pro-Castro organization, Steve agreed. The group, Fair Play for Cuba, attracted little attention in its first years of existence. Nationwide there were a few thousand members.[2] Following the Kennedy assassination in 1963, Fair Play for Cuba was linked to Lee Harvey Oswald. At that time the organization became infamous and its membership evaporated. Steve recalls his own role as especially nominal. He was the titular faculty sponsor for Berkeley Fair Play for Cuba in 1960–1961.[3]

Meanwhile Steve's work on the Generalized Poincaré Conjecture had opened up higher dimensional topology. Smale followed with a number of interesting results, culminating in June 1961 with his spectacular *h*-cobordism. The *h*-cobordism theorem may be thought of as an inter-interdisciplinary result between the fields of algebraic topology and differential topology. The algebraic topology equivalence of *h*-cobordism had been introduced, under a different name, by René Thom at the 1956 Symposium in Mexico City. A differential topology equivalence is the seemingly stronger diffeomorphism. In 1960 the tools of algebraic topology were sufficiently developed to permit progress on the classification of manifolds by *h*-cobordism. However, the emerging field of differential topology was stymied by its lack of facility for showing that manifolds were diffeomorphic. Milnor framed a question as to whether *h*-cobordism was actually equivalent to diffeomorphism.[4] Subsequently he produced a counterexample, resolving the problem in the negative. Smale's *h*-cobordism theorem refined Milnor's question, adding an additional algebraic topology condition.[5] Steve's result was again valid in the environment of dimension five or larger. The ramifications were overwhelming, and included the Higher Dimensional Poincaré Conjecture as a corollary.

The *h*-cobordism theorem remains one of the fundamental results in topology, but Smale's article[6] is seldom studied. To understand how this came about, consider the historical context. Smale's deep theorem appeared and evoked a great deal of interest among researchers and students in the field. The original source was difficult to penetrate. At Princeton, Milnor customized the development of the material in presentations to his seminar class. Milnor was a superb expositor. Notes from his lectures were often passed informally among mathematicians. When Milnor published his *Lectures on the h-Cobordism Theorem* in 1965,[7] the mathematics became more accessible, despite the 107 page length of the proof. The book was a valuable service and became the preferred vehicle for learning the theory.

Another impact of the book was to separate Smale's name from the result. Milnor's book, rather than Steve's paper, became the standard reference for the *h*-cobordism theorem and the Generalized Poincaré Conjecture. How were subsequent generations of mathematicians to become aware of Smale's contribution? His name never stuck to the theorem, as is sometimes the case with major results such as Morse theory. To learn of Smale's role required following the thread back to Milnor's preface, where the proof was properly attributed.

Once again, Smale had obtained a huge result, and he was frustrated by the failure of the mathematical community to fully acknowledge his contribution. Unlike the Higher Dimensional Poincaré Conjecture, there was no dispute that the *h*-cobordism theorem was Steve's theorem. The lack of linkage between his name and these theorems can be explained by the presence of Milnor's book and the absence of a sense of history among mathematicians. Smale saw less benign factors, largely associated with his subsequent departure from topology.[8]

> *I did make a very clean break with topology in '61. I just said that I thought the questions connected with dynamical systems were more interesting. That's the most clean break in mathematics I've made. Certainly topologists felt—some topologists, at least, and probably many—that I wasn't a topologist anymore and that I was sort of, you know, dismissable. I understand that, because by leaving the field I sort of implied that maybe it wasn't so interesting. You know, people took it that way.*

Had Smale betrayed topology, and was there a passive aggressive reaction among his former colleagues in failing to recognize his contributions? Such questions are difficult to assess. Steve did indeed leave topology. Any abrupt transfer is unusual in mathematics, attesting yet again to the vertical nature of the discipline. No doubt there were a variety of reactions among topologists, including feelings of betrayal, but also relief. Smale had knocked down some large problems, leaving fruit on the ground to be gathered by others.

The switch from topology to dynamical systems involved a substantial risk. Topology was a hot subject and Smale had joined Milnor as its newest stars. If Steve had remained in topology he was assured of a high status. His future work and lectures were certain to capture the attention of the topological brahmins, but there was also a niche for just proliferating his work on the Higher Dimensional Poincaré Conjecture and the *h*-cobordism theorem. In 1961, Smale's vision for dynamical systems was not even a mathematical subspecialty. Steve had written a couple of papers and had some conjectures. If he succeeded in pushing through his ideas, there was no guarantee that the theorems would attract an audience. However the challenges posed by the vertical nature of mathematics were mitigated by two factors. Steve was not a novice in dynamical systems in that he had already obtained the horseshoe and other results. Moreover, his recent topological work had employed tools from dynamical systems. Steve was working on new questions, but could take advantage of some of the techniques that he had honed for topology.

The transfer from topology to dynamical systems coincided with another change in Steve's career. His rapid ascent up the academic ladder consisted of only one year at the associate professor rung. Following Steve's arrival at Berkeley, the University of Chicago and Columbia University weighed in with offers of full professorships for the next year. Berkeley moved to counter with a promotion to the same level.

Smale was in an enviable position. He was assured of becoming a full professor in a first class mathematics department at the extraordinarily young age of 31, but at which university? The Smales had just bought their first house, giving Berkeley the advantage of inertia. Both Steve and Clara liked the Bay Area, and it was a nice environment for their children. Clara wanted to stay.[9]

The competition was Columbia. Few people are indifferent toward New York and Steve had been attracted to the city since his college days, living in Greenwich Village during the summer of 1950. Personally and professionally, Smale had a choice between two terrific opportunities. In the end, the tiebreaker was money. Columbia promised to double his salary from $9,000 to $18,000, while Berkeley responded with $13,000. Steve went to the higher bidder, joining the Columbia faculty for the fall term of 1961.[10]

B. Full professor at Columbia

Dynamical systems was now the focus of Steve's research. In 1961 it was a mathematical wilderness, lacking established problems such as the Poincaré Conjecture. Part of Steve's mission was to create an agenda for development. While he had succeeded in doing outstanding mathematics in the relative isolation of Brazil, interaction with experts was valuable in framing new questions. Steve was connected with many of the world's leading practitioners. Chicago, Princeton, and Berkeley were prime stops on the lecture circuit, and Steve had attended major conferences in Europe and Mexico. However, some of the best work on dynamical systems had occurred in the Soviet Union, and most Soviet mathematicians were restricted to travel behind the Iron Curtain.

Prior to beginning his teaching at Columbia, Smale traveled to the Soviet Union, lecturing on dynamical systems in Moscow and at a conference in Kiev. For the first time Steve interacted with D. V. Anosov and his other Soviet counterparts. Considering Smale's cosmopolitan background and his reluctance to indulge in hyperbole or superlatives, the trip made quite an impression.[11]

> *After Kiev I went back to Moscow where Anosov introduced me to Arnold, Novikov, and Sinai. I must say I was extraordinarily impressed to meet such a powerful group of*

four young mathematicians. In the following years, I often said there was nothing like that in the West.

The concept of structural stability, for lower dimensions, had originated in the Soviet Union in the 30's.[12] Through his conversations with Peixoto, Smale had identified the higher dimensional generalization, and made some guesses about its characterization. The horseshoe had forced Steve to adjust his ideas. To capture structural stability required inclusion of Morse–Smale, the horseshoe, and perhaps some other dynamical features. In the Soviet Union, Smale shared his thoughts and conjectures, as he grappled to further understand the concept. It was a profound intellectual exchange.

Back at Columbia, Smale taught his graduate course on dynamical systems. One day a precocious sophomore wandered into his office to obtain a clarification on the definition of structural stability. Mike Shub was not enrolled in Smale's class, but he had learned of the concept in discussions with more advanced students. As could be expected, Smale's answer was way over Shub's head, but the meeting served as an introduction for a significant relationship.[13]

For the spring of 1962, Steve took a leave of absence from Columbia. The Smale family was off to Lausanne. Steve continued his work on dynamical systems, and began to move off in a new mathematical direction. However, dynamical systems was the topic of his lectures in Lausanne and other parts of Europe. He was making progress in reconciling the horseshoe with his original vision, but there were serious obstacles ahead.

In August at the Stockholm International Congress of Mathematicians, Steve reported on the state of his work. Dynamical systems was still in its formative stage, but Smale boldly revealed his unproven ideas. Taken in the context of future developments, the write-up of the address[14] provides a fascinating glimpse into the workings of a great mind. There were extraordinary insights, including even Steve's doomed attempt at a post-horseshoe rehabilitation of his model for a generic, structurally stable system. Jacob Palis aptly described the paper as "an unfinished painting with several superposed sketches."[15]

At the Stockholm Congress, Sinai updated Smale on the progress that the Soviets had made since their discussions of the previous year. In particular Anosov had obtained a significant new structural stability theorem. A parallel development of dynamical systems was proceeding in the Soviet Union. Ironically, just as the seeds were beginning to germinate, Smale's own mathematical interests were again shifting.

In Lausanne, Steve had begun to look at infinite dimensional problems. Recall that n-dimensional Euclidean space is given by the set of n-tuples of real numbers. Analogously one might consider the collection of all (or some subset of) infinite sequences as an infinite dimensional space. The notion of a manifold can also be extended to infinite dimensions. Important examples of infinite dimensional manifolds are given by sets of functions between two spaces, such as the set of diffeomorphisms from a sphere onto itself. To establish the manifold structure is a major task.

For the fall term of 1962, Smale was scheduled to teach graduate differential topology. A student enrolling in the course might have expected the syllabus to begin with n-manifolds and build up to the h-cobordism theorem. Mike Shub, then a junior, was among the large audience on the first day when Smale announced that the material would follow his current, rather than previous, interests.[16]

> *My friends advised me to try to take Steve's graduate course in differential topology since he was a famous topologist who had proven a great theorem. In those days Columbia College didn't have much of an undergraduate mathematics curriculum. Many math majors vied in taking graduate courses which we were frequently hopelessly unprepared for. I enrolled for Steve's course. The very first day, he arrived and announced that*

> *the course would be about infinite-dimensional differential topology because that was where the most interesting work was to be done. In the class the first day was Sammy Eilenberg and other luminaries of the Columbia math department. Steve began by defining the derivative in Banach space. He didn't quite get it right, and the class degenerated as various of the luminaries shouted out suggested corrections.*

As Shub struggled with the ambitious task of absorbing infinite dimensional topology, Steve continued to get hung up on the details of his lectures. At times, other undergraduates corrected Steve's mistakes, causing Shub to actually giggle at the incongruous scene.[17] How could a senior know more than a great mathematician? Shub discussed the matter with Serge Lang, a Columbia professor who had recruited Smale. Lang explained to Shub "that while the undergraduates were locally correct, Steve was almost always locally wrong but globally correct."[18] In other words, Steve saw the big picture and his errors were never fatal. While this was undoubtedly the case, Steve's struggles with the details showed an inadequate preparation for his lectures and indicated that he regarded teaching as a low priority.

In late October, a differential topology class was unexpectedly cancelled. Perhaps Steve had encountered last second difficulties in organizing the lecture. At the next meeting, Ralph Abraham substituted for Smale. Steve soon returned and resumed teaching. Had he been ill or possibly attending a conference? Shub was not given an explanation.[19]

The actual reason for Steve's absence was motivated by political, rather than health or professional factors. It occurred at an important moment in the Cold War. On October 22, President Kennedy spoke to the nation. In a grave tone he revealed the findings of recent reconnaissance flights over Cuba. Missile sites, with nuclear capabilities, were under construction. To force the dismantling of the missile bases, the United States was initiating a naval blockade of Cuba. Kennedy directly confronted the Soviet Union and threatened nuclear war: "It shall be the policy of this nation to regard any nuclear missile launched from Cuba against any nation in the Western Hemisphere as an attack by the Soviet Union on the United States requiring a full retaliatory response upon the Soviet Union."

As Kennedy and Soviet Premier Khrushchev went one-on-one, there was a widespread fear that the confrontation would escalate into a nuclear holocaust. The concerns were well founded. The world was on the brink of a war between the two superpowers. Kennedy was close to approving an invasion of Cuba, but he was unaware that there was already a nuclear capability on the island.[20] Khrushchev backed down at the last moment, agreeing to remove the missiles. War was averted, but slight changes in timing or decision making on either side produced scenarios of nuclear engagement. What was Steve's initial response to the Cuban Missile Crisis?[21]

> *I reacted strongly to the growing threat that atomic war could start any day. I became intensely angry at Kennedy, being aware that the United States already had missiles located on the Soviet border in Turkey. When I became convinced that Soviet missiles were en route to Cuba, I became angry at Khrushchev as well. It didn't make sense to die in a nuclear war due to the insane militarism of the two countries.*
>
> *So, Clara and I with Nat and Laura packed a few of our belongings and started driving to Mexico!*

Thirty years later Steve reconstructed his analysis. He saw a "reasonably good chance that there would be an atom bombing of America by Russia." To stay was to risk his life and to leave was to risk his job.[22] The Columbia administration was unlikely to be impressed by Steve's logic in abandoning his teaching duties. Prior to leaving, Steve informed Abraham and Lang of his plans. The two colleagues were supportive, and undertook to cover Smale's class. As the Smale family approached the Mexican border, the missile crisis abated. When Steve phoned Columbia and learned that he was still a member of the faculty, he decided to fly back and resume his teaching, leaving Clara to drive the children. Shub and virtually everyone else at Columbia were unaware of the circumstances of Steve's absence. Who could have guessed?

Steve's foray into infinite dimensions lasted a couple of years. During that period he produced two important generalizations from the finite dimensional theory.[23] First, he (and also Richard Palais independently) extended Morse theory to the infinite dimensional setting, for maps satisfying what is now known as the Palais–Smale condition.[24] The other result involved the Morse–Sard theorem, a remarkable finite dimensional analysis result, circa 1940. Smale identified a class of functions on infinite dimensional manifolds for which the Morse–Sard conclusion remained valid.[25] Once again, Steve had the mathematical power and independent insight to see the huge (even infinite dimensional) picture while the thinking of others was limited by the finite dimensional hypotheses of the classical theorems.

Smale's mathematical meandering from topology to dynamical systems to infinite dimensions was unusual. What was remarkable were the theorems that he produced in each field. The leading mathematics departments continued to solicit his services. Shortly after Smale's arrival at Columbia, Milnor delivered a Princeton job offer, but Steve declined, having just settled in at Columbia.[26] Berkeley remained interested. They lost Steve to Columbia over money. For the 1964–65 academic year, Berkeley was able to match his salary of $22,000. Steve decided to return to California. With the geographic move came another mathematical shift. Steve was thinking about dynamical systems again.

Mike Shub and another Columbia mathematics student were beginning graduate school at Berkeley in 1964. For Shub, Smale was just part of the attraction. The Berkeley campus had acquired a reputation as a great place. The cross country move was facilitated by an offer from Steve. The students drove the Smale Peugot to California while the family flew. The Columbia contingent arrived in June, and Steve initiated a dynamical systems seminar.

C. Free Speech Movement

As Smale returned to Berkeley in the summer of 1964, the university administration was formulating a policy to restrict political activity at the entrance to the campus. The ensuing reactions led to the formation of the Free Speech Movement (FSM), a seminal event in the political and cultural transformation that occurred on university campuses in the sixties. The student activity of the FSM began in the middle of September and concluded in December. During these three months a small, disparate group of student radicals successfully united and won the support of the university community in a battle against the administration. The students' victory went against a conventional wisdom that portrayed students as powerless and subject to the whims of their university's administration.

One of the early impacts of the FSM was its contribution to the Berkeley antiwar Vietnam Day Committee (VDC), created the following spring by Jerry Rubin and Steve Smale. While Steve's role in the FSM was peripheral, an understanding of the FSM is a prerequisite for a study of the VDC. The legacy of the FSM extended well beyond Berkeley and included the student revolts at Columbia in 1968 and then at Harvard. The 1969–70 academic year marked the greatest defiance on university campuses, especially following the tragic shootings at Kent State and Jackson State. Throughout the country, previously apathetic students demonstrated against the Vietnam War and in support of civil rights, while strongly asserting themselves in curricular and social reform at their own institutions. It was a turbulent time on college campuses.[27]

The FSM came into existence as a contentious reaction to a new university policy limiting the nature of political activity. The Berkeley campus has a complex history of restrictions on political expression and where it may occur. The most frequently used entrance to the campus was located at the south central border, via Telegraph Avenue. Traditionally, students were permitted substantially more freedom in the area just outside (south of) this entrance. The accommodation began in the 1930's when University of California President Robert Sproul reacted to a left wing student movement by establishing a ban on campus political activity. At that time the entrance was located at Sather Gate. Following Sproul's decision, political solicitation moved to the city-owned south side of the gate. There the University tolerated the activity, which declined in subsequent years. The prohibition on campus politics persisted and Democratic presidential candidate Adlai Stevenson delivered a 1956 campaign speech from an automobile parked south of the Sather Gate entrance.

When the University expanded southward in the early sixties, Sather Gate was enveloped into the campus interior and the entrance shifted to the intersection of Bancroft Way and Telegraph Avenue. Political solicitation and activity then moved to card tables set up on the sidewalk at the new south boundary entrance. Meanwhile university regulations underwent several modifications. By 1963 communist and other controversial speakers were permitted on campus, provided the sponsoring organization could navigate a number of obstacles. These included a 72 hour advance notification and a program providing balance and a tenured faculty moderator.

Civil rights actions were becoming more prominent in the Bay Area. During the 1963–64 academic year, demonstrations in support of black employment were conducted at Mel's Drive-In, Lucky supermarkets, the Sheraton-Palace Hotel, automobile dealerships, and the Oakland Tribune. The tables at Bancroft and Telegraph were strategically located for solicitation of funds and recruitment of demonstrators, both essential to the civil rights effort. As Berkeley students participated in these activities, with some going to the southern states during the summer, they learned organizational methods which they would subsequently transfer to the FSM. Most prominent among these students were Jack Weinberg and Mario Savio.

Jack Weinberg was born in 1940 in Buffalo, New York. At the age of 21 he transferred to Berkeley and received a mathematics degree in January 1963. He continued in the graduate program the following semester while becoming involved in the Congress of Racial Equality (CORE). That summer he traveled the South, spending a month in Sumter, South Carolina, working with black activist students at Morris College. Weinberg recalled that "through a combination of challenge and dare, I and some other students, on my second night in town, went to a Holiday Inn and we were refused service, and I was arrested." His subsequent activities received further scrutiny from the authorities, and Weinberg eventually left Sumter as part of a negotiated settlement.[28]

When Weinberg returned to Berkeley in the fall of 1963, mathematical scholarship quickly gave way to civil rights work. Dropping out of graduate school, he became a full-time civil rights activist. As the chair of the campus CORE chapter, he organized and participated in demonstrations, frequently adding to the arrest section of his resumé. Rather than return to the South in the summer of 1964, Weinberg remained in the Bay Area, maintaining a high civil rights profile.

Other Berkeley students traveled to Mississippi to participate in the Freedom Summer of 1964. Among them was Mario Savio, a machinist's son, who had transferred to Berkeley the previous year.[29] In a 1965 interview[30] Savio described his route into activism. He learned of the Sheraton-Palace demonstrations from a leaflet he received at Bancroft and Telegraph. At first he was undecided whether to make the trip to San Francisco, but the topic arose at a small party, and Savio spontaneously joined his friends. To obtain transportation, they proceeded to a location adjacent to campus where vehicles had been organized for that purpose. At the Sheraton-Palace, while deciding whether to risk arrest, Savio realized he had found his niche. "You sit there looking at your finger, and there comes a point at which you realize that you have lifted your finger. I was arrested there. There were 160 people arrested, and I was one of them." While in jail he learned of the Mississippi Summer Project. Prior to his southern departure, Savio was chosen to chair the campus Student Non-Violent Coordinating Committee (SNCC). In Mississippi he worked on voter registration in Holmes County and taught in McComb.

Savio returned to Berkeley for the 1964–65 academic year, profoundly affected by his experiences. Weinberg looked forward to the return of the students, expecting a growing participation in civil rights activity. While both student leaders had paid their dues in the field, only Weinberg appeared to have acquired the technical and verbal skills required for his position. Many questioned whether Savio had the prerequisites, ironically doubting his capabilities as a speaker.

The student leaders' tactical experience was dwarfed by that of Clark Kerr, president of the University of California and an expert on labor negotiations. The University of California system included Berkeley and several other campuses, each directed by a chancellor. The chain of command in the California system consisted of the regents ceding power to the president who in turn was above all of the chancellors. In 1958 Kerr, a former industrial relations professor, had risen from Berkeley chancellor to the prestigious position of university president. With the baby boomers and Great Society approaching, Kerr described his vision of a *multiversity* leading the *knowledge industry.* Here vast numbers of students would be educated while faculty produced research sponsored by the government, all meeting the demands of industry and society.[31] Under Kerr's leadership the University of California underwent extraordinary expansion with the Berkeley enrollment increasing from 18,728 in 1960 to 25,424 in 1964. While Kerr's standing as an educational administrator flourished, students became disenchanted with the impersonality and large classes of the multiversity.

Politically, Kerr was a liberal Democrat. In 1964 Berkeley had a small number of radicals, while there was a delicate balance between the liberals and conservatives.[32] The FSM would dramatically strengthen the radical left and increase their rift with the liberals. In time Berkeley radicals would view the "L" word with contempt comparable to that displayed by George Bush and Newt Gingrich in the eighties and nineties.

The chancellor of the Berkeley campus was philosophy professor Edward Strong, but Kerr would be the University point man in the FSM negotiations. Another key administration figure was Vice Chancellor for Student Affairs Alex Sherriffs, a conservative psychology professor. During the summer of 1964, Sherriffs moved to shut down the tables at Bancroft and Telegraph. Whether Sherriffs acted on his own or was pressured is unclear, but the vice chancellor did push the decision in a series of meetings with midlevel administrators.[33] There were several possible motivations behind the ban on the tables. Businesses were feeling the impact of the civil rights activity organized at the tables. From a public relations standpoint, Sherriffs was appalled by the spectacle of entering the campus through a gantlet of scruffy solicitors.[34] Finally, there was ambiguity as to the ownership of the protest venue and the applicability of various statutes and regulations. Other administrators suggested a more deliberate course, but Sherriffs was adamant about removing the tables.

At Sherriffs' direction, Dean of Students Katherine Towle issued a letter, dated September 14, 1964, prohibiting card tables, solicitation, and advocacy at Bancroft and Telegraph, effective September 21, 1964. Many student organizations relied on this location and negative reaction to the new policy was immediate and widespread. A remarkably diverse coalition of clubs quickly formed with the goal of obtaining a more feasible policy. Adopting the name United Front, the group represented civil rights, political, and religious organizations. The broad umbrella of the United Front included socialist groups as well as the Young Democrats and Youth for Goldwater.[35] While the name would be shortlived, the theme of operating as a single group would persist as both a powerful weapon and challenge for the movement. Negotiations of the United Front achieved only minor revisions in the administration's position. Meanwhile, both Kerr and Strong went on record defending the no solicitation policy. At this stage there was an impasse, but the administration had every reason to be confident of its ability to sustain its position.

As the students tried to negotiate, they also sparred with the new regulations, setting up tables at Bancroft and Telegraph. On September 28 at a scheduled university meeting, Strong announced the third, but still unsatisfactory to the students, iteration of the regulations. Unaware of Strong's plans to address the issue, the United Front organized a nearby rally and march to picket the university gathering. For their involvement in the activities Savio and Art Goldberg (a civil rights worker and leader of the local political party Slate) were cited by the administration. That day and the next, the United Front escalated their defiance by setting up tables at the old Sather Gate venue. The administration met this challenge by confronting the "table-manners," each of whom retreated, excepting Sandor Fuchs who was cited.

September 30 provided a foreshadowing of the subsequent events. The students increased their resolve as five were cited for refusing to abandon tables at Sather Gate. Each was summoned to appear before a dean later that afternoon. At the appointed time the five appeared, along with Savio and hundreds of others. The dean refused to meet with the protesters and an impromptu sit-in began in Sproul Hall, an administration building. Neither side had foreseen this event, but each was faced with a dilemma as to formulating the degree of reaction. The administration announced the indefinite suspension of all eight cited students. Savio was emerging as a leader, but could the 21 year old, with a history of stuttering, effectively inspire the crowd? The answer was immediate as Savio suddenly transformed into Patrick Henry. Listeners were moved by his emotion and words, for the first of many times. The machinist's son invoked a machine metaphor which ingeniously wove ridicule of Kerr's own knowledge industry metaphor with the students' alienation toward the impersonality of their institution and the events of the day.[36]

> *You've gotta be a part; part of a machine. Now, every now and then, the machine doesn't work. One of the parts breaks down. And in the case of a normal, regular machine, you throw that part out; throw it out and you replace it. Well, this machine, this factory here, this multiversity, its parts are human beings. And, sometimes, when they go out of commission, they don't simply break down, but they really gum up the whole works. That's what we're all doing here. We've kind of gone out of commission. We won't operate according to the way the parts of this machine should operate, and the machine started to go out of commission. But the remedy is the same! In the case of a regular machine, in the case of this machine, you throw the parts out! And that's what they decided to do. That's what the statement says. They're an indefinite suspension, I presume that's close to the words he used, of those students who weren't good enough parts, who didn't function well enough.*

After Savio's oration captured the crowd, there was further discussion with a minor administrator who was in no position to negotiate demands. By early morning the students were faced with deciding whether to continue the sit-in or move to a new front. They wisely saw that little more could be gained then in Sproul Hall, and decided to adjourn and reconvene with tables at Sather Gate later that day.

The previous night in Sproul Hall served as a pep rally for October 1, pumping everyone up for a confrontation over the tables. The students decided to up the ante by refusing to personally identify themselves when confronted, but rather provide a long list of names of individuals acknowledging participation in the activity. Jack Weinberg and CORE conspicuously made their statement, setting up a large table constructed from a door. The CORE table stood out among many, and it was then the administration's turn to respond. Weinberg recalled the subsequent sequence of events:[37]

> *They came up to me. They asked me for my ID. I refused to give it. They said if you don't leave, we are going to arrest you. The deans went off to get the police to arrest me. I got up and started making a speech. I was drawing a crowd as they came to arrest me. I had already drawn several hundred people around me. So when they came to arrest me, I went limp. They carried me to the police car which had pulled up by the steps. By the time they got me into the police car there were people sitting down all the way around it. You know there are always 10 people who swear they were the first. I'm sure they all did it without anybody else's lead because it was a pretty standard thing. As soon as they got me into the police car it was surrounded. That was right at noon when the whole student body flows past there. It was an incident.*

Quite an incident it was, with the police car immobilized by a mass of defiant bodies. Eventually Savio climbed up onto the car to speak. He was followed by others in a sort of mega–group therapy session. It was an exciting scene that drew Smale and thousands from the community. First there was the real time drama of Weinberg and the car. The car served both as prison and prisoner, but, moreover, it was a stage where speakers vented their rage against the University. Many in the audience validated these emotions while others were appalled. Part of the excitement involved the uncertainty as to when and how it would end.

The sit-in began at noon on Thursday afternoon. It continued throughout the night and on Friday with Weinberg remaining in the blockaded car. For the first day Kerr refused to negotiate, but the scheduled Parent's Day on Saturday called for cloture. The president wanted to avoid a police crackdown on the demonstrators. Late Friday afternoon, Kerr agreed to meet with a delegation of students.

The negotiations pitted an experienced labor mediator against an unprepared, disparate committee of students, including Savio and a representative of the Young Republicans. Despite a great deal of friction between many of the parties, an agreement was reached. Concerning the short term, the protest would end and Weinberg would be booked but not prosecuted. The issues of the suspended students and campus political activity were deferred to committees. The terms of the agreement were drafted and signed.[38]

Thirty-two hours after Weinberg's arrest, Savio climbed onto the police car for the last time and announced the settlement. The crowd dispersed and Weinberg was booked and released on his own recognizance. Of course there remains an obvious question. The optionally equipped police car did not contain a lavatory and 32 hours is too long to abstain. Investigation of this thread indicates the difficulty in reliance on memories. Weinberg recalled that he was permitted to give and receive objects through the car window.[39] In this manner he was able to receive fluids and, when necessary, urinate into a cup which was passed to a friend for disposal. With regard to bowel function, Weinberg maintains that he quickly concluded that the siege might be protracted and took a decision to refrain from solid food. Despite offers from the police to permit him a comfort trip, Weinberg declined for fear that he would not be allowed to return to the car.

David Goines was one of the five students who had been cited on Wednesday. In the chapter of his book, "Jack goes to the can," Goines describes an agreement in which Weinberg was allowed to use the student union facilities. The police were concerned that this would provide an opportunity for Weinberg to escape, while the students were worried that he might not be allowed to return. According to Goines (and corroborated by others), Weinberg was chaperoned by two representatives from each side, a solution providing protection to the students, police, and the car. Did Weinberg go or not? There are two sincere, but contradictory, recollections.

The settlement with Kerr did not really resolve any of the pre–police car issues, and the students quickly met to organize. In lengthy meetings over the next few days, they selected the name Free Speech Movement and developed its structure. To maintain the broad-based coalition of the United Front, the FSM would have an executive committee with representatives from each organization. The 50 person Executive Committee elected a smaller steering committee to handle most matters. The composition of the Steering Committee was more radical, with Savio and Weinberg exerting a strong influence. Another member was Bettina Aptheker, the daughter of a prominent CP official and a party member herself. Although Bettina did not publicly disclose her CP affiliation, her last name was a more than sufficient connection.

Organizing a successful movement required more than just a chain of command. The masses needed information and direction. This was accomplished by rallies and brochures, which in turn required sound equipment, writing, typing, printing, and distribution. A variety of "centrals" developed to perform the myriad infrastructural tasks, staffed by cadres of volunteers. It wasn't seamless, but many students had found meaning and happiness by participating in the FSM subculture.

As the deal with Kerr was implemented, the FSM quickly became enmeshed in a variety of fronts. A vital element of the agreement was the separate committees to address political activity and the suspended students. Disputes over interpretations as to the staffing of these committees led to allegations of bad faith by both sides. The students refused to accept the administration stacked committees, eventually forcing compromises. The result was a faculty appointed ad hoc committee to make a discipline recommendation on the eight students. To study political advocacy there was an 18 person committee with six representatives each from the faculty, administration, and students. A major victory for the FSM was a parliamentary agreement that the committee would act as a troika, requiring a majority of each of its constituencies to put forward a proposal.

The students were excited at the prospect of press coverage, but appalled by the allegations they read of communist and off-campus influence. Jack Weinberg was the only nonstudent among the leaders, and he was a Berkeley alumnus who had been enrolled in graduate school one year earlier. Bettina Aptheker was a stealth communist and others had left wing sympathies, but how was that relevant to this movement? When a reporter pressed Weinberg to reveal who was really behind the FSM, he recalls: "I was trying to find words to express that this was an autonomous expression of *us.* And I thought he was trying to get at what deep, dark plot was behind it. I said we have a saying in the movement, 'that we don't trust anyone over thirty.' ... There was a generational divide."[40] Weinberg's off the top of the head remark, that the FSM was a "youth movement" rather than a tool of the Old Left or anyone else, became substantially more famous than its source.

Early October was spent formulating the shape of committee tables. For the next month the FSM supported a moratorium on direct action, allowing the committees to pursue their missions. Meanwhile there were other formidable challenges. For the FSM to succeed, it was crucial that they remain the sole bargaining agent on behalf of the students. Given the diversity of the Executive Committee and the centrifugal force to which they were subjected, this was no small task. However, any splits would have given Kerr the opportunity to manipulate the factions.

One potential competitor might have been an organization of graduate students. This large Berkeley group included the teaching assistants, who performed a substantial portion of the undergraduate instruction. In the university status hierarchy the graduate students fit somewhere between the faculty and the undergraduates, thinking of themselves as junior faculty but treated as senior students. Prior to 1964 the graduate students lacked sufficient organization even to nominate a representative for the FSM Executive Committee. In October the free speech issues led to the formation of the Graduate Coordinating Council. Over the next month the GCC gained structure and sought to determine its role. During this period Steve Weissman, a first year graduate student in Latin American history, rapidly ascended to a leadership role in the GCC and FSM, chairing meetings and serving as a liaison between the FSM, graduate students, and faculty.

Weissman was a bright guy with an astute knowledge of politics, particularly its history and dynamics. He grew up in Tampa, Florida, and was active in Jewish youth organizations. As state president of the B'nai B'rith AZA, Weissman became skilled at leading meetings. In 1957 he had a brief abortive undergraduate experience at the University of Pennsylvania Wharton School of Business. Returning home, Weissman obtained a BA from the University of Tampa night school. While working on his MA in Russian History at the University of Michigan, Weissman intersected with Tom Hayden and the formation of the Students for Democratic Society (SDS). The Cuban missile crisis prompted Weissman to join the new radical group.[41]

As with Savio and Weinberg, Weissman was attracted to the civil rights cause. Following the completion of his master's thesis on Russian relations with Cuba, Weissman planned to begin a stint as a civil rights worker in the South. The position failed to materialize, and Weissman transferred to Berkeley, arriving on campus in late August of 1964. His earliest FSM memory involved sitting around the police car. When a friend helped organize the GCC, Weissman followed along. Although

Weissman was not among the first GCC representatives selected for the FSM Executive Committee, his facility for chairing a meeting was soon recognized, and he became a leader of the GCC. To Weissman, his role was clear. "The trick was to keep graduate students comfortable enough to be in the FSM without forming an alternative group that was for the same ends, but with different means."[42] The ends and means conundrum would also be a problem for many faculty members.

By early November it was apparent that the troika committee could not produce recommendations on campus political activity. While there was unanimity among the constituencies on the need to relax the campus restrictions on political activity, there remained one major obstruction to an agreement. The University was intent on insulating itself from any association with illegal activity, but the civil rights movement relied on a tradition of nonviolent civil disobedience, sometimes leading to arrests and convictions. Could campus speakers and table-manners advocate and solicit individuals to engage in such activity? What if the original intent was benign or ambiguous, but the demonstration led to violations? The University wanted to exercise its authority in these situations, but the students insisted that they were only subject to constitutional restraint and punishment. To Savio and Weinberg, with their civil rights motivation, this was the core issue for the FSM. Others were willing to settle for tangible gains or continue negotiations, while some did not fully appreciate the distinctions.

This was a period of great contention among the FSM as they pondered whether to resume direct action. Disagreement among the Executive Committee over how to react was exacerbated by the fact that the decision making dynamics of the Steering Committee had evolved to a point where a more radical subset was determining policy. After a failed coup attempt, the Steering Committee was reconstituted with a larger role for the graduate students. A close vote of the Executive Committee supported the resumption of direct action.[43]

Tables reappeared at the Sather Gate venue, and then, a few days later, the ad hoc faculty committee released their findings on the eight suspended students. The report was sharply critical of the administration's handling of the matter. It recommended immediate reinstatement of the six table-manners with their disciplinary records upgraded from suspension to censure. For Savio and Goldberg, the committee suggested that the term of their suspensions essentially consist of time already served.[44]

Of course the committee's standing was merely advisory. The administration's next move came in conjunction with a regents' meeting on November 20. The response was delivered in three stages. First, the regents accepted a recommendation from Kerr that stiffened the faculty suggested sanctions against the students, but did not add any future suspension time. Regarding the free speech issues, the regents approved a liberalization of the rules. The new policy would permit fund raising and solicitation in selected areas to be announced in stage two by the administration (and subsequently included Bancroft and Telegraph). While representing a major gain over the September policies, the new policy did not allow advocacy of illegal actions such as sit-ins, considered an essential weapon by the civil rights movement.[45]

The first two stages by the administration were masterful. With the suspensions they had asserted their power by demonstrably toughening the faculty recommendations, but the nature of the increase was not especially provocative. On political activity, they had conceded everything, except a complex issue which only the radicals would feel was worth a fight. To Weinberg and Savio, however, this was the *raison d'etre* of the FSM.

The FSM was again split over how to respond. Savio and Weinberg advocated direct action to boost the speech issues. The leadership narrowly approved a Sproul Hall sit-in to take place on November 23. Weissman, who had become an influential member of the Steering Committee, was opposed. He felt that the administrative actions failed to meet the aggrievement threshold required to sustain such a protest. The challenges facing the movement were evident at a rally preceding the sit-in. A few hundred participants entered Sproul Hall, some enthusiastic but others confused or ambivalent. The Steering Committee voted again, reversing their previous decision by one vote. The sit-in was aborted. In retrospect it was a wise move, but it is easy to understand the disappointment of the partisans. It appeared that there was insufficient support to counter the administration's moves and closure was likely on the FSM issues. With Thanksgiving and Christmas vacations approaching, the FSM seemed moribund.

Looking then at the scorecard, both the FSM and administration could spin their gains or lament their losses, but it would have been difficult to declare a loser or winner. However, unknown to the FSM, there was still a third remaining stage to the previous administration-regents actions. A standard technique employed by universities is to announce unpopular news over vacations. Hence, spring break is when tuition increases are often unveiled. Over the Thanksgiving weekend Savio, Goldberg, and two others received letters from Chancellor Strong, instituting disciplinary action for their post-September 30 activities, i.e., the police car sit-in.[46] Whatever the reasoning, the letters revived the FSM cause, validating the student suspicion that the administration was vindictive and acting in bad faith. Hundreds had participated in the police car sit-in, but two months later four were singled out for punishment.

The November 23 sit-in had failed for lack of a well focused stimulus. Now there was an issue (complete with martyrs), but what was the best response? Two possibilities were either another sit-in or a strike. There was sufficient outrage among the student body to provide the critical mass for a sit-in. The FSM elected to pursue this course, mobilizing its considerable logistical skills for the enterprise.

Meanwhile the notion of a strike had intrigued GCC leaders for some time. Weissman viewed a strike as a penultimate step in an FSM victory.

"The only way we could win was to get the TAs* to go out on strike. If we created a situation, the faculty would find a way to support us."[47] However, the TAs would require a strong provocation before participating in a strike. The GCC's early strategy was to keep the graduate students under the FSM tent until the administration supplied the provocation. As the sit-in approached the choreography became clear.

1. FSM sit-in
2. Police crack down on students
3. Outraged TAs organize strike
4. Faculty find solution which is favorable to FSM.

These were four large steps whose order was crucial, but GCC preparations for the strike proceeded in parallel with those for the sit-in. The first three steps followed the script. On December 2 thousands attended the FSM rally conducted outside Sproul Hall. Savio, whose passionate oratory was so crucial to the movement, returned to the machine metaphor for his most memorable words:[48]

> *There is a time when the operation of the machine becomes so odious, makes you so sick at heart, that you can't take part; you can't even tacitly take part, and you've got to put your bodies upon the gears and upon the wheels, upon the levers, upon all the apparatus and you've got to make it stop. And you've got to indicate to the people who run it, to the people*

* teaching assistants

who own it, that unless you're free, the machine will be prevented from working at all!

Joan Baez then performed and over 1000 entered Sproul Hall, singing "We shall overcome." Mike Shub was among the protesters. For several hours, people were permitted to enter and leave Sproul Hall. During this time Smale mingled among the protesters. In the evening access was restricted with Shub and about 1000 others remaining. Smale and Baez had both departed.

It appeared that Kerr had lost control of the campus, and Governor Edmund "Pat" Brown elected to intervene. Brown was in communication with the deputy district attorney of Alameda County, Edwin Meese III. Meese was stationed on the Berkeley campus and recommended sending in the police to end the demonstration.[49] The governor issued the order early in the morning. The decision was not unexpected by the FSM. As the police approached, Jack Weinberg and other civil rights veterans instructed each floor's occupants in techniques of passive resistance. The FSM planned to protract the arrest phase into the day, in order that the campus might witness their peers being dragged out of the building. It took the 367 police officers 13 hours to clear the building and arrest 773 participants.[50] The horrifying spectacle brought support for the FSM and condemnation of the administration.

Although the FSM was delighted with the impact and magnitude of their action, there was concern that the jail time might create a vacuum of leadership on the campus. There was an urgent need to organize bail and the forthcoming strike. As the arrests proceeded, several participants prevailed upon Steve Weissman to make an escape. Eventually Weissman climbed down a rope from a second story window. As Moe Hirsch, Steve Smale, and other faculty assisted students with bail, Weissman lit the fuse on the previously planned strike. The first three stages were complete and both sides looked to the faculty for support.

The large Berkeley faculty was among the most distinguished in the academic world, a status based on their scholarly research. To achieve this acclaim required considerable devotion to their work, and permitted little time or inclination for campus affairs. Most faculty were not significantly impacted by the early FSM activity, but a few tried unsuccessfully to mediate the dispute. A popular position was to agree with the goals of the FSM, disassociate from the means, and remain in the background. Smale and Hirsch were unusual in that their support of the FSM was without qualification.

Steve Weissman recalls the two mathematics professors as valuable alternative role models for the graduate students.[51] These students were susceptible to pressure from the faculty who controlled their stipends, as well as serving as thesis advisors and job references. There was little future in academe for graduate students who alienated the faculty in their department. As the campus chaos increased in late November, the FSM threatened the very existence of the University, and became persona non grata with some of the faculty. Graduate students had a great deal at stake in deciding their loyalties. To keep them comfortable in the FSM camp, Weissman needed a prominent professor to publicly endorse the movement, offering 100 percent support. Hirsch and Smale came forward, and provided the graduate students hope that other faculty would come around.[52]

When the campus erupted with the arrests and strike, teaching and research were impacted, and the faculty was thrust into the *real* campus world. Kerr met with a group of department chairs, fashioning a proposal that would be presented the following Monday at a convocation in the University's Greek Theatre. Meanwhile the Two Hundred, a group of more liberal faculty, sought a solution compatible with the FSM demands. On Tuesday the faculty would meet to decide their position.

The FSM was determined to have some role in the Monday Convocation. When Savio requested permission to speak at the gathering, he was refused. He then asked that he be allowed to make an announcement of a followup FSM meeting. A second rejection did not deter Savio who declared his intention to make the announcement. The film *Berkeley in the Sixties* captured Savio's appearance of torture and disbelief, as he listened to Kerr's address. When the meeting concluded, Savio walked to the microphone. Before he could speak, Savio was dragged off the stage by two police, one of whom was pulling him by the tie. Consider the scene: the president has just presented his free speech vision to the university community, and then the icon of the Free Speech Movement is mugged as he tries to speak. The momentum abruptly shifted to the FSM.

As 1000 faculty attended the Tuesday meeting, the discussion was piped outside to thousands of students. The motion of the Two Hundred passed overwhelmingly, supporting the FSM position on advocacy. The faculty had moved under the FSM tent. Ten days later the regents joined the bandwagon and the administration was defeated. Strong was replaced the following month and Kerr was critically weakened.

During those last three months of 1964 an extraordinary event had occurred on the Berkeley campus. A small group of students battled and defeated a powerful university administration. In the process they prompted a university community to examine and alter the traditional roles of students, faculty, and administration. Subsequent years would continue the process on other campuses, but it began at Berkeley in the fall of 1964.

While the FSM was a popular movement that benefited from the efforts of thousands, there were three people whose names and contributions most defined what occurred. They were Jack Weinberg, Mario Savio, and Steve Weissman. Thirty years later Weissman, who had organized the graduate students and represented the FSM to the faculty, provided the following analysis of their respective contributions:[53]

> *Jack was really key to the whole thing because of his understanding of direct action and how to really use it. He was really brilliant at it. Mario gave the movement its personality and its public face and in many ways things would not have happened the way they happened without him, but he also created our biggest problems. He had an ability to reflect what was going on or what a lot of people were feeling, except he would do it in a huge way. You would see him do what I considered to be the wrong thing up publicly, and everyone would go with him, and everyone would get disappointed by the answer and then come back, and he would get up and take a new position, and everyone would be with him … I don't know of any movement on predominantly white college campuses where the leader had the personal importance he did in that way, but in terms of what actually happened, I think Jack was more important, because of his knowing how to push the other side by using very simple direct action tactics. My role was much more behind the scenes. I ended up chairing all the meetings and rallies because I knew how to chair meetings … The meetings were quite open. We didn't railroad, but believe me I knew who to call on first and who to call on last. I'm accused of such and I'm guilty of that.*

And whom did Weissman call on last? "Whichever way I wanted it to go."

The Lone Ranger of the Antiwar Movement

06

A. Free speech after the FSM

After the regent's vote in December, the mission of the Free Speech Movement was complete. When the students returned to the Berkeley campus in January, their energy shifted to the legal defense of the hundreds arrested in the Sproul Hall sit-in. The county elected to press charges, and the legal proceedings continued through the summer. Each of the leaders received jail time, while the followers got fines and/or probation. For those on probation, a subsequent offense would land them in prison. It appeared that the government's strategy was to make an example of the leaders and to motivate the rank and file to remain out of trouble.

By March 1965 the free speech boundaries had already been tested. A young man named John Thompson had been attracted to Berkeley by the permissive culture. On March 3, he sat in the Bancroft-Telegraph area, holding a sheet of paper on which he had written the word: FUCK. Mario Savio wandered past and observed there was ambiguity as to which part of speech Thompson wished to denote. This prompted Thompson to add the parenthetical information: (verb). Eventually Thompson was arrested, leading to a small scale protest demonstration.[1] The ensuing cause became known as the Filthy Speech Movement, and included a controversial student magazine, the SPIDER. FSM leaders were confronted with a dilemma. While they generally agreed that Thompson's act should be protected speech, this was not the issue on which they wanted to push the free speech envelope. In any event the cause was short-lived. One consequence was a resignation ploy by Kerr and acting Chancellor Meyerson, who were caught between the speech issues and the regents.[2] The administrators responded to the pressure by announcing their intention to resign, forcing a grudging vote of support from the regents. Despite this mandate of sorts, the episode revealed the instability of the University of California leadership. Meanwhile, the new free speech environment was about to be tested by the birth of an anti-Vietnam War movement.

The speeches of Isaac Deutscher, Norman Mailer, and others at the May 1965 Berkeley Vietnam Day demonstrated that campus speech had entered a new phase. In organizing the massive teach-in, Steve Smale and Jerry Rubin adroitly exploited the administration's impotence in the post-FSM environment. As cochairs of the Vietnam Day Committee, Smale and Rubin then introduced and pushed nonviolent civil disobedience as a tactic for the antiwar movement. When the FSM fought for concessions on speech and advocacy, the leaders expected that the immediate application would involve the civil rights cause. It was the Vietnam Day Committee, however, that was the true beneficiary. To follow these events requires some additional perspective.

B. Vietnam Day

The mood of the country in 1964 was substantially different than that in FSM-inspired Berkeley. It was post-McCarthy and pre-Vietnam. Americans were strongly influenced by the Cold War and nuclear fears. John Kennedy had inspired a country that was not yet cynical about its leaders. There was a conventional wisdom that "we are the good guys and the commies are the bad guys."

Foreign policy in the Kennedy and Johnson administrations was designed by men whose views were shaped during World War II and the Cold War. They saw a world that had been saved from the grasp of Adolf Hitler, and then faced the pervasive threat of communism. Most significantly, they subscribed to the Domino Theory espoused by Eisenhower in 1954.[3] Under this model, the countries of the world were arranged as rows of dominoes, subject to the monolithic force of communist domination. If South Vietnam were to fall, then it would trigger a chain reaction knocking down the entire Southeast Asian row, increasing communist power and threatening United States security. While this simplistic notion of foreign dynamics neglected nationalistic movements within countries and overlooked the obvious tension between communist powers, it did pervade the thinking of the Kennedy and Johnson administrations. Robert McNamara was secretary of defense and one of the primary architects of Southeast Asian policy. Originally a subscriber to the Domino Theory, he has recently recanted his views.[4]

Vietnam posed a difficult policy problem, and presidential advisors were often divided and rarely confident in their recommendations.[5] In 1964 South Vietnam possessed a weak internal government, subject to challenges from within and from the North. It was an axiom of American foreign policy that the South Vietnamese domino must remain upright. To maintain its stability, the United States sought to bolster the South and discourage the North. In 1964 there were 16,000 United States military advisors in Vietnam. Meanwhile the CIA was providing support for South Vietnamese actions in the North.[6] Despite these efforts the South continued to lose ground, and the Johnson administration considered overt military support. Possible options included bombing and ground troops.

There was a legal ambiguity as to whether the president could take such actions unilaterally. In any event Johnson recognized the political wisdom of obtaining congressional support, and seized an opportunity in August. During that period South Vietnamese boats, under United States supervision, engaged in missions to attack North Vietnamese targets. On August 2, 1964, the American destroyer

Maddox was conducting electronic surveillance in the Gulf of Tonkin, when it was attacked by North Vietnamese torpedo boats. Two days later, during stormy weather, the *Maddox* reported a new attack. Later that day, follow-up investigation cast doubt as to whether an attack actually did occur. (Questions remain today.) Despite the difficulty in obtaining conclusive confirmation, Johnson ordered retaliatory raids, and appeared on national television to announce this action and his intention to seek legislation. Absent from his remarks was any mention that the North Vietnamese had received some provocation and that there was doubt as to the validity of the second attack.

Public response was enthusiastic as Americans were incredulous that a third world country would challenge the United States. Three days later Congress passed the infamous Gulf of Tonkin Resolution, providing the president a blank check to undertake military action in Southeast Asia. In the entire Congress, the only negative votes were registered by Senators Wayne Morse and Ernest Gruening. Although administration representatives assured the Congress that there would be no abuse of the new authority, Johnson would repeatedly betray the pledge.[7]

In late 1964, opposition to United States involvement in Vietnam was virtually nonexistent. The radical Students for Democratic Society reluctantly decided to sponsor an April protest march, but their only immediate support came from some elements of the far left.[8] During this period the administration realized that the South Vietnamese government was on the verge of collapse. While this information was withheld from the public, advisors debated policy changes. Suggestions varied, but one constraint was becoming clear. If you accepted the Domino Theory, which the administration did, then American security required an escalation in involvement.

February 1965 was a significant month in the evolution of the United States military involvement in Vietnam. Just as the Johnson administration was moving toward the initiation of bombing attacks, there was a guerrilla raid on a South Vietnamese military base in which eight American soldiers were killed. The president ordered retaliatory air attacks, and issued a statement stipulating the limited nature of the action. Despite enthusiastic public support for Johnson's response, some opposition began to develop. Women Strike for Peace picketed the White House, fourteen protesters were arrested at the United Nations, and on some college campuses there were calls for United States withdrawal from Vietnam. At Yale 600 students attended a meeting at which four professors advocated this course.[9]

Meanwhile Johnson authorized Operation Rolling Thunder, a program of regular bombing of North Vietnam. In making this decision the president consulted with advisors and selected members of Congress, but there was no congressional or public debate, indeed there was no White House announcement. To protect the air bases, Johnson approved the deployment of marine battalions, as the ground mission was upgraded from that of advisory. Operation Rolling Thunder represented a major policy change. While previous actions could be spun as defensive, advisory, or retaliatory, the United States was now unquestionably embarked on an open-ended offensive military campaign against another country. Although the Gulf of Tonkin Resolution provided Johnson with this authority, he was silently taking the country into a war. Announcement of the bombing campaign came from the American embassy in South Vietnam, one week after its initiation.

Lyndon Johnson had campaigned for president in 1964 as the peaceful alternative to Barry Goldwater, whom he had portrayed as a warmonger. Even some of Johnson's campaign supporters felt betrayed by his actions on Vietnam. As the new phase of American involvement in the War began, the first nontrivial antiwar response followed in March. Typically, opponents to the Johnson policy advocated a negotiated settlement in Vietnam, leading to a coalition government. This approach had been rejected by the Johnson administration because they projected that it would lead, in the medium term, to a communist government. While Johnson had no analysis to indicate that bombing would produce a different result (in fact he had serious doubts[10]), it would at least delay the outcome.

At the University of Michigan some faculty sought a new form of protest that could incorporate the university setting into its method. They devised the notion of a "teach-in."[11] The original model consisted of a one day moratorium on regular classes, with lectures replaced by lessons on Vietnam. The concept was supported by 49 faculty members and immediately became a source of controversy with the university administration and state legislature. Eventually the plan was revised so as to take place overnight, rather than pre-empt classes. Thus from 8 pm on Wednesday, March 23 until 8 am the following morning, several thousand Michigan faculty and students participated in the first teach-in.[*]

A significant impact of the Michigan teach-in was its proliferation to campuses throughout the country. Other colleges followed the template, perhaps adding their own customizing touches. Despite Johnson's effort to avoid discussion of his Vietnam policy, the dynamics of a spring college fad thrust the topic into the United States intellectual forefront. At Berkeley, an anthropology professor quickly organized a campus meeting for the Wednesday evening of the Michigan teach-in. Steve Smale was one of six faculty members who spoke to the 1000 attendees. Smale called for termination of the air strikes.[12]

At the beginning of April, President Johnson approved additional ground troops, designating their mission as active combat.[13] This time the White House announced the increase, but denied that it involved any shift in policy. The acknowledged incremental escalations in

* The word "teach-in" is now part of common usage. American Heritage Dictionary defines it as "An extended session, as on a college or university campus, for lectures and discussions on an important and unusually controversial issue."

American involvement together with the popularity of the teach-ins increased interest in the scheduled SDS march. While the main protest was set for Washington, there was a parallel event in San Francisco. To increase turnout and effectiveness, SDS sought endorsements from other organizations. Smale, as the chair of the Political Affairs Committee of the Berkeley Faculty Union, succeeded in obtaining the union's sponsorship as well as the following statement:

> *The Berkeley University Teachers Union condemns the recent air strikes by U.S. military forces against North Vietnam, which have greatly increased the risk of global war. It deplores the failure of our government to justify adequately our involvement in Vietnam and urges the President to avoid further escalation of the war and to seek a negotiated settlement immediately.*

The *New York Times* reported that the Washington participation in the SDS march exceeded 15,000.[14] Steve noted that while the San Francisco march attracted a sizable turnout[15] that the strictly controlled organization was reflective of the Old Left. Even signs required the approval of an organizing committee.[16]

The idea of a Berkeley teach-in originated with a young couple, Barbara Gullahorn and Jerry Rubin. For Rubin, later one of the most prominent radicals and characters of the sixties, the teach-in provided his first opportunity for leadership and notoriety. Gullahorn recalled how it all began as a romantic reconciliation.[17]

> *At the end of an argument, we started talking about the teach-in that had happened at Michigan; and then, why didn't we try to do a teach-in at Berkeley. Then I can remember holding hands and walking to see Ralph Gleason. It was sort of a way of doing something together. It had a certain personal element.*

Gullahorn had arrived in Berkeley in 1960 as an undergraduate at the University. Politically she was a liberal who became inspired by John Kennedy's vision of the Peace Corps. Gullahorn majored in political science and planned to enter the Peace Corps following completion of her degree. Meeting Rubin profoundly altered her outlook.[18]

Rubin moved to Berkeley in January 1964, ostensibly to begin graduate study in sociology. He was 26 years old. Full of energy and seeking his niche, Rubin was intrigued by political and social issues. Naturally he was drawn to the Berkeley civil rights protest culture, and quickly transferred from academe to activism, becoming a regular on the picket lines.[19]

During this period, Rubin and Gullahorn met via a mutual friend. The Kennedy-liberal woman and Marxist man began arguing over politics. The intense discussions led to romantic involvement. As Gullahorn began to examine her perspectives in a new light, Rubin seized an opportunity to visit Cuba. With her boyfriend's influence limited to an occasional phone call, Gullahorn reverted back to her liberal mode, returning home to her mother in Palo Alto.[20]

Inspired by Fidel Castro and Che Guevara, Rubin returned to California late in the summer.[21] When confronted with an immediate choice between Berkeley activism and his girlfriend, Rubin followed her to Palo Alto. There he moved into the Gullahorn household, an awkward environment where the young couple attempted to reconcile their differences. Meanwhile the FSM began in Berkeley without Rubin. "I don't think he has ever forgiven me for, having in a sense, kept him from the birth of the Free Speech Movement" recalled Gullahorn.[22] Rubin did, however, persuade Gullahorn to return to Berkeley where the couple became active in the FSM. When the big Sproul Hall sit-in occurred, the position of the more moderate Gullahorn prevailed. The couple remained on the outside while others had their radical credentials validated by arrest. Jerry missed out again.

The Vietnam teach-in provided a new opportunity. Rubin and Gullahorn decided first to sound out Ralph Gleason, a *San Francisco Chronicle* music critic who was prominent in the radical culture. Gleason greeted the idea with enthusiasm and made nominations for the program. Others offered suggestions as well and the concept quickly gathered momentum.

The upcoming teach-in would provide a real test of the new campus free speech policy, and a professor could be a valuable ally in obtaining use of university facilities. At an early stage of the planning, Rubin and Gullahorn solicited Steve Smale in his office. None of the three could remember the precise sequence of events that brought them into contact for the first time, but it seems likely that Steve's contribution to the April march would have had a direct or indirect impact. Gullahorn recalls that Steve received them with a combination of bemusement and graciousness.[23]

Smale was approaching a major juncture in his life. At 34 he appeared to have it made. A highly paid professor at a first rate university in the idyllic Bay Area setting, he had already achieved an enviable standing among mathematicians. Most people would have been quite content with this situation, and continued down the professional path. However, just as Smale had previously reacted to the Korean War and Cuban Missile Crisis, Rolling Thunder provoked a new outrage. He was especially concerned that Vietnam might lead to a war involving the United States and Soviet Union.[24] As chair of the Political Affairs Committee of the Berkeley Faculty Union, Steve had some standing to push the protest.

Rubin and Gullahorn's visit succeeded in enlisting more than an ally. Smale became a full partner who would play a major role in shaping the direction of the project and its sequels. That Steve took on this responsibility reveals several aspects of his perspective and approach. At that time the Johnson administration policy on Vietnam had the overwhelming support of the American people. While there had been a large number of teach-ins and a few demonstrations, there was no national antiwar organization nor any serious prospect of altering the president's position. Smale elected to lead an unpopular cause, not only believing that it was important, but that there was a possibility of having an impact. Just as with the Poincaré Conjecture, it was an audacious risk of time and energy. In addition he was investing his reputation in an enterprise from which faculty typically maintained some insulation from students. Recall that with the FSM, faculty played supporting and conciliatory roles, but never served on the Executive or Steering Committees. In the university caste system, faculty and students were not peers. However, to the amazement of both Rubin and Gullahorn, Steve joined the meetings at their apartment, participating as a true equal.

A Rubin–Smale symbiosis quickly formed. Despite their contrasting backgrounds and personalities, Steve and Jerry developed a strong rapport and maintained a great deal of communication, presenting a united front at the meetings. They tended to agree on most things, including the priority of producing a large spectacle and the important role that the media served in obtaining that end. They became co-chairs of the Vietnam Day Committee (VDC), organizer for the Vietnam Day teach-in beginning on May 21. Following Steve into the VDC were his wife Clara, colleague Moe Hirsch, and student Mike Shub.

The VDC confronted several contentious issues. While the campus was an obvious candidate for the Vietnam Day venue, some suggested that an alternative location might provide more credibility. To Steve and Jerry the logic was simple. They wanted to maximize the size of the crowd. This would not only permit the program to reach the most people, but press reports of the attendance would then have the greatest nationwide impact. Why not have Vietnam Day in the most convenient location to attract the tens of thousands of potentially sympathetic students? The campus location was selected, but that raised additional concerns as to whether the university administration would accommodate the event. Here Smale took charge as administrative liaison. His strategy was to hold off on any formal request for facilities and proceed with the planning as if the campus would be placed at the VDC's disposal. Meanwhile he obtained the endorsement of the faculty union.

Another volatile issue involved the selection of speakers. Leftist history in the United States is replete with turf battles, particularly between Marxists and Trotskyists. Smale felt that the San Francisco march had been handicapped by a narrow ideological perspective. For Vietnam Day he envisioned an inclusive program with liberals, all flavors of leftists, and even advocates of current administration policy. When this view prevailed, the task remained to identify and invite individuals. Steve recalled the process of putting together the program.[25]

> *We didn't spend much time on analysis and theory. We knew what we wanted and had some good instincts about how to obtain it. Therefore, our mode was one of continually doing things, all kinds of things, which would make Vietnam Day into a bigger and sharper antiwar protest. Jerry and I did, however, talk for many hours about how to deal with various problems that came up.*
>
> *It was a lot of fun to work in the Vietnam Day Committee. We didn't think in terms of duty. It was more of an exciting challenge: How to make something that would be the greatest teach-in, the biggest Vietnam War protest. How to make Johnson cringe.*
>
> *For example we would sit around in Jerry's apartment making suggestions for speakers. It was like a competition to see who could propose the biggest, most provocative names. There was practically no limit to our ambitions; Bertrand Russell, Fidel Castro, the ex-president of the Dominican Republic, Norman*

Mailer, U.S. Senators, Jean-Paul Sartre, and so on. Then we would call them up then and there, or send telegrams. And sometimes it worked and invitations were accepted.

But there was another important aspect to sending out these invitations. It helped gain the attention of the media …

Financing the cosmopolitan assortment of speakers was another matter. If all went well, Vietnam Day would be well attended and there would be an opportunity to solicit funds at the actual event. However, obtaining advance contributions on a scale to support the ambitious program was not possible. What then was one to tell invited speakers from the east coast about their airfare? The organizers elected to offer an optimistic spin to the prospect of raising sufficient funds for reimbursement of expenses.

Initially the teach-in was planned to last 24 hours. As opponents of the War began lining up to speak, the organizers shifted to a 30 hour span, and then longer. A variety of entertainers were scheduled so as to provide a festive element to the long proceeding. While the overwhelming focus of the teach-in was on opposition to the Johnson policy, efforts were made to include some administration supporters in the program. That these attempts largely failed was not surprising. It was reasonable to expect an audience that was at least adversely disposed toward this position, if not outright hostile. Additional difficulties were the result of the State Department's rapidly changing strategy on public relations. Despite President Johnson's efforts to obscure his Vietnam actions, the Michigan and other teach-ins had generated interest in the issue. In a late April speech, Secretary of State Dean Rusk attempted to blunt criticism with an arrogant statement directed at the faculty involved in the teach-ins. "I continue to hear and see nonsense about the nature of the struggle [in Vietnam]. I sometimes wonder at the gullibility of educated men and the stubborn disregard of plain facts by men who are supposed to be helping our young to learn—especially to learn to think."[26]

In early May, Smale and Hirsch responded by inviting Rusk to speak at Vietnam Day. At that time the State Department was formulating plans to confront their critics, and agreed to supply two representatives who would be named at a later date. Teams of midlevel government staffers with Vietnam experience were then being dispatched for appearances at several midwestern universities. Following a particularly contentious encounter at the University of Wisconsin, the State Department concluded that the so called "truth teams" were not producing the desired results.[27] They then backed out of Vietnam Day.

Two Berkeley political science professors, Eugene Burdick and Robert Scalapino, were invited to present their views, known to be supportive of the Johnson policy. Both declined, citing the stacked program of opposition speakers and noting the several entertainment interludes on the agenda. Burdick described Vietnam Day as "a protest masked as an ideological circus." Scalapino was more outspoken and raised disciplinary issues, castigating Smale and Hirsch.[28]

> *The May 21 meeting on the Berkeley campus is symbolic of the new anti-intellectualism that is gaining strength today.*
>
> *A few individuals, most of whom would not dream of treating their own disciplines in this cavalier fashion, have sponsored a rigged meeting in which various ideologues and entertainers are going to enlighten us on Vietnam.*
>
> *Only a handful of the performers have ever been to Vietnam or made any serious study of the problems. The objective is propaganda, not knowledge.*

Scalapino's attack received quite a bit of local press prior to the teach-in. Smale, Hirsch, and Rubin countered, accusing Scalapino of slander and cowardice. Embedded in the rhetoric was the issue as to whether mathematicians might constructively address an important question of foreign policy. The vertical nature of mathematics certainly makes it unlikely that political scientists or politicians could prove significant theorems, but Congress and the executive branch do determine the National Science Foundation budget. On the other hand Vietnam issues were generally accessible, despite their obfuscation by the president. Passage between mathematics and social science is not symmetric.

As the VDC invited speakers and put out publicity for the massive campus teach-in, there was no assurance that the University would permit the event to take place. Certainly it would have been impossible in the pre-FSM period one year earlier. Although Rubin and Gullahorn feared the university administration might subvert their plans, they deferred to Smale who was confident that he could exploit the recent shift in campus power. With the exception of the Filthy Speech Movement, the new campus speech rules were untested, and open to interpretation. Smale knew "that the UC administration, by imposing university rules, could have structured the protest to make it innocuous. To meet this, we developed our own program, and planned it on the campus before the requisite permissions had been obtained. To some extent, the authorities were presented with a *fait accompli,* giving them little choice."[29]

The FSM had seriously weakened President Kerr and forced Chancellor Strong out of office. With the FSM immersed in its own legal proceedings, there was no intersection between the initial VDC leadership and that of the FSM. However, if acting Chancellor Martin Meyerson had placed any substantive restrictions on Vietnam Day, there was a genuine likelihood of reviving the FSM cause. From Meyerson's perspective, at least Vietnam Day was directed at the national administration rather than that of the University. On May 11, ten days prior to the start of the teach-in, Smale and Rubin made their first formal request to the chancellor. Included with their plans and needs was a bold statement. "Of course we require that all the usual regulations regarding limitations on the distribution of literature, fundraising, speaker's approval, and other political activity be suspended for this period."[30]

Smale's plan succeeded as Meyerson conceded on the central issues. In order to accommodate the large crowd predicted by the VDC, the administration approved the requested venue and even agreed to make some desired physical modifications (e.g., removal of a fence), with reimbursement after the event. Equally important was the waiver of a requirement for advance approval of speakers. There would be free speech. Smale had driven a truck through the opening created by the FSM. Meyerson did, however, draw the line at the VDC request for the cancellation of classes.

When May 21 finally arrived, Rubin and Smale delivered the teach-in that they had advertised. The traditional evaluative measures of such events include duration and attendance. At Berkeley they even surpassed the 33 hour schedule. Newspaper estimates placed the crowd at 10,000 during its peak moments, but a significantly larger number attended some part of the proceedings. Fundraising efforts recouped all the expenses.

By more substantive criteria, the program provided speakers of greater prominence than its predecessors. In addition to Norman Thomas, Benjamin Spock, and Senator Ernest Gruening there was the passionate pro-Marxist speech of historian Isaac Deutscher and the anti-Johnson oration of novelist Norman Mailer. The profanity in Mailer's intense personal attack on the president led a local radio station to suspend its live broadcast of the teach-in. The maverick journalist I. F. Stone, who did not complete college himself, made a big impression. He gave two lectures, receiving a standing ovation both times. Stone was a tireless researcher with a disdain for the inside source approach.

Moreover, he had the ability to distill his vast knowledge into lucid arguments that ridiculed administration policy.

There was a strong anti-administration ambience at the teach-in. Two chairs were mockingly placed on the stage, designated for the representatives that the State Department had withdrawn earlier. Two other supporters of administration policy did appear, but they were unsuccessful in persuading the crowd.

Among the most enthusiastic attendees was Marilyn Milligan, who had arrived at Berkeley in 1963 for postdoctoral work on animal behavior. Like other Berkeley people during this period, experiences outside the laboratory "transformed" her life.[31] In her first year she participated in some of the civil rights activity. Year two began with the FSM. From the police car sit-in through her December arrest in Sproul Hall, Milligan was a loyal FSM supporter. She still recalls her disappointment when the abortive sit-in was called off. The rallies provided an excitement and meaning that were absent from her work. Milligan was further along the academic path than most participants, and struggled at dividing her time between work and the FSM. Vietnam Day provided another diversion. Milligan was enthralled by the scene, remaining overnight and enlisting to participate in future activities.

C. Troop train protests

With Vietnam Day, Smale and Rubin had pushed the teach-in genre as far as it could go. In Berkeley, they had expanded awareness of the Vietnam War and created hostility and skepticism toward the Johnson policy. To alter that policy would require a proselytization far beyond the Bay Area. Vietnam Day, the grandest of all teach-ins, had only made page 26 of the *New York Times.* Any serious effort to stop the War would require new approaches.

The Vietnam Day Committee was originally created with the purpose of organizing Vietnam Day. With the success of the teach-in, the organizers realized that Vietnam Day had the potential to be a first step in an antiwar movement, rather than a single shot at raising Berkeley awareness and an entry in the teach-in section of the Guinness Book of Records. Marilyn Milligan and other FSM alumni were volunteering their support to the already existing VDC. Two questions faced the leaders:

1. Would Smale and Rubin continue the VDC?
2. How do you stop a war?

In planning their post-Vietnam Day lives, Smale and Rubin were in substantially different positions. As Gullahorn recalled, "Jerry always had high hopes for greatness in all things he did."[32] Cochairing the VDC had provided Rubin his first moments in the limelight. Continuing the organization offered the prospect of furthering the cause and his own notoriety. Without a career, there was no competing channel for his considerable energy.

Smale faced another choice between career and cause. For a month he had focussed on the VDC, finding excitement and fulfillment. Smale had also obtained a major dynamical systems result.[33] He had produced a 4-manifold on which structural stability was not generic, deflating his original vision for the field and raising new questions. The summer was approaching. Academic scholarship does not require a day to day accounting of effort, leaving each individual substantial freedom in organizing their schedule. However, despite the frequent public perception that university faculty take a vacation for the entire summer recess, most research mathematicians find that the period provides the best opportunity for the total immersion required for deep research. With few distractions from students and administrative demands, it is normally a time of intense thought on mathematical problems. If Smale continued his leadership level in the VDC, it would likely undermine the valuable research opportunity available during the summer months.

Whether or not notoriety was a factor for Smale, as it was for Rubin, prospects existed with both mathematics and the VDC.

At Vietnam Day, Yale historian Staughton Lynd had proposed massive nonviolent civil disobedience as the next step in opposing the War, citing the Oakland Army Terminal as a target.[34] Rubin recalled meeting with Smale after Vietnam Day. To Steve the next step was clear: close down the Oakland Army Terminal.[35] Just as with proposing big names for Vietnam Day, when it came down to discussing the first major plan for civil disobedience to oppose the Vietnam War, Smale and Rubin seemed to build on each other. Both had the audacity to think big, very big.

This was a major moment in the history of protest against the Vietnam War. Smale and Rubin were starting a serious movement to stop the War. The SDS had had this opportunity after the April protest, but had failed to seize it. If Rubin and Smale had not begun in late May 1965, others would undoubtedly have taken the action after the later escalations. However, it was the Smale/Rubin symbiosis that actually took the steps to implement Lynd's scenario of an internal revolt against the War. Moreover, they believed that they could make it work.

Smale and Rubin drafted a call to action, committing themselves to an expansion of the VDC. Timing was important. The announcement was made in the afterglow of Vietnam Day, but any major action would necessarily await the return of the students in the fall. "The Vietnam Day Committee is planning another protest meeting on October 15 with massive civil disobedience on October 16. All peace and political groups, nationally and even internationally, are being asked to support these days of protest."[36] As their collaboration continued, a major unifying thread was their ideas on how to use the media to facilitate their ends. Smale knew that television images of protest could cause people to examine their government's policy. As a former reporter, Rubin had good instincts for dealing with the press and how to attract their interest. Together they viewed the media as a vital resource. This was a profound change from the somewhat adversarial perspective of the FSM.

The VDC did not hibernate over the summer. While organizing support for the weekend of October 15 was the long range goal, there were opportunities to attract attention in the intervening period. Smale reflected:[37]

> *The first such demonstration occurred when Lyndon Johnson visited San Francisco June 25; this was a chance that couldn't be missed. The man most responsible for U.S. aggression in Vietnam was coming to our territory. It turned out, however, that I had already made plans to go to the Hawaiian Islands with my family for a short vacation. But during that sojourn in Hawaii, I was anxious about the anti-Johnson activity taking place in San Francisco.*
>
> *On my return I learned that the demonstration had been successful; however, I couldn't help feeling, as I read the news stories, that a great opportunity had not been fully exploited by the VDC. "If I had been there ... "*

Several thousand turned out to protest, picket, and march. The numbers were good, but the demonstration was mundane. If the VDC had devised some novel spectacle or confrontation, it would have been the hook for widespread stories by the press entourage accompanying the president, rather than another generic picketing note.

Vietnam Day brought people and money into the VDC. To facilitate operations, a house was rented near campus. The VDC was a nonhierarchical organization with lots of open meetings and votes. Planning for the October event, as well as other initiatives, was delegated to a rapidly developing committee system. National and international committees solicited support for parallel demonstrations.

There were also Lawyers, Finance, Defendants, and Commando. Other groups addressed the various infrastructure needs of a large gathering.

Another arm of the VDC, led by Marilyn Milligan, sought to organize the community against the War. She went door to door, mostly in West Oakland where people were willing to listen to her talk about Vietnam. For the postdoctoral scientist "it was very exciting. I found that I couldn't wait to get up in the morning and go start knocking on doors. It was one of the most interesting things I've ever done in my life."[38] To complement the canvassing, there were occasional neighborhood rallies, including musical performances by VDC volunteer Joe MacDonald, who would later achieve notoriety as the leader of the group Country Joe and the Fish.

Despite the summer education efforts of the VDC, the president's Vietnam policy retained overwhelming support among the American people. A July poll registered a 65 percent approval rating for Johnson's handling of the War. While the White House's optimistic spin was succeeding at home, word from the front was discouraging. Military commander William Westmoreland recommended that a doubling of troops was necessary, but not sufficient, for the American effort. As the staff debated the future troop commitments, public opinion remained hawkish with only 11 percent supporting the VDC position:[39]

47 percent more troops

19 percent keep the present number

11 percent take troops out

23 percent not sure

In July 1965 there were 81,400 United States troops in Vietnam and 509 had been killed in action (less than 1 percent of the eventual number). Johnson approved the troop increase, sensing that he was sinking into a quagmire.[40] Again there was no congressional debate as the president invoked the Gulf of Tonkin Resolution.

The War was escalating with no end in sight. Both Johnson and Smale knew that public opinion can be fickle in such matters. As Johnson continued to obscure his actions, the VDC sought to attract attention. For the October demonstration there was support in Japan. The Japanese connection came about at the instigation of the French mathematician Laurent Schwartz, a Fields Medalist and political activist who was visiting Japan in 1965 when he learned of the VDC from Smale.

To increase national support, Jerry Rubin was dispatched to Washington. There he pitched the October International Days of Protest to other activists, seeking to form a national coalition.[41] While he did achieve some success in his diplomatic initiative, Rubin missed out on more exciting developments in Berkeley.

On August 4, the VDC learned that a train would pass through Berkeley on the following day, transporting soldiers to Oakland. This information inspired the suggestion of a new form of protest—confronting a troop train. Smale quickly embraced the concept. He saw it as a flamboyant expression of defiance with the potential to attract national attention to the VDC. Time was short and the VDC house, under Smale's leadership, began to mobilize for the protest. They were operating at two serious disadvantages over the Johnson protest. There was little lead time, meaning that posters and the other traditional publicity methods were ineffective in obtaining demonstrators. Furthermore, the trains lacked a press entourage.

Obtaining a critical mass of protesters required an immediate word of mouth campaign. Clara played a crucial role in this respect as she fired up the Women Strike for Peace "phone tree" (Clara called a list of names, each of whom called their separate list of names, etc., efficiently reaching the entire membership). In attendance at the demonstration were women with children and older women, delighting Smale who wanted to obtain the most from television and press images. To maximize coverage, Steve made calls to alert the media.

The subsequent front page article in the *San Francisco Chronicle* estimated that there were 150 demonstrators.[42] The pictures portrayed several of the protesters blocking the tracks as the train approached at 10 mph. If the tactic succeeded and the train stopped, then the VDC would board and lobby the soldiers. However, when the train maintained its speed, each demonstrator was confronted with a decision. It was a tense moment as the protesters withdrew at the last instant, barely averting serious injury. Nevertheless, antiwar activity had entered a new phase. Steve recalled his feelings as the first troop train demonstration ended:[43]

> *There was rage and alienation at the whole establishment, exhilaration from the events, and excitement anticipating the challenge of the next days lying ahead of me. There was also pleasure from our success. What was this success? We certainly didn't prevent any troops from going to the front lines. I believe that the shock of a militant confrontation constituted an important victory. It was a confrontation in which U.S. soldiers going to Vietnam were put in the center.*
>
> *Moreover, it was a confrontation fraught with uncertainty. Up to that time the antiwar demonstrations were orderly; even in the civil disobedience of the traditional pacifist movement, all parties knew just what to expect. Now no one knew what was going to happen. The spokesmen for the Army and the police clearly showed this concern in their statements. With the extended news coverage of this and subsequent troop train clashes, the country was beginning to see the cost of the War in terms of the threat to its institutions. The stakes in the fight between the government and the protest movement had been raised.*

> *The troop train ambush gives an example of how a few people with the right action at the right time can create an enormous effect. There was no violence, no injury, and no arrest. Nevertheless, the willingness to take these risks was vital to the success.*

As the crowd departed, Smale urged them to return on the following day, when other troop trains were scheduled. To maximize turnout it was desirable to have accurate information on when the trains would reach Berkeley. The VDC quickly set up an intelligence network with scouts positioned at cities up the line, reporting train information to the VDC house. Meanwhile, the Berkeley police were formulating their own strategy to escort the trains.

On the next day there were two trains and twice as many demonstrators. Both sides had upgraded their communications, employing bullhorns and walkie talkies. When the train arrived, it slowed to a walking pace, as a wedge of police formed in front. The police led the engine for several blocks, picking up the pace as they progressed. The VDC was unprepared for the tactic and the train successfully made its escape. For the second train the scene was largely repeated, except that two VDC members positioned themselves ahead of the wedge. This produced the incongruous image of two protesters (one carrying a banner) leading the police and train down the tracks.

While the VDC was harassing trains, the president was signing the Voting Rights Act. Enactment of this historic bill into law was the lead story in newspapers the following day. In the *San Francisco Chronicle* the headline was juxtaposed with a picture of the troop train protest, rather than the traditional shot of the president handing a pen to a civil rights luminary. The *New York Times* carried the VDC on page 3.

Next the VDC moved into the mainstream political arena, petitioning the Berkeley City Council to stop the troop trains. At a Council meeting, Smale urged passage, drawing an analogy to Germans who ignored Nazi atrocities during the Second World War. An outraged conservative councilman, John DeBonis, retorted "You should not even be teaching at the university! You are out here aggravating a riot! I've been wanting to tell you off for a long time."[44] Smale even failed to receive the support of the liberal Council members, as the vote affirmed the train route.

In leading efforts to block troop trains and comparing the United States government to that of Adolf Hitler, Smale was pursuing a radical path that was drawing a great deal of flack, from all directions. DeBonis expressed the feeling of many people on the right. Their shock at Smale's actions was exacerbated by their expectations of behavior by a mathematics professor. During this period Steve received a number of death threats, both by mail and phone.

It was difficult to be indifferent to the troop train action. Some liberals were attracted to the VDC while others were appalled. Smale, whose goal was to build an exponential growth in the antiwar movement, was elated by the results. On balance, the troop train activity had significantly increased VDC membership and interest. The movement continued to build momentum and controversy.

Among the radicals, support for the troop train demonstration was not unanimous. Both Jerry Rubin and Steve Weissman returned to Berkeley after the action had already begun. While Rubin shared Smale's enthusiasm, Weissman had reservations. He viewed the troop train notion as a "brilliant tactic," but expressed concerns as to the manner in which it was structured.[45] These concerns reflected a significant difference in the leadership philosophies of the VDC and the FSM, involving the roles and responsibilities of leaders and followers.

Most FSM decisions were made by the Steering Committee. If bodies were needed for demonstrations or civil disobedience, both Weissman and Weinberg felt a personal responsibility to guarantee that each follower made their own decision to participate, and that the action proceeded as advertised. They viewed this good faith standard both as a moral imperative and as a necessary ingredient in bringing people into the movement. Weissman was especially influenced by old family stories of duplicitous CP leaders manipulating the actions of their followers.[46] To avoid such scenarios in their demonstrations, the FSM made full advance disclosure of the risks, and relied on monitors to maintain the boundaries of their scripts.

In between the leadership models of the Old Left manipulation and FSM responsibility, Smale adopted a *laissez faire* approach. VDC decisions were made in large meetings that were open to anyone, even informants. Smale was an independent thinker with his own methods of analysis, supplemented by a willingness to undergo risks. In policy discussions, Steve made his views clear, but refused to argue. If others elected to follow, that was fine. However, he encouraged independence. Unlike with the FSM, VDC members were given considerable latitude in customizing their protests. Smale assumed that everyone made their own decisions, and he did not feel a strong responsibility for the actions of his followers. In practice, he was extremely influential in the VDC and, up to this point, his views had generally prevailed.

Consider the troop train protest and the conflict between Smale and Weissman. There was a genuine danger to the participants, that included children and young people. Things could quickly escalate into a chaotic situation, placing individuals in jeopardy of the train. Weissman urged that the demonstration be structured so that participants could make an informed decision, rather than a reactive one. Smale felt less of a burden in this respect, seeing the dangers as evident. Furthermore he had reconciled the ends and means conundrum. To stop a terrible war required building a mass movement. Sustaining the necessary momentum involved physical risks that might lead to the injury of some of its partisans. "The U.S. military machine was killing thousands of Vietnamese with napalm and other bombs. To risk a life in a confrontation with that military machine wasn't unreasonable, it seemed to me."[47] Whatever guilt Smale did feel was assuaged by the depth and danger of his own personal involvement. He was at the tracks and he was the one receiving the death threats.

Milligan and other VDC members saw themselves as willing to die for the cause. What impact would a death or serious injury have had on the movement? Some saw quite harmful consequences, with a backlash damping further participation. However, the shootings at Kent State had the opposite effect. In the case of the troop train demonstrations, a death would certainly have generated more publicity, perhaps promoting the story to the front page of the *New York Times.* Moreover, the movement would have then had a martyr.

At the next troop train demonstration, on August 12, the leadership of the VDC was tested. Early that morning Smale, Rubin, and two others received telegrams from the superintendent of the railroad. The long text of the message detailed a number of criminal laws and conspiracy statutes, along with penalties that might be applicable in another troop train action.[48] While Smale did not wish to be arrested, he accepted the risk and was undeterred by the threat.

About 500 protesters gathered at the train station in the morning. Due to the threats on Smale's life, a Berkeley detective was detailed to accompany Steve, along with another for Rubin. The close proximity of the detectives to the leaders inhibited their discussion of tactical maneuvers. This was particularly frustrating since, for the early morning, the police effectively contained the demonstrators from the tracks. However, the lack of discipline among the VDC paid off, as a rank and file member made a break through the police lines. Others picked up on the cue, and then ran along the tracks and across backyards. When the train approached, two miles of track were lined with protesters.

The railroad went on the offensive, as the engineer activated a special device that generated a menacing spray of steam. However, many demonstrators were unintimidated, and police were forced to remove them one by one. Marilyn Milligan remembers "standing on the tracks and the train coming and a policeman pushed me off the tracks. I can still see that train bearing down. I don't know what would have happened if that policeman hadn't ... I don't think I would have."[49]

When the gantlet succeeded in slowing the train, protesters jumped onto its side, grasping steel handles. Although they failed to gain access to the soldiers, newspaper and television images showed people clinging to the train. One woman managed to travel for nearly a mile, despite repeated attempts by the police to remove her. There were a few minor injuries and arrests.

This was the last significant troop train protest. The VDC soon shifted its focus to other actions, while the Army managed to find routes that avoided Berkeley. Nevertheless, the defiant train actions dramatically initiated a new phase in opposing Vietnam policy. Smale analyzed the impact:[50]

> *The troop train demonstrations escalated the antiwar protest. We were warning Johnson and the government that if they continued the War, then the fabric of American society would be threatened. Eventually that became the key reason in Congress, in the government, and in the military to end the War. The War was causing too much division among Americans and even America's institutions were becoming insecure.*

At this stage, however, the VDC had not even persuaded their congressional representative. Jeffrey Cohelan was a liberal Democrat who supported President Johnson's Vietnam policy. In late August the VDC picketed the congressman's office. The following year they would produce a peace candidate to oppose Cohelan in the election.

Another opportunity for a VDC statement arose when General Maxwell Taylor arrived in San Francisco for a speaking engagement.

Taylor, having previously served as Chairman of the Joint Chiefs of Staff and United States Ambassador to South Vietnam, was a special advisor to the president on Vietnam. The VDC had an opportunity to confront one of the primary architects of the War. Two months earlier Smale had missed the Johnson visit and believed that the demonstration had suffered from his absence. For the Taylor protest Steve had his chance to guide the proceedings. The VDC sign crew geared up, pursuing a Nazi theme, depicting Taylor as a war criminal on a wanted poster. Smale viewed the poster as too extreme, but did not obstruct its use.[51]

When Taylor arrived at the Fairmont Hotel on August 24, there were 100 VDC people to greet him, including Smale, Rubin, Gullahorn, and Weissman. They surrounded his limousine, as one protester lay prone on the hood. The *San Francisco Chronicle* described what transpired when hotel manager Dick Swig attempted to escort Taylor to his room.[52]

> *But the demonstrators began crowding into the elevator too and jammed the entrance, so Swig and the general got out and strode across the lobby with the demonstrators in hot pursuit and walked up the stairs to the executive offices on the mezzanine where they locked themselves in.*
>
> *About 50 demonstrators made it up to the offices with their signs and their wooden barbed-wire barricade, splashed with red paint and bearing the legend: "Is This Freedom?"*
>
> *They broke into a clapping, foot-stomping chorus of "We Shall Not Be Moved" at the top of their lungs and Ben Swig, the owner of the hotel, appeared and got into a heated political debate with Jerry Rubin, cochairman of the protest committee.*

Not surprisingly, Rubin and Swig were unable to resolve their differences and the police arrived to free the general. The demonstrators were warned that they faced arrest. The VDC decided that a token group should be arrested, and it remained to select the victims. Neither Rubin nor Weissman had been arrested in the FSM crackdown, and both felt the need to validate their credentials. For Rubin, arrest was an obligation of leadership, but Weissman had an additional motivation that was especially compelling. He was scheduled for a draft physical two days later. Being in jail offered an attractive conflict.[53]

Smale, at this stage, was unwilling to solicit arrest. Gullahorn had similar feelings, enraging Rubin who eventually persuaded his girlfriend to join him.[54] A total of five people were arrested. They went limp and were carried away, while Smale urged the remaining protesters to continue the "haunt." Moe Hirsch bailed Gullahorn out of jail and Weissman prevailed upon his comrades to remain in prison for a couple of days. The VDC announced that the men refused bail as part of a protest.

On his second day in San Francisco, Taylor delivered a speech while the VDC picketed outside the hotel. Beyond the publicity, the haunt achieved some measure of success, as General Taylor departed one day earlier than planned, exiting the hotel via a freight elevator and service entrance.[55] Later that year the arrested protesters were sentenced to 30 days in jail.

D. March to Oakland

The medium range goal of the VDC was the October 15–16 International Days of Protest. Rubin's August Washington trip led to the formation of the National Coordinating Committee to End the War in Vietnam.[56] The name seemed to say it all. With an office in Madison, Wisconsin, the group promoted nationwide efforts for October 15 and beyond. There would be parallel demonstrations in other cities, both nationally and internationally.

In the Bay Area, support was building for the march to the Oakland Army Terminal and the subsequent civil disobedience. The troop train and Maxwell Taylor demonstrations had kept the VDC in the headlines, despite the small number of participants. With classes resuming in September, more bodies would soon be available. A natural transference was occurring from the FSM to the VDC. Vietnam would be the campus issue for fall, 1965.

Forcing the Johnson administration to change its Vietnam policy was a hard problem. As with the Generalized Poincaré Conjecture, intractability was more attraction than deterrent for Smale. During the summer of 1965 he worked full time on the VDC. It was exciting for him to watch the organization take shape and grow. Steve enjoyed the interactions with Rubin, Milligan, and others. At times he considered leaving mathematics permanently. That would have been fine with some of the Berkeley faculty.

The Berkeley faculty included a large number of liberals who opposed Johnson's Vietnam policy, but applauded the Great Society domestic initiatives. They objected to the rhetoric and actions of the VDC, particularly the Nazi metaphors and use of civil disobedience. An economics professor, Carl Landauer, was disturbed that the public might construe the VDC position as representative of the entire Berkeley faculty, rather than what he viewed as a small minority. To correct this impression, he and a history professor circulated a criticism of the VDC, seeking signatures of their colleagues. The draft was dated September 9.

On that same day in San Francisco, Smale was a principal speaker at a VDC press conference to publicize the October protest. Steve promised civil disobedience, outlining some of the actions contemplated for the Oakland Army Terminal. Reaction was sharp among Berkeley faculty with 250 signing the Landauer petition. The attack became more personal when the chancellor's office informed Smale that some of the faculty were seeking his termination.[57]

It appeared that the October protest might provide the grounds for Smale's dismissal. Both Steve and Jerry volunteered to participate in civil disobedience, anticipating arrest. The VDC Defendants Committee was composed of people who expected to be arrested on October 16. At their meetings, the Committee explored tactical options, as well as planning a legal defense. Smale wanted maximal publicity from the demonstration, and the judicial venue provided another opportunity. This approach was a major departure from the legal strategy of the FSM. After the Sproul Hall arrests, the FSM deferred to their lawyers. The attorneys cooperated with the government in an attempt to reduce prison time for their clients. Of course the FSM had already won the political issue. Since nobody believed that the October protest would stop the Vietnam War, the VDC planned to incorporate the arrest aftermath into their strategy. To maintain visibility, the Defendants committed to refuse bail and remain in jail. Political considerations, rather than expediency, were to motivate the legal decisions.

The Defendants wanted to put the War on trial, even if that led to harsher punishment.

The original protest schedule called for a large campus teach-in on Friday, October 15, and a Saturday march from Berkeley to the Oakland Army Terminal. There the Defendants planned to block access to the base, committing at least a misdemeanor offense. Other options involved lobbying the soldiers to disobey orders. Possible approaches to gain access included climbing fences and dropping leaflets from airplanes. There was a greater legal risk from these more aggressive acts, and the philosophy of the VDC leaders was to permit individuals to select their own action.

As the VDC welcomed an influx of supporters in September, they maintained their democratic decision making process. Typically, long discussions preceded votes. Marilyn Milligan remembers a several hour debate over whether a flyer would read *Stop the War in Vietnam* or *Stop the War in Vietnam Now.*[58] More substantive issues were decided in the same fashion. When policy is determined by debate and vote, status and eloquence have a strong influence. This was always the case in the VDC, but September marked the emergence of significant new personalities.

Prior to that time, the vision of Smale and Rubin had guided the VDC. They presented a united front at meetings, having developed their strategy and reconciled their few differences in advance. The charismatic Rubin was the out-front partner, who usually presented their position. Although Smale attended the interminable meetings, he tended to remain in the background, picking his spots to make a point. Outside the meetings Steve and Jerry worked on building the organization. During the first several months they gained respect and devotion from the VDC workers.

However, Rubin and Smale's influence diminished as the VDC expanded in September. With more speakers and participants, Smale became less conspicuous at meetings. Rubin was overshadowed by the presence of Weinberg and Weissman, both articulate speakers with a loyal following, earned from their work on the FSM.

A division developed in the VDC, split between the FSM and original VDC leadership. Joining the FSM faction was Frank Bardacke, a Berkeley graduate returning from a year in Uganda. Bardacke was a likable person with strong left wing credentials. He had been on the front lines in the past and would continue his activism in the future. Bardacke had missed out on the FSM, much to his dismay. Now the VDC provided an opportunity to protest American policy in Vietnam, a cause in which he had been involved even prior to the Gulf of Tonkin.[59]

The primary point of contention between the FSM and VDC leaders involved the use of civil disobedience at the Oakland Army Terminal. To Rubin and Smale, the action was an essential step in their program to stop the War. Massive civil disobedience was front page and network news material. It would provoke thought about the War and attract more supporters. While the FSM leaders were quite willing to participate in civil disobedience themselves, they were reluctant to pursue a course in which they might lead others into unanticipated chaos.

Aside from the philosophical differences between the two factions, there was an important personal element. Weinberg and Weissman did not trust Rubin. They viewed him as an opportunist who would exploit others without regard for their well being. Moreover, the meeting dynamic led Weinberg and Weissman to believe that the civil disobedience was being pushed by Rubin, rather than a Rubin/ Smale partnership.

It came down to a hawk (Rubin, Smale) and dove (Weissman, Weinberg, Bardacke) argument. Rubin and Smale wanted to take aggressive action. The doves were trying to bring people into the movement incrementally. They felt a moral responsibility to protect their followers from taking a large step that they might subsequently regret. Neither side fully appreciated the motivation of the other. As Smale watched helplessly, the eloquent doves moderated the stance of the VDC. Confrontational civil disobedience was discarded. The roads would not be blocked and no fences climbed. There would be a march to Oakland with a "teach-out" on a vacant field adjacent to the base. Monitors would preclude freelancing. While arrests were possible, no action was intended to provoke them. The VDC had departed from the course charted by Smale and Rubin.

Despite his disappointment, Smale continued his efforts to build support for the October protest. Another creative action, such as the troop train or Maxwell Taylor protest, was desired to keep up the VDC's publicity profile. Governor Brown provided a nice target. There were bitter feelings on the Berkeley campus toward the man who had authorized the Sproul Hall crackdown on the FSM. An opportunity arose when Brown rebuked a state Democratic official for criticizing United States policy in Vietnam. The VDC initiated a poster campaign against the governor and planned a protest at his San Francisco residence.

Smale drove to San Francisco to organize the demonstration. Accompanying him was a new VDC volunteer named Sam Angeloff. On the surface Angeloff was an interested, competent worker. Actually, he was a *Life* magazine reporter who had infiltrated the VDC to research a story about the antiwar movement.[60] One of the subtexts of Angeloff's subsequent article was the ease with which one could become involved in the VDC. Anyone was welcome, and there did not appear to be any secrets or security.

Smale and Angeloff located the Brown residence and then proceeded to San Francisco State College to recruit protesters. While previous VDC efforts had proven that a small number of people could obtain headlines, this demonstration posed a real challenge. Angeloff wrote "After two hours the best we could do was 20 people, four of them newsmen. We drove back to the governor's home and got out of our five cars to divide up our picket signs, taking too much time and feeling rather silly."[61]

After a short time a neighbor informed the picketers that Brown had not lived in the house for five years. The address was Brown's voting residence, but his daughter was the actual occupant. Although the Governor missed the demonstration, it did provide some contribution to his granddaughter's celebration of her third birthday.[62] The two hour picket did receive a little press, but it was not one of the VDC's greater successes.

One of Smale's favorite VDC ventures was the War Dogs episode.[63] The army was training dogs, for military use, at the edge of Tilden Park. This large recreational area in the Berkeley hills was a popular location for both adults and children. When the VDC learned that the dogs had invaded their turf, they attempted to gain some publicity while harassing their enemy, the military. However, with its elaborate committee system, Smale was not aware of all the initiatives. In particular there was a VDC Commando unit that engaged in what might be called *dirty tricks.* The Commandos produced signs that read:

CAUTION

ARMY WAR DOGS

IN THIS AREA

Do Not Leave Raw Meat Exposed

Keep Children and Pets Within Sight

If Dog Approaches Do Not Move

Wait for Handler

A military style font was selected for the lettering, and the Commandos posted the bona fide looking warning throughout the park. It was just the type of prank that Smale loved, but Steve had no knowledge of the signs when he visited the park director to complain about the dogs. At the mention of war dogs, the director became agitated and described the object of his distress. Earlier he had observed a bus unload a group of boy scouts for an outing in his park. However, he was puzzled when the troop immediately returned to the bus and departed. Investigating the site, the director discovered the sign. Since he had not posted it, he assumed that the army was responsible for terrorizing park patrons. Who else could have produced the signs? The director was especially annoyed that the army commandant denied any knowledge of the outrage. When Smale saw the sign, he immediately inferred the role of the Commandos, but the director remained baffled, despite Steve's laughter. When Smale recalled the story thirty years later, he began to laugh hysterically.

In late September, the Oakland Army Terminal protest schedule was modified again. The seven mile march to Oakland was moved up to Friday night, October 15, following the campus teach-in. Assuming they reached the base, the teach-out would continue through the following day. VDC publicity emphasized their opposition to civil disobedience. They were merely exercising their rights of free speech and assembly. The original VDC philosophy of provoking arrests had shifted to the classic FSM tactic of asserting constitutional protections that the authorities might reject. If a crackdown occurred, the demonstrators were the aggrieved party.

Quite a number of authorities had some degree of jurisdiction over the weekend protest. There were the city governments of Berkeley and Oakland, the county of Alameda, the state and federal governments, the University of California, the Oakland Army Terminal, and the Oakland Port Commission (which controlled the venue of the teach-out). Behind the scenes, officials jockeyed for political cover.

On October 1 the county district attorney wrote Governor Brown that "there is in this situation potential danger of riot and mob violence."[64] He requested the assistance of the California National Guard and Highway Patrol and suggested that the governor urge President Kerr to block the university portion of the demonstration. If Kerr did not act, the letter threatened that the state might be civilly liable for the consequences. The governor forwarded the confidential message to Kerr who relayed it to Roger Heyns, the new Berkeley chancellor. In an accompanying letter Kerr passed responsibility to Heyns.[65]

The president did not withdraw completely, offering further advice in a note to Heyns on the following day. "In Monterey yesterday evening I had the opportunity to talk with the governor and others. October 15 and 16 are taking on significance not only as press and TV lures, but also as political testing grounds. The test will be how well law and order is preserved and by whom." Kerr suggested that Heyns consult with counsel. He also reiterated his desire that Kerr himself and the governor remain insulated from the problem.[66] Heyns did obtain an opinion from the university legal officer, advising that they were protected from legal culpability for actions by the VDC.[67] With the VDC publicly eschewing civil disobedience, Heyns approved the campus portion of the protest.

The cities and county were opposed to the protest, and had tried to pressure the state government. It was a volatile political issue and the buck was passed from the governor down to the president of the university and then to the new chancellor of the Berkeley campus. However, if Heyns obstructed the demonstration, he invited confrontation. With his approval, the Berkeley teach-in and march staging proceeded without impediment. That shifted the burden back to the local authorities who vented their frustration in an October 14 letter to the chancellor.[68] On the same day, Berkeley and Oakland rejected the VDC requests for parade permits. In other developments, National Guard units were called up and the Oakland Port Authority declared their property off limits for the teach-out.

The VDC reaction was that denial of a parade permit was prohibition of free speech. They filed appeals before federal judges, but declared their intention to march, regardless of the legal outcome. If they did manage to begin the parade to Oakland, the seven mile route provided the police with lots of opportunities for intervention. With so many contingencies, the VDC elected a steering committee to make real time decisions. Both factions were represented among the nine members that included Bardacke, Weinberg, and Weissman as well as Milligan, Rubin, and Smale. Rubin was barely elected.

On the day of the march, a federal judge declined the VDC's request for protection against city enforcement of the parade ordinances. The state attorney general took a different view, discreetly urging local officials to permit the march.[69] Given this admonition and the difficulty in confining the demonstrators to the campus, the Berkeley city government reluctantly allowed the march to proceed without a permit.[70] While Berkeley was willing to make accommodations, Mayor

John Houlihan of Oakland refused to concede an inch. The stage was being set for a confrontation at the Berkeley-Oakland border.

It was a Friday and the weather was perfect for the protest, but Steering Committee member Steve Weissman was ill with a high fever. Realizing the physical demands of the weekend, Weissman sought pharmaceutical assistance from a sympathetic left wing doctor. He received a shot that replaced the fever with euphoria. Weissman felt great, but he was in no position to make decisions.[71]

When 10,000 people appeared on campus for the evening march, Smale was gratified by the fruition of his past several months of effort. This was step two in his program. The first step was the large turnout for Vietnam Day. A teach-in, however, is a passive act. By marching to Oakland, all of these people were making an active statement of opposition against the War. To some other members of the Steering Committee, the turnout, itself, was enough to declare the demonstration a success.

The plan called for the march to begin at the conclusion of the Berkeley teach-in talks. Monitors were to direct the crowd, with the Steering Committee at the front of the pack. However, the scenario of control was pre-empted at the start. Impatient demonstrators began the march while Weissman and Weinberg were still on the stage. When the leaders learned the crowd was already on its way to Oakland, they frantically rushed to assume their positions. As Berkeley police cleared the way in their city, Oakland police massed at the border.

Two conflicts were developing. It was evident that the Oakland police planned to use all of their force to prevent the march from proceeding into their city. Would the VDC turn around, walk into a bloodbath, or select some intermediate course? The decision was in the hands of the Steering Committee. A variety of intermediate contingencies were discussed at meetings prior to the march, including a sit-in or a flanking maneuver. With 10,000 people approaching the border, the VDC leadership was divided. Smale advocated a sit-in, but Weinberg was opposed. Thirty years later they recalled their differing analyses and goals.[72]

Smale wanted "to go up to the Oakland border. Not cross it where police were on the other side. Not to cross, not to provoke, but go up to it, stop, stay there, stay there as long as it took, months." Smale savored the scenario of a protracted stand-off, forcing the police into an untenable position. He felt that television and press images of the sit-in would generate support for the movement.

Weinberg operated from a different perspective. "Looking back, I had a more small scale vision. I had a poorer understanding of national politics, of how things are playing in the media. I had a very strong sense of the need that people should know what they are getting into. Activity is something that I wanted and wanted to promote, but nonviolent philosophy of people choosing to do this." For Weinberg, there were problems with the sit-in concept. The crowd was beyond the control of the leaders and some were likely to confront the police rather than sit-in. In his judgement this scenario would suck many novice demonstrators into an unbargained for battle with the police, involving numerous injuries. Finally, he incorrectly inferred that Rubin, rather than Smale, was the architect of the strategy. Jerry was "how do you set up a situation, step back, and let it blow." Weinberg felt responsible for the welfare of the demonstrators, and he would not permit Rubin to selfishly manipulate them to achieve his own ends.

Smale had a different assessment of the situation and his responsibility as a leader. "Leadership would stop and set an example." The marchers "can decide for themselves. They knew the dangers of the march had been in the newspaper for weeks and days before that. The dangers of the Oakland police and all that was not a secret. My feeling was that marchers, they can decide for themselves if they want to undergo these dangers." Smale conceded that the situation was unstable and there might have been flurries between police and demonstrators, but he disagreed with Weinberg on their likely magnitude and the acceptability of the risk.

Milligan and Rubin sided with Smale. Bardacke and the dazed Weissman deferred to Weinberg whose position prevailed on a 5-4 vote. With the rejection of the sit-in, the marchers were directed on a course parallel to the border. A left turn down a subsequent street led to Oakland and a right turn back to Berkeley. Some of the marchers shouted "LEFT," but Weinberg was in control and determined to avoid a violent confrontation. Following a right turn, the crowd headed for a park in downtown Berkeley where festivities continued through the night.

On Saturday afternoon, the VDC made another attempt to reach Oakland. Smale joined a few thousand marchers who, again, were met by police at the Oakland-Berkeley border. This time the impasse was broken by a third party. Several members of the Hell's Angels motorcycle gang charged through the Oakland police lines, attempting to intimidate the marchers. When the Berkeley police intervened, a few injuries occurred. It was a stunning spectacle that altered the mood of the antiwar demonstration. A short time after the incident, the crowd dispersed.[73]

The International Days of Protest reached a large number of cities including Los Angeles, Chicago, Brussels, and Tokyo as well as the college towns of Boulder, Madison, and Ann Arbor. In New York, over 10,000 protesters marched down Fifth Avenue, while 500 people filled London's Trafalgar Square.[74] In late May, Smale and Rubin had initiated an effort to establish an international campaign to stop the War. The movement was off the ground.

Despite the success of Smale's foray into political action, he was demoralized by the turning back of the march. The Steering Committee vote, decided by the narrowest of margins, was confirmation that the VDC had shifted in the direction of the "political people." With some degree of bitterness, Steve realized that the organization was not going to be what he hoped it would be. Smale had lost faith in the VDC's future.[75] No contract bound Steve to the VDC. Unlike some others in the movement, he had an attractive alternative. Smale was still a Berkeley mathematics professor with prodigious research ability. As abruptly as he had left mathematics for the VDC, Smale switched back after the march.

For Weinberg and Bardacke, Steve's departure was barely noticeable. Bardacke was on the winning side of the vote, but recanted his decision a few weeks later. In subsequent years he was on the losing end of many battles, but learned to savor the victories and follow his political ideals through the defeats. He could not understand why Smale left the VDC.[76]

While Rubin also felt that Smale "overreacted," to him the loss of his cochair had a profound impact. Smale had shaped Jerry's thinking.[77] They were partners in leading the VDC, and each would have been less effective without the other. After the VDC attained a level of success, their leadership and course were threatened. Smale's reaction was to bail out. Rubin felt betrayed and anticipated a further decline in his own influence.

The VDC continued, but Smale's assessment was correct. Civil disobedience was repudiated. When their petition for a November march into Oakland was rejected, the VDC obtained relief from the courts. With the aid of an injunction, they did reach Oakland on November 20, but the destination was a park rather than the Oakland Army Terminal.[78] The size of the demonstration was half that of October 15, when interest in the VDC may have peaked.[79]

New areas of contention were developing. VDC member Bob Scheer was contemplating a candidacy for Congress in the Democratic primary. Scheer was an intellectual journalist who had visited Vietnam in 1964. His monograph *How the United States Got Involved in Vietnam* was a VDC-recommended primer on the issues. Should the VDC support Scheer and work within the political system? Certainly Scheer had politically correct views on Vietnam, while the incumbent, Jeffrey Cohelan, was a target of the VDC. However, the Democratic Party was an anachronism to many of the radicals. Embracing Scheer's candidacy acknowledged a legitimacy of the capitalist system.[80]

One of the original precepts of both the VDC and FSM was that of a radical movement under a large tent. Communists and Trotskyists had a long history of competition and animosity, but Smale and Rubin succeeded in bringing both groups into the VDC. With the Scheer candidacy the ideological conflicts erupted. The VDC split with Scheer gaining the larger faction. "After the communists and liberals left the VDC to work for Scheer, the VDC fell under the control of the Trotskyists and then slowly disintegrated."[81]

Scheer waged an aggressive grassroots campaign. Despite strong support for Cohelan from the Democratic establishment, Scheer obtained a surprising 45 percent of the vote. The antiwar movement lost the election, but they made an impression. It would be Cohelan's last term. In 1968 antiwar candidate Ron Dellums defeated Cohelan and began his long tenure in Congress.

On April 8, 1966, the VDC headquarters was destroyed by a bomb. Four people were injured. It was the symbolic *coup de grace* for an organization that was in disarray. Teach-ins, marches, and election campaigns could not sustain a radical group. In May, only 1000 people attended a teach-in to mark the first anniversary of Vietnam Day.[82] Rubin concluded "I think the VDC may have played itself out."[83]

The success of the International Days of Protest raised hopes for national and international antiwar movements. The National Coordinating Committee to End the War in Vietnam (NCC) was the *de jure* organizer for the United States. Further development of the NCC was anticipated by delegates to a 1965 Thanksgiving-weekend convention in Washington, DC. Making the trip as representatives of the VDC were Milligan, Rubin, Weinberg, and Weissman.

Milligan had backed Smale on the sit-in at the October 15 march. Losing the vote and seeing Smale leave were major disappointments for her, but Milligan quickly reconciled her own involvement, and rededicated herself to the VDC. She was excited by the prospect of attending the NCC conference in Washington. Naively, Milligan expected to bond with others who shared her focus on ending the War. Political infighting, however, was the central theme of the conference. The battle between the Marxists and Trotskyists began with a struggle for the microphone at the opening session, and then continued throughout the weekend.[84] Milligan's excitement at attending the conference quickly turned to disillusionment. This time she did not recover. In December she went on a vacation to Mexico with her husband. Milligan was pregnant by January, and the VDC was no longer a priority.[85]

E. Vietnam Protest after the VDC

By the spring of 1966, the VDC and NCC had disappeared. The War continued to escalate. In March 1967 there were 470,000 U.S. troops in Vietnam under the authority of the Gulf of Tonkin Resolution. When the commander of the United States Military in Vietnam, William Westmoreland, requested 200,000 additional troops, Secretary of Defense Robert McNamara decided the cost was too great. In advising the president against the increase, McNamara wrote:[86]

> *The Vietnam War is unpopular in this country. It is becoming increasingly unpopular as it escalates—causing more American casualties, more fear of its growing into a wider war, more privation of the domestic sector, and more distress at the amount of suffering being visited on the noncombatants in Vietnam, South and North. Most Americans do not*

> *know how we got where we are, and most, without knowing why, but taking advantage of hindsight, are convinced that somehow we should not have gotten this deeply in. All want the war ended and expect their president to end it. Successfully. Or else.*

Opposition to the War was growing. Robert Kennedy and Martin Luther King were among the high profile critics of the president. National antiwar coalitions were forming with names such as *Student Mobilization Committee* and *National Mobilization Committee.* On April 15, 1967, over 300,000 participated in a New York march against the War. Administration officials had become persona non grata on college campuses and were "haunted" at public appearances. In many cases family members, friends, and even passersby lobbied them against the War.[87]

A Washington antiwar protest in October 1967 had elements of *déjà vu* from Berkeley two years earlier. Jerry Rubin was the project director, though Abbie Hoffman replaced Steve Smale as his partner. Original plans involved a march to the Capitol. Rubin decided that the Pentagon was a better destination. Years later he acknowledged that his choice of the Pentagon was influenced by his collaboration with Smale on the Oakland Army Terminal.[88] As with the VDC, there were disputes over the inclusion of civil disobedience. The negotiated plan called for a peaceful demonstration at the Lincoln Memorial, followed by a march to the Pentagon where individuals were free to create their own mode of civil disobedience or select from an à la carte menu. Estimates for the Lincoln Memorial crowd ranged from 50,000 to 100,000 with 20,000 to 35,000 proceeding to the Pentagon where 683 were arrested.[89] In many respects it was the protest that Rubin and Smale had originally conceived for Oakland. This time the marchers reached their objective, there was civil disobedience, and massive publicity (including Norman Mailer's Pulitzer Prize winning *Armies of the Night*). Reaction to the protest was mixed, but it certainly made an impression. Defiance, as well as the threat of violence, was on the rise.

By the time of the Pentagon protest, McNamara had concluded that the War was hopeless.[90] When he recommended an alternative course, Johnson found him a new job as president of the World Bank. Replacing the messenger did not lessen Johnson's difficulties. In February 1968 the trusted CBS News anchor Walter Cronkite editorialized for withdrawal.

There was no keener observer of political dynamics than Lyndon Johnson. The president watched in agony at the rapidly declining support for the War and the increasing opposition. Incumbent presidents were virtually assured of renomination by their party. Little known Senator Eugene McCarthy challenged Johnson and shocked the country by drawing 42.2 percent of the vote in the New Hampshire primary. A few days later Robert Kennedy declared his candidacy. Johnson got the message. In a nationwide address on March 31 he observed, "there is a division in the American house tonight." He concluded with the startling announcement, "I will not seek, and I will not accept, the nomination of my party for another term as president." The War had taken its biggest casualty and the antiwar movement claimed a victory.

The next domestic battleground was in Chicago at the August 1968 Democratic convention. Thousands of protesters arrived, many looking for action. Jerry Rubin was among the leaders promoting militancy. Again he got it. Television dramatically covered a series of violent confrontations between police and demonstrators. It was as big a story as the politics. Not all of the Chicago protests led to bloodshed. The Democratic Convention coincided with an American Mathematical Society conference in Wisconsin that Smale attended. About 50 to 100 mathematicians went to Chicago to make their statement. Smale directed them on a protest march. They passed the police without incident. With some pride, Smale recalls they were the only group that got close to the convention center.[91]

Vice President Hubert Humphrey won the Democratic nomination, but the victory was Pyrrhic. The convention alienated the supporters of Humphrey's opponents. During the campaign, Johnson gave Humphrey little slack to separate himself from the president's unpopular Vietnam policy. Meanwhile, his opponent, Richard Nixon, boasted of a "secret plan" to end the War. Nixon won in a close election.

With a change in administration, there was a short hiatus in antiwar activity. Eventually, Johnson's War became Nixon's War, and the new president was accountable for the daunting problem that he inherited. Nixon undertook a course to decrease the domestic pressure for peace on himself while simultaneously motivating the North Vietnamese to negotiate. To accomplish the former he enacted a sequence of troop reductions, announced one at a time, over the following four years. In contrast to the fanfare associated with the withdrawals, there were covert bombing escalations and offensives, intended to send the opposite message to the enemy.

During this period the antiwar movement faced a number of challenges. Nixon's strategy was somewhat effective domestically, while internal problems continued to plague the movement. There were disputes over the choice of various forms of civil disobedience, as well as political and personal differences among the leaders. Meanwhile, the Nixon administration employed many of its resources to monitor and disrupt the peace movement. At times the tactics ranged from abuse of power to illegal acts. When Daniel Ellsberg released the Pentagon Papers, the White House created a clandestine unit, known as the Plumbers, who broke into the office of Ellsberg's psychiatrist. A later mission of the Plumbers resulted in the Watergate episode.

Despite the efforts of the administration, the peace movement had an enormous impact on United States politics and culture. There were rallies of extraordinary magnitude, notably in November 1969 and April 1971. Although the antiwar movement had its ineffective periods, it possessed a resilience that refused to let up on political leaders. For eight years Vietnam War opposition rachetted upward. Deliberations on War policy involved consideration of the reactions it might provoke among the protesters.[92] Vietnam was the biggest political issue of the day, and it was an overwhelming aspect of college life, particularly following the killings at Kent State. On January 23, 1973, the United States agreed to withdraw its remaining troops from Vietnam. The War ended with 58,000 American and 2,000,000 Vietnamese deaths.

For most of the Vietnam War, Steve Smale was doing mathematics and Jerry Rubin was attracting attention with outrageous anti-establishment antics. Rubin became notorious as a leader in the antiwar culture, while Smale is barely mentioned in historical treatments of the movement.[93] Rubin characterized his VDC cochair as "the forgotten man of the successful movement to stop the War."[94] What was Smale's imprint on the antiwar movement? Consider his six months with the VDC, in its context as the start up of an eight year operation that transformed public opinion in an unprecedented manner.

At the beginning of the bombing in 1965, Americans tended to blindly support their country's foreign policy. A small segment of the public did oppose the Vietnam actions, and Staughton Lynd made the radical suggestion of generating massive civil disobedience to change the government policy. It was Smale and Rubin who took the step of implementing Lynd's idea. Their vision was for an exponential growth of dissidence. To attract support they devised creative demonstrations and enticed the media to publicize their cause and provoke thought. For several months they were successful, but it is even more remarkable that they had the audacity to try. While the VDC did not succeed in stopping the War, Smale and Rubin blazed a trail for others. Especially notable was their timing (early), promotion of inclusiveness (all flavors of leftists), their practice of exploiting the media as an asset to the cause, and the introduction of new tactics (troop trains). In retrospect Rubin paid tribute to Smale's vision:[95]

> *In six months he laid out the whole direction of the antiwar movement. He was almost like the Lone Ranger. He came in on his horse and gave us the message, and then dropped the silver bullet and went off.*
>
> *What happened in Berkeley with the Vietnam Day and with the troop train protests … was the script for the whole decade. Steve Smale wrote that script, and then he left the others of us to carry it out.*

The Summer of 1966

07

A. Another Berkeley subculture

The Berkeley campus impact of the Free Speech Movement extended beyond the new freedoms on political activity. Previously, there had been a sharp demarcation between faculty and graduate students. In the post-FSM environment, such class distinctions became less formal, and a more collegial relationship emerged. Some professors had difficulty adjusting to the new campus culture, but this was not the case for Smale and Hirsch. Both had already accepted a peer status in sitting around Jerry Rubin's apartment, planning Vietnam Day.

When Smale shifted his energy back to mathematics in late 1965, he was well positioned to assume a new role as faculty mentor. Berkeley had a large number of talented mathematics graduate students, each needing a thesis advisor. Smale was not only a politically correct choice, but he also possessed the currency which was most vital to a graduate student: good problems. Steve had launched a major program for the development of dynamical systems, offering enough interesting conjectures and questions to sustain a small army of mathematicians. Hirsch and a young faculty member named Charles Pugh joined the steady stream of graduate students enlisting each year. They quickly developed a subculture within the mathematics department, facilitating discussion and interaction. Somehow the excitement of mathematical discovery comfortably integrated with the social and political transformation occurring around them.

Among the first graduate students to join the group were Mike Shub, Jacob Palis from Brazil, and Nancy Kopell. All would go on to distinguished mathematical careers of their own. However, for the last seven months of 1966, their early progress was challenged as Smale was temporarily away from the Berkeley campus. During this period the students benefited from each other, as well as Hirsch and Pugh who generously gave of their time in guiding Smale's charges. Steve's own plans called for a summer in Europe, followed by a residency at the Institute for Advanced Study. The European phase included mathematical sojourns and conferences, as well as travel with his family. The highlight would occur in August when Steve was one of the eminent mathematicians invited to deliver a plenary address at the International Congress of Mathematicians in Moscow.

B. The Nobel Prize and the Fields Medal

In 1897 Zurich hosted the first International Congress of Mathematicians. Over 200 attended the three day meeting that was to feature lectures by four outstanding mathematicians. Henri Poincaré was scheduled to deliver the first talk, but illness necessitated his paper being read by a substitute.[1] The second Congress, three years later in Paris, extended the format to an entire week. Again there were a few invited *plenary* addresses to the entire gathering, separated by smaller sessions broken down by field. This structure of the International Congress of Mathematicians remains today. The meetings have occurred every four years since, excepting the war periods and a one year delay of the 1982 Warsaw meeting due to the political upheaval in Poland. Other more specialized or localized mathematics conferences occur frequently, but none possess the scope and prestige of the ICM.

The number of plenary speakers has increased to approximately fifteen per Congress, but the honor of delivering these hour lectures remains among the highest status available to a mathematician.

Ironically, the most influential lecture in the history of mathematics was a nonplenary talk at the 1900 Paris ICM. In his address "Mathematische Probleme" and a subsequent paper, David Hilbert of Germany described 23 unsolved problems, setting an agenda to challenge mathematicians for the new century.

During the early twentieth century, Europe was the center for mathematical research. Promising American scholars often traveled across the Atlantic for graduate and/or postdoctoral work, studying under the European masters. Naturally, the ICM tended to select venues that were convenient to its audience and the plenary speakers. The emerging North American school finally attracted the 1924 meeting, originally scheduled for New York but actually held in Toronto.

The Canadian mathematician J. C. Fields organized and raised the money for the Toronto ICM. Fields received his Ph.D. from Johns Hopkins in 1887, one of the foremost American programs of his day. After a short period of teaching in the United States, Fields went to Europe and further refined his mathematical knowledge. During his ten years abroad, Fields developed an important friendship with the influential Swedish mathematician, Gösta Mittag-Leffler. In 1902 Fields accepted a position at the University of Toronto, where he remained until his death.[2]

Fields' time in Europe coincided with the founding of the Nobel Prizes. When the wealthy Swedish scientist Alfred Nobel died in 1896, the world was shocked to learn the substance of his will. Nobel left most of his estate to endow the annual award of prizes to those who "have conferred the greatest benefit on mankind."[3] He designated five separate disciplines for recognition: physics, chemistry, physiology or medicine, literature, and peace. While the will was short on implementation details, the $40,000 magnitude of the prizes dwarfed the salaries of academicians. Prestige quickly followed the wealth, and the prizes became the most coveted awards of the twentieth century.

Following the Toronto ICM meeting, Fields sought to remedy the omission of mathematics from the Nobel list. Due to Fields' successful fundraising, the 1924 Congress had produced a surplus. Fields managed to direct the money toward a prize in mathematics, which he conceptualized in a memo just prior to his death in 1932.[4]

> *It is proposed to found two gold medals to be awarded at successive International Mathematics Congresses for outstanding achievement in mathematics ...*
>
> *It would be understood, however, that in making the awards while it was in recognition of work already done it was at the same time intended to be encouragement for further achievement on the part of the recipients and a stimulus to renewed effort on the part of others ...*
>
> *One would here again emphasize that the medals should be of a character as purely international and impersonal as possible. There should not be attached to them in any way the name of any country, institution or person.*

Fields designated additional endowment funding from his estate, permitting stipends to accompany the medals. At the 1932 ICM in Zurich, the concept was approved and a committee was designated to select the honorees for the following Congress. In the opening session of the 1936 Oslo ICM, King Haakon VII presented the first awards.[5] The winners were 29-year-old Helsinki born Lars Ahlfors of Harvard University and 39-year-old Jesse Douglas of New York and the Massachusetts Institute of Technology. The Fields Medal, as it became known despite the intentions of its creator, featured a profile of Archimedes.

Both the Fields and Nobel winners receive a gold medal and accompanying honorarium. Mathematicians describe the Fields Medal to laypeople as the "Nobel Prize of mathematics." This characterization (similar to a college calling itself the Harvard of some region or genre) contains a great deal of information. The prestige of the Nobel is appreciated by everyone, but the standing of the Fields Medal is restricted to the world of mathematics.

The iconification of the Nobel Prize among science awards owes to its seniority and the lottery-magnitude of its stipend. When the first Nobels were awarded in 1901, physicist Wilhelm Röntgen and chemist J. H. van't Hoff each received a payment in excess of $40,000 in the currency of the time. Seventy years later, mathematicians were taking home a $1500 stipend with the Fields Medal. Although the financial reward of the Fields Medal failed to measure up to that of the Nobel, among the mathematical community it was a prize of comparable honor.

Another contrast between the Fields and Nobel awards has been the age of the recipients. While Fields made no mention of a specific age requirement, he did suggest that the medal both reward prior work and encourage future achievement. In interpreting this intention, every Fields selection has designated recipients who were under the age of 40. Nobel's desire was to reward the greatest past accomplishment. Röntgen and van't Hoff were 56 and 49, respectively.

Why is there no Nobel Prize in mathematics? It is a natural question, especially for a mathematician who annually observes, perhaps with some envy, the fuss surrounding the announcement of the Nobel winners. The question provides the title for a short article by Lars Gårding and 1962 Swedish Fields Medalist Lars Hörmander.[6]

While the author of this book was in graduate school, he was told that the Swedish mathematician Mittag-Leffler had had an affair with Nobel's wife. It is a fairly common experience for American mathematics graduate students to hear this story from another student or a faculty member, possibly at a precolloquium gathering, party, lunch, office discussion, or even in a classroom.[7] Despite the fact that Nobel never married, the legend continues to spread as part of the socialization of mathematicians. In a field with little in the way of prurient substance, it is easy to understand the persistence of the myth. To a young (typically male) mathematician, defensive of the nerdish image of his chosen profession, there may be some satisfaction in believing that less glory is available because of the sexual exploits of an ancestor. However, it is a profession with an extraordinarily high standard of verification. Inadequately supported mathematical claims are frequently the object of ridicule, and the Nobel's wife story is *trivial* to debunk.

Gårding and Hörmander refer to the above story as the French-American version. When a mathematical statement is shown invalid, one might try to adjust the hypothesis or conclusion so as to be consistent with the data. For example, in some tellings of the story, there is a triangle with an unnamed woman who is not specified as Nobel's spouse. Other formulations omit the woman, but maintain that Nobel rejected mathematics in order to deprive Mittag-Leffler of an opportunity at the award. Gårding and Hörmander refer to this as the Swedish version.[8]

The theory that Nobel's action was motivated by animus toward Mittag-Leffler is difficult to prove or disprove. Gårding and Hörmander state that there is no basis for the theory. Crawford reached the same conclusion following her intensive study of the establishment of the Nobel Prize.[9] A statement from J. L. Synge, an associate of Fields and the executor of his estate, suggests that Fields accepted the allegation. "Perhaps I should insert here something that Fields told me and which I later verified in Sweden, namely, that Nobel hated the mathematician Mittag-Leffler and decided that mathematics would not be one of the domains in which Nobel Prizes would be available."[10]

Considering the historical context of these stories, the Swedish version may well have originated out of the political battles waged by Mittag-Leffler. In 1881 Mittag-Leffler joined the faculty of the

Högskola in Stockholm, three years after its founding. Mittag-Leffler, who was well connected in the European mathematical community, set out to boost the profile of mathematics at his new institution. Two of his most successful ventures occurred in the areas of publication and personnel. Shortly after his arrival at the Högskola, Mittag-Leffler began a mathematical periodical, *Acta Mathematica*, which rose to become among the most important outlets for the dissemination of mathematical research. Under Mittag-Leffler's influence, the Högskola attracted some of Europe's leading scholars to long and short term residencies.[11] The Russian mathematician Sonya Kovalevsky joined the faculty in 1884.[12] When Kovalevsky contemplated returning to St. Petersburg, Mittag-Leffler solicited the wealthy Nobel, among others, for funds to endow a chair for Kovalevsky at the Högskola. That Mittag-Leffler made such a request in 1890, suggests the absence of enmity at that stage. In examining the correspondence, Crawford concluded "The polite letters that Nobel and Mittag-Leffler exchanged during their lifetime seem to belie the assumption of a rift between the two."[13] Nobel declined the request, and Kovalevsky died the following year.[14]

> *The building up of the Högskola was an arduous process that caused a series of conflicts culminating in the mid-1890s. Since these conflicts are thought to have influenced Nobel when he formulated his last wishes as well as to have affected the manner in which these wishes were carried out, they deserve brief mention here. The underlying cause of the disagreements at the Högskola were the conflicting views of two factions as to the basic aims of the institution. One faction, led by Mittag-Leffler, consistently supported Lehrfreiheit, that is, the view that the Högskola should be devoted to free learning and research at the highest level and not concern itself with exam or degree requirements. The other, led by Pettersson and including Arrhenius and Bjerknes, wanted to see the Högskola develop into a full-fledged university with*

> *the right to grant degrees in the same range of subjects as the state universities. The election of Pettersson as rector in 1893, followed by that of Arrhenius, who held this post from 1897 to 1901, were significant victories for this group. They were won, however, only after head-on confrontations with the old guard led by Mittag-Leffler. Pettersson's reelection as rector early in 1895 only occurred after a drawn-out fight during which the Högskola for a short while, partly as a result of Mittag-Leffler's machinations, found itself with three rectors, two of whom had been duly elected by the council of teachers at two successive meetings, the third having been appointed temporarily by the governing board. This episode was followed only four months later, in May 1895, by a row over the appointment of Arrhenius to the newly created chair in physics. In an effort to block Arrhenius, Mittag-Leffler succeeded in having Lord Kelvin called in as one of the three experts judging Arrhenius's competence. Kelvin's opinion was predictably negative, but was transmitted to Mittag-Leffler in a personal letter rather than, as called for by the rules of appointment (sakkunnighetsinstitutionen), in a formal opinion addressed to the Högskola. This made it possible for Pettersson, the new rector, to disallow it ...*
>
> *In Stockholm's small intellectual community, where practically everybody knew someone in the press, these conflicts received considerable attention from the newspapers, which made much of "the trouble at the Högskola."*

The political struggles at the Högskola peaked between the drafting of the penultimate and final versions of Nobel's will, dated March 14, 1893, and November 27, 1895, respectively. In the earlier version, the Högskola was among the beneficiaries with a modest bequest.[15]

The final will omitted the Högskola entirely, but it was only one of several changes.

Nobel, the inventor of dynamite, held a deep interest in the movement toward world peace. In early 1893 he broached the idea of a peace prize in a letter to Bertha von Suttner, one of the most influential people in his life. Two months later Nobel produced a will which stipulated 20 percent of his estate for various named individuals including relatives, 16 percent for specified institutions including the Högskola, and 64 percent to endow a prize fund. While the leaders of the Högskola quarreled, Nobel further developed his notion of prizes. These ideas were implemented in the final will, along with drastic revisions in the allocations. Each individual was left a specific amount, summing to substantially less than the one fifth share of the previous version. All of the remainder was to endow prizes in the five areas of physics, chemistry, physiology or medicine, literature, and peace. No institution was included, except to administer the prizes.[16]

When the contents of the will were revealed after Nobel's death, it was a shock to virtually everyone, including his relatives and even his personal assistant Ragnar Sohlman who was designated as one of the two executors.[17] Who could have guessed? There was nothing like the Nobel Prize then in existence, and Nobel had been discreet about his plans, except with Bertha von Suttner. Von Suttner occupied a unique role in Nobel's life. They had met when von Suttner sought a job many years earlier. Nobel was strongly attracted to her, but she was already involved in a complicated relationship that soon led to marriage. Nobel remained fascinated by von Suttner and they did maintain a correspondence. Von Suttner founded a peace organization and authored the novel, *Lay Down Arms!* In 1905 she was awarded the Nobel Peace Prize.

Why is there no Nobel Prize in mathematics? The immediate answer is that it was not among the five areas designated by Nobel in his final will nor, as with economics, was it added later. Why then was mathematics excluded from the list? Nobel clearly wanted to make a big splash and promote those activities which "have rendered the greatest service to mankind." Under these criteria, peace and medicine were obvious choices. Given his own scientific work, it is likely that discoveries in physics and chemistry would have grabbed his attention. Finally, literature had an important aesthetic value to Nobel. Since he was so secretive about his plans, one can only speculate about his motivation, or lack of it, with regard to mathematics. Gårding and Hörmander assert that "the thought of a prize in mathematics never entered Nobel's mind."[18] Whether or not this was the case, it seems likely that mathematics would have ranked, in priority, behind the other five. Too many prizes might have diluted their impact, and cases could be made for other disciplines as well. Simply put, there was no compelling reason to include mathematics.

However, the Swedish and French-American explanations did originate somehow, and do continue to circulate. The contentious environment of the Högskola was ripe for allegations of blame.[19]

> *Although it is not known how those in responsible positions at the Högskola came to believe that a large bequest was forthcoming, this indeed was the expectation, and the disappointment was keen when it was announced early in 1897 that the Högskola had been left out of Nobel's will of 1895. Recriminations followed, with both Pettersson and Arrhenius letting it be known that Nobel's dislike for Mittag-Leffler had brought about what Pettersson termed the "Nobel flop."*

Thus the predicate of the Swedish version had an early origin, and was employed by Mittag-Leffler's opponents to explain the disinheritance of the Högskola.

Mittag-Leffler's posture toward the award winners was hardly that of a scorned party. Each year he hosted a dinner in honor of the laureates.[20] Moreover, Mittag-Leffler was intimately involved in the selection process for physics and chemistry. Nobel had left these decisions in the hands of the Swedish Academy of Sciences to which he and Mittag-Leffler both belonged. There were three stages in the selection process: nomination, review by standing committees of the Academy, and consideration by the entire Academy. Although Mittag-Leffler never served on either the physics or chemistry committee, he did manage to exert his influence on each of the three stages. He was especially active in choreographing nomination campaigns, even attempting to disguise his involvement from his enemies.[21]

Mittag-Leffler's greatest efforts were on behalf of Poincaré. The absence of a Nobel Prize in mathematics was no deterrent to Mittag-Leffler. Poincaré's work covered a broad spectrum, ranging from abstract topology to mathematical physics. Mittag-Leffler began a tactical campaign to obtain a Nobel Physics Prize for Poincaré.[22] At the turn of the century, the Academy physicists were disposed toward the applied side of their discipline. To reach Poincaré required a significant shift in emphasis. Mittag-Leffler planned a series of moves to steadily push the theoretical envelope. The first successful step in his program was the 1902 award shared by H. A. Lorentz. The effort ended with Poincaré's death in 1912, at which point he had the distinction of receiving the most nominations of any nonwinner.[23]

We hypothesize the following chronology for the Nobel–Mittag-Leffler stories. The omission of mathematics from Nobel's list reflected the priorities of the benefactor. When the Högskola was passed over, Mittag-Leffler's opponents seized the opportunity to place the blame on him, fabricating the notion of Nobel's animus. As Mittag-Leffler plotted to manipulate the physics awards toward mathematicians, the story's consequence evolved, changing Nobel's retribution from the Högskola to mathematics. Stipulating this Swedish version, it is easy to imagine the story being spruced up in numerous retellings, eventually leading to the French-American version with Nobel's (nonexistent) wife.

C. International Congress of Mathematicians in Moscow

The original Fields plan called for two medals to be presented at each of the quadrennial ICMs. Following the inaugural award in 1936, there was a 14 year hiatus in the Congresses, due to World War II and its aftermath. During this period many leading European mathematicians migrated to the United States.[24] Europe's loss was America's gain, and the center of mass of the mathematical world abruptly shifted westward. In 1950 the ICM resumed in Cambridge, Massachusetts, where the second Fields Medals were presented. Below is a list of the subsequent ICMs, along with the Fields Medal winners, their age, country of birth, and institutional affiliation.[25]

1950 Cambridge, MA
Laurent Schwartz, 35, France, University of Nancy
Atle Selberg, 33, Norway, Institute for Advanced Study

1954 Amsterdam
Kunihiko Kodaira, 39, Japan, Princeton University
Jean-Pierre Serre, 37, France, College de France

1958 Edinburgh
René Thom, 34, France, University of Strasbourg
Klaus Roth, 32, Germany, University of London

1962 Stockholm
Lars Hörmander, 31, Sweden, Stockholm University
John Milnor, 31, USA, Princeton University

Smale was fortunate to have had early mathematical contact with several of the Fields Medalists. In large measure this was via his thesis advisor Raoul Bott. It was Serre's work that was the topic of the topology seminar at Michigan. During Smale's last year as a graduate student, he met Serre while visiting his advisor at the Institute for Advanced Study in Princeton. Thom learned of Smale's thesis results directly from Bott. At the Mexico City conference in 1956, Steve met Thom, who was also on his way to the University of Chicago for the following term. At Chicago, Smale attended Thom's lectures on transversality. While that mathematical tool played an important role in Smale's future work, there was also a vital interaction. At that time Steve was a fresh Ph.D. embarked on his first major result, and Thom was about to win the Fields Medal. These were two young mathematicians who possessed an abstract mathematical power beyond most people's comprehension. Thom appreciated Smale's ideas. That interest provided direct support to Smale as well as boosting his profile at Chicago.

Milnor was also at Mexico City. Just 25 years old, that year he published the result that won him the Fields Medal in 1962.[26] When Steve went to the Institute for Advanced Study in 1958, he attended Milnor's lectures at Princeton. Consider then that among the six Fields Medalists in the 1954–62 period of Steve's thesis and early professional work, Smale was intimately acquainted with the ideas of three of them, immediately after their discoveries.

Smale wanted his own Fields Medal. With the higher dimensional Poincaré Conjecture in 1960 and the *h*-cobordism theorem in 1961, Steve knew that he had the credentials. He thought 1962 "was the logical time I would have gotten the Fields Medal."[27] The Stockholm Congress was also a logical time for Milnor who, like Smale, had profoundly influenced the course of topology. Other mathematical areas offered their candidates as well, and a committee was charged with the task of selecting just two winners. It was a purely qualitative problem. Considering all the mathematicians in the world under the age of 40, who had done the best mathematics? How do you compare seminal results in topology, as well as with those in analysis or algebra? The committee chose Milnor and the Swedish analyst Hörmander.

Not winning was a tremendous blow to Smale. He attributed his loss, in part, to mathematical politics.[28] Comparisons between Milnor and Smale were inevitable. They were the two brilliant, young, American topologists with the most outstanding theorems. Unlike Smale, Milnor's mathematical gifts became apparent at an early stage of his life. The New Jersey native attended Princeton University and then joined its faculty. With the Institute for Advanced Study and the University, the city of Princeton had a special prestige in the mathematical world. Milnor had the more influential backers. Steve had followed an unconventional path to the top and was deprived of full credit for his work on the Poincaré Conjecture. It was not the sort of profile that attracted unanimous enthusiasm from the mathematical establishment.

With the Nobel Prize, Steve would have had another opportunity the following year, but the next Fields Medals were to be awarded in 1966. That year Steve's European summer tour began in Paris. The French left were organizing a Vietnam protest against the United States. It was called Six Hours for Vietnam and followed much of the Michigan teach-in format with speeches, movies, songs, and breakouts into parallel seminars. Among those endorsing the rally were Jean Paul Sartre and Fields Medalist Laurent Schwartz.[29] Schwartz had a long record of political activity, having opposed his country's policies in Algeria and Vietnam. When Smale sought to globalize the Vietnam protest in 1965, he solicited Schwartz' support, receiving an enthusiastic response. While Smale's role as an antiwar leader had evaporated since the Oakland march of the previous October, he remained a prominent opponent of the War. Steve had already planned to spend the summer in Europe with a vigorous schedule of mathematical endeavors. When Schwartz invited him to speak, Smale decided to begin his time abroad with the Paris rally. Joining Steve on the platform were officials from North Vietnam and China. Smale recalled the experience.[30]

> *The Salle de la Mutualité—I remembered that great old hall on the left bank for the political rallies I had attended there fifteen years earlier. Now there were several thousand exuberant young people in the audience, and I was at the microphone. My French was poor and since my talk could be translated I decided to speak in English. I still wasn't at ease giving nonmathematical talks; even though I had scribbled out my brief talk on scratch paper, I was nervous.*
>
> *There was a creative tension in that atmosphere that inspired me to communicate my feelings about the United States in Vietnam. As I was interrupted with applause my emotional state barely permitted me to give the closing lines: " … As an American, I feel very ashamed of my country now, and I appreciate very much your organizing and attending meetings like this. Thank you." Then I was led across the stage to M. Vanh Bo, the North Vietnamese representative in Paris; we embraced and the applause reached its peak.*

A reporter covering the rally observed:[31]

> *Indeed Professor Smail* [sic] *of the University of California got almost as big a hand as Monsieur Vanh Bo of North Vietnam. Such is the sentimentality of the left. Even the chant that followed Smail's speech—"U.S. go home! U.S. go home!"—was obviously a left-handed tribute.*

After the Paris speech, Steve resumed his mathematics mode.[32] A few days later he and René Thom drove to a conference in Geneva. While en route Thom, a member of the Fields Committee for the 1966 ICM, disclosed that Smale was to receive the award in Moscow. Steve had attained the status he deserved and craved. It was an extraordinary feeling. His place in mathematical history was secured.

Clara and the children joined Steve in Switzerland. The Smale family remained in Geneva until the end of July. During that time Steve was in residence at the University of Geneva except for a week at a Bonn mathematics conference. Smale had a remarkable capacity to effectively immerse himself in his current interest. The previous summer he was completely devoted to stopping a war. In Geneva he was doing mathematics, and the conditions were highly favorable. The University had a fine mathematics department, providing stimulating interaction. There were no teaching or administrative responsibilities. The locale offered nice settings for family outings. Finally there was the satisfaction that he was soon to receive the Fields Medal.

When Steve was a child, his father enjoyed taking the family on automobile excursions around the country. Now Steve had the opportunity to explore Europe with his own family. The Fields Medal presentation was scheduled for August 16, the opening day of the ICM. The Smales rented a camper and, during the first two weeks of August, drove from Geneva to Athens, camping and touring Yugoslavia and Greece along the way. Smale planned to fly from Athens to Moscow on August 15, participating in the ICM while his family remained in Greece.

Unbeknownst to Smale, his whereabouts were a hot topic in the Bay Area. In early August the House Committee on Un-American Activities (HUAC) sought to subpoena Steve for testimony later in the month. Some members of Congress were disturbed by the troop train demonstrations and other activities, which they construed as aiding military opponents. Pending legislation was directed at prohibiting such activities. The HUAC scheduled hearings on the bill, issuing subpoenas to various antiwar leaders.

In the fifties, the prospect of subpoenas provoked dread among members of the Old Left. Recipients, such as Chandler Davis, suffered blacklisting and jail. At that time Steve's leftist activities narrowly escaped the public scrutiny of the HUAC. Now, as the former cochair of the VDC and leader of the troop train protests, he was too prominent. Moreover, the HUAC always had a thing for college professors. Subpoenas were issued for Smale, Jerry Rubin, and several others. The HUAC was soon to learn that they had lost their power of intimidation. The first indication was the posture of the Berkeley community toward the subpoenas. Among the radicals, subpoenas were coveted as recognition of one's standing. When Stew Albert lamented his failure to attract the solicitation of the HUAC, Jerry Rubin diagnosed his condition as "subpoenas envy."[33]

Serving Steve was problematic. At the time he was somewhere in Yugoslavia, on a roundabout course to the ICM in Moscow. Investigation by the *San Francisco Examiner* turned up a few facts which they interpolated with a fictional motive. The story ran under the headline "UC Prof Dodges Subpoena, Skips U.S. for Moscow."[34]

> *Dr. Stephen Smale, University of California professor and backer of the Vietnam Day Committee and old Free Speech Movement, is either on his way or is in Moscow, The Examiner learned today.*
>
> *In leaving the country, he had dodged a subpoena directing him to appear before the House Committee on Un-American Activities in Washington.*

Steve was, in fact, headed to Moscow, and he did not receive the subpoena. However, he was unaware of its existence and there was absolutely no linkage between the recent subpoena and the long planned trip to the Moscow ICM. Mathematical colleagues rallied to Smale's defense. The following day an article in the *San Francisco Chronicle* set the record straight, chiding the *Examiner's* "somewhat exaggerated published report." However, it is the nature of such corrections that they fail to completely erase the original distortions.

Back in Athens, Clara dropped Steve at the airport on August 15, and drove off with the children. He was ticketed on a flight to Moscow, excited about the Fields Medal ceremony of the following day. The euphoria was interrupted by a trip into a bureaucratic twilight zone. Customs detected some irregularity with Smale's passport.[35]

> *Slowly I began to understand. When we had come across the Greek border, customs had marked in my passport that we were bringing in a car. (The government was concerned that we would sell the car without paying taxes.) Now the Greek officials were not letting me leave the country without that car. The customs officials were adamant.*

Unable to produce the car, Smale was prohibited from boarding the only Athens-Moscow flight of the day. Imagine Steve's frustration. A last moment technicality was depriving him of the greatest moment of Fields acclaim. Eventually Steve obtained the assistance of an American embassy official who successfully intervened on his behalf. Smale was on a plane the following day, and the timing was tight.[36]

> *I arrived late in Moscow and rushed from the airport to the Kremlin where I was to receive the Fields Medal at the opening ceremonies of the International Congress. Without a registration badge the guards at the gate refused me admission to the palace. Finally, through the efforts of a Soviet mathematician who knew me, I obtained entrance and found a rear seat. René Thom was speaking about me and my work:*
>
> *"si les oeuvres de Smale ne possedent peut-etre pas la perfection formelle du travail definitif, c'est que Smale est un pionnier qui prend ses risques avec un courage tranquille;"*

Steve had just missed the opportunity to formally receive the award, but he did hear Thom's tribute to his risk taking. An anonymous gift permitted the award of four Fields Medals in 1966. The other winners were:[37]

Michael Atiyah, 37, England, Oxford University

Paul Cohen, 32, USA, Stanford University

Alexander Grothendieck, 38, Germany, University of Paris

Moscow was an interesting selection as host city. It was the home of some of the world's best mathematicians, but the severe travel restrictions imposed by their government prohibited many of them from attending congresses and conferences in the west. The 1966 Congress was a truly international gathering. Over 4000 mathematicians participated in the technical program at Moscow University, smashing the attendance records from all previous Congresses. Especially notable was the large Soviet delegation.

The communist government repressed various forms of intellectual inquiry. Earlier that year the cases of two Soviet writers, Yuli Daniel and Andrei Sinyavsky, had attracted international attention. Satirical works by these men were smuggled out of the country and published in the West under pseudonyms. The authors were indicted and tried for the crime of producing "anti-Soviet propaganda, harmful to the Soviet people."[38] In February 1966 Daniel and Sinyavsky were convicted and sentenced to five and seven years of forced labor, respectively.

Smale had visited Moscow and Kiev five years earlier, when he lectured on the horseshoe and sketched some of his ideas for dynamical systems. During the visit he interacted with a number of young Soviet mathematicians and was struck by their prodigious talent. In 1966 the development of dynamical systems was hitting its stride with independent investigations underway in Berkeley and the Soviet Union.

At the Congress Smale was able to renew his discussions with the Russians. The stimulation was not restricted to mathematical endeavors.

Despite the Daniel–Sinyavsky case, there was a degree of freedom in the country. Smale, through his Russian mathematics friends, became connected with a network of anticommunists.[39] Among this group of writers, artists, mathematicians, and other intellectuals there was contempt for Soviet policies. Not only did they express their solidarity for Daniel and Sinyavsky, but they supported the United States' intervention in Vietnam. Smale was shocked to encounter the latter manifestation of anticommunism.[40]

During the previous year Smale and Laurent Schwartz had collaborated to internationalize opposition to the Vietnam War. Now Smale had moved to the background, but Laurent Schwartz continued the effort. The Moscow ICM provided an opportunity to petition mathematicians from around the world. A Japanese mathematician conceived the International Appeal Against the War in Vietnam and Schwartz pushed the idea. Prior to the Congress, Schwartz enlisted the help of Chandler Davis (HUAC Class of 1954). Davis' vita included being fired at the University of Michigan, 6 months of jail time, being hired by the University of Toronto, and his own organization of antiwar efforts. He was already planning to attend the ICM, excited about the opportunity of meeting Vietnamese mathematicians. Davis liked the concept of a Moscow petition, recalling "the idea was to initiate the Appeal at the Congress, to agree on the wording, and collect tons of signatures."[41] Following the Congress, the document was to circulate around the world, gathering opposition to the War. There was some discussion of seeking endorsement of the resolution by the Congress as a whole, but the notion was quickly rejected. The structure of the program was not conducive to such initiatives and there appeared little prospect for success.

To conduct the Appeal at the meeting required some infrastructure, such as a meeting room and duplication machinery. Securing these facilities in a totalitarian state was problematic. Although the Appeal's mission was in line with Soviet policy, there was a reluctance to permit political activity outside of rigid government control. I. Petrovski, president of the Congress and rector of Moscow University, refused a request for a room by a Schwartz–Davis led delegation. Mimeograph machines were even further out of the question. Access to these vital propaganda tools was severely restricted.

With the assistance of some Soviet mathematicians, Davis and Schwartz secured a meeting room. This time they did not request official sanction, but Davis was certain that the authorities were aware of the plans, and elected to look the other way. Invitation was by word of mouth and attracted about 80 people. Schwartz chaired the meeting, and there was discussion of the appeal's wording.[42]

It remained to print and duplicate the petition. In 1966 Moscow, there were no Kinko's. The only mimeograph facilities were under government control. Already rejected by the Soviets, Davis solicited the help of another government. At the North Vietnamese embassy, Davis typed and the Vietnamese mimeographed.[43]

Smale was among the proponents and signatories of the statement condemning United States actions in Vietnam. There were other political activities at the Congress, including petitions associated with Smale's HUAC subpoena. Timing constraints limited the effectiveness of these efforts. While Schwartz had circulated preliminary appeals prior to the Congress, few participants were even aware of the subpoena in advance. Smale, himself, learned of it on the plane to Moscow, receiving the information from a mathematician who boarded during a stop in Hungary.

The antiwar efforts of Schwartz, Davis, and Smale were acknowledged in a moving gesture. They were the guests for a banquet hosted by four Vietnamese mathematicians at the Congress. The dinner took place in a dormitory where the Vietnamese prepared their own native cuisine. Mathematicians from different countries often overcome cultural barriers by sharing their scientific interest. Here the bond was much stronger. There was a devotion to a common cause that somehow was multiplied with the mathematics. The banquet was a memorable experience.

The Vietnamese conceived the dinner, but the guests devised a complementary agenda, planning to issue a press release afterward. However, the banquet set in motion a chain of events that led to a greater political statement, spotlighting Steve. Smale recalled:[44]

> *At the banquet, I was asked to give an interview to a reporter named Hoang Thinh from Hanoi. I didn't know what to say and struggled with the problem for the next day. I felt a great debt and obligation to the Vietnamese—after all, it was my country that was causing them so much pain. It was my tax money that was supporting the U.S. Air Force, paying for the napalm and cluster bombs. On the other hand, I was a mathematician, with compelling geometrical ideas to be translated into theorems. There was a limit to my ability to survive as a scientist and weather further political storms. I was conscious of the problems that could develop for me from a widely publicized anti-U.S. interview given to a Hanoi reporter in Moscow. In particular, I knew that what I said might come out quite differently in the North Vietnamese newspaper, and even more so when translated back into the U.S. press.*
>
> *This was the background for my rather unusual course of action. On the one hand, I would give the interview; on the other hand, I would ask the American reporters in Moscow to be present so that my statements could be reported more directly. I would give a press conference on Friday morning, August 26.*

At the ICM Smale faced the problem of whether to give an interview to Thinh, the Moscow correspondent of the North Vietnamese Press Agency. His analysis and solution exhibit parallels to his mathematical style. As a mathematician Steve liked to undertake *hard* problems which had resisted the attacks of others. Frequently he succeeded in finding an innovative solution, employing techniques from completely different areas of mathematics. After the fact, his methods seem natural, but nobody else thought to apply them. While other mathematicians were failing with conventional techniques, Smale understood the problem in a new fashion, and blazed his own trail. To succeed in this manner requires audacity and profound mathematical power.

When confronted with the interview request, most people would have seen two choices: decline or accept. Steve was inclined to accommodate the reporter, and, in the process, make his own political statement. However, there were serious qualms. Smale was a person who had lost faith in communism a decade earlier, and he never trusted his own country. He was sufficiently pragmatic to recognize the dangers of distortion from North Vietnamese and American filters, especially when composed in that order. A third party observer at the interview might have provided some insurance, but the above sequence could still play out. Furthermore, Smale wanted to direct his future energy toward mathematics rather than clarifying his politics. Neither conventional solution was appealing, but Smale was rarely constrained by convention. How many people would have elected to conduct a press conference in Moscow in 1966? The thought process did not require the intellectual power of his mathematical work, but it was a creative solution.

The logistics of holding a press conference presented a major problem. In Moscow the emphasis was on government control. Recall that Davis and Schwartz were unable to officially obtain a meeting room for discussion of the appeal. Even when there was an expectation of shared goals and interests, the government permitted no slack for free expression. Smale ruled out any official request, but there was concern that the Soviets might treat the press conference as a subversive act.

Steve wanted to create a situation that maximized his physical safety and minimized the potential for distortion of his remarks. These priorities called for an open venue, a diverse audience, and short notice. The steps of the University were easily accessible to ICM participants. In addition to Hoang Thinh, Smale invited the American and Soviet press, as well as various friends and colleagues.

The next move was up to Thinh. Smale's unilateral change in the arrangements, from interview to press conference, shifted the dynamic from a North Vietnamese production to a Steve Smale show. If Thinh were to participate in a public demonstration, unauthorized by the Soviets, there was potential embarrassment for both the North Vietnamese and USSR governments. Late Thursday, Thinh responded. He declined to participate in the press conference, rather, providing his list of questions and requesting written responses.

Thinh's interview overture had begun the whole process. With his withdrawal, the press conference might have been cancelled. However, Smale was determined to go ahead with his condemnation of United States' policy in Vietnam. A press conference in Moscow was in the best tradition of the Rubin–Smale-led VDC. An audacious act to attract the press, who would then communicate the message.

With the press conference scheduled for the following morning, Steve worked on the message.[45]

> *As I wrote down my words attacking the United States from Moscow, I felt that I had to censure the Soviet Union as well. This would increase my jeopardy, but, having just received the Fields Medal and being the center of much additional attention because of HUAC, I was as secure as anybody. If I couldn't make a sharp antiwar statement in Moscow and criticize the Soviets, who could?*

That night Steve discussed his ideas with Chandler Davis who recalled Smale's demeanor: "He was really happy about the opportunity. His attitude was of trying to take the most advantage of it by getting things absolutely clear."[46] Davis cautioned that the attacks on the Soviet Union and the United States might cancel each other out, blunting the criticism of American policy in Vietnam.

On Friday morning, a small group gathered on the steps for the press conference. Chandler Davis, Moe Hirsch, Bob Williams, and Berkeley department chair Leon Henkin were among the mathematicians joining the American reporters in attendance. Other mathematicians passed by on their way to the final day of the Congress. Steve had constructed the situation so as to afford protection. If the Soviets forcibly terminated the press conference, the American reporters were on the scene to report a big story. A more diplomatic effort was unlikely to stop a veteran of VDC maneuvers. Just prior to the opening, a woman from Novosti, a Soviet press agency, requested a private interview. Smale agreed, but deferred their meeting until after the press conference. Then he read the following statement, handwritten on two sheets of paper.[47]

> *This meeting was prompted by an invitation to an interview by the North Vietnamese Press. After much thought, I accepted, never having refused an interview before. At the same time, I invited representatives from Tass and the American Press, as well as a few friends.*
>
> *I would like to say a few words first. Afterwards I will answer questions.*
>
> *I believe the American military intervention in Vietnam is horrible and becomes more horrible every day. I have great sympathy for the victims of this intervention, the Vietnamese people. However, in Moscow today, one cannot help but*

> *remember that it was only 10 years ago that Russian troops were brutally intervening in Hungary and that many courageous Hungarians died fighting for their independence. Never could I see justification for military intervention, 10 years ago in Hungary or now in the much more dangerous and brutal American intervention in Vietnam.*
>
> *There is a real danger of a new McCarthyism in America, as evidenced in the actions of the House Un-American Activities Committee. These actions are a serious threat to the right of protest, both in the hearings and in the legislation they are proposing. Again saying this in the Soviet Union, I feel I must add that what I have seen here in the discontent of the intellectuals on the Sinyavsky–Daniel trial and their lack of means of expressing this discontent, shows indeed a sad state of affairs. Even the most basic means of protest are lacking here. In all countries it is important to defend and expand the freedoms of speech and the press.*

Following his statement, Smale responded to the questions submitted by Thinh urging immediate withdrawal of United States forces from Vietnam. "At this point, a middle-aged Russian woman hastened up to the professor, touched him on the right arm and told him that Vladimir G. Karmanov, the secretary general of the mathematics conference wanted to see him immediately for an "urgent" discussion."[48] The second interruption, more intrusive than the first, was deflected in the same manner. Smale agreed to see Karmanov, following the completion of the press conference. A short time later, Smale had fielded the remaining questions. The press conference was over, and Steve moved on to his commitments with the two Soviet women. Karmanov's representative was the more assertive, ushering Smale to meet the secretary general at his nearby office. Following along was an entourage that included the other woman, curious reporters, and concerned mathematicians.

Despite the purported urgency of the meeting with Karmanov, the conversation was devoid of substance, mathematical as well as political. Karmanov played the role of a director of tourism. He presented Steve a coffee table book, featuring pictures of the Kremlin and text in German. Instead of asking about the press conference, Karmanov was anxious to learn what sights Steve had seen or might wish to see, making available an automobile and guide. The meeting ended amiably with Steve declining the tour.

What followed is a bit murky. Steve reluctantly agreed to accompany the Novosti representative to an unspecified location for their interview. He was then escorted, at a fast pace, to a waiting car, as Soviet bodyguards brushed off the press and mathematicians. Steve recalled: "I felt pressured and a little scared. But all the while I was treated not just politely, but like a dignitary. It was hard to resist."[49] The scene was out of a Hollywood movie script. Steve's automobile drove off at high speed with the American reporters in hot pursuit. Where were they taking him? Hirsch "thought the worst" when Steve was put in the car.[50] The excursion ended at Novosti headquarters where there was another hour of small talk, but no interview. Finally Steve was returned to the Congress at his own request.

The Congress was concluding, but there were questions for Steve from reporters and his very relieved friends. What happened and what were the agendas behind the two meetings? The press conference placed the authorities in a quandary. Unaware of what Steve planned to say and unable to prevent it diplomatically, his statement was a worst case scenario. The first step in damage control was to isolate Smale from the reporters and prevent further mischief. Firmness replaced finesse in the trip from Karmanov to Novosti. The process bought some time during which the problem may have been considered at a higher level. In any event, no further action was taken. Smale returned to his hotel room late that night, without incident. The following morning he made his scheduled 7:00 am flight to rejoin Clara and the children in Athens.

The Saturday *New York Times* carried the story, with accompanying photograph, at the bottom of the front page. The well written article provided accurate coverage of Smale's press conference and adventure in Moscow. However, the headline and picture caption gave a false impression. In the photograph, Steve was shown between a pair of larger men. The caption "CALIFORNIAN GETS ESCORTED TOUR: Dr. Stephen Smale, center, being led to car in Moscow after his news conference was halted when he criticized the Soviet Union" was juxtaposed with the headline:

American Critical in Soviet—Briefly

Together, the picture and headline implied that the Soviets had ended the press conference prematurely. The portion of the story appearing on the front page, while correct, was also consistent with that inaccurate portrayal. To learn the complete story, one needed to read the continuation on page 13. The *Washington Post* headline, "Russians Halt Speech by U.S. Professor," was followed by the sentence: "Soviet authorities ended a press conference abruptly yesterday when an American professor, who is a leader of the anti-Vietnam-war movement, compared U.S. "aggression" with Russia's brutal intervention in Hungary."[51]

Smale had included the American press in order to secure both his safety and the accurate dissemination of his message. While both were achieved, the papers failed to correctly convey the actual sequence of events. As a result, the distorted impression prevailed.

Although the news appeal of the story might have been enhanced by the image of the Soviets shutting down Smale's speech, the historical value was in the substance and setting of the statement. It was a remarkable speech. The world was polarized in communist and anticommunist camps. Smale saw parallel tragedies in each system that were concealed from their respective constituencies by Cold War denial. He paired the United States' aggression in Vietnam with the Soviet invasion of Hungary. The efforts of the HUAC were compared with Daniel–Sinyavsky and communist repression. All of this conveyed in two lucid paragraphs by an American in Moscow.

D. The HUAC hearings

On August 16, the day that Smale received the Fields Medal, a HUAC subcommittee began hearings on *Bills to make punishable assistance to the enemies of U.S. in time of undeclared war.*[52] The legislation had been introduced by the subcommittee chair, Representative Joe Pool of Texas. Pool organized the hearings into two stages. First was an investigative phase to establish the problem. Among those scheduled for testimony were informants, radicals, and government officials. Topics included the troop train demonstrations and efforts to provide medical support for the North Vietnamese and Viet Cong armies. Once the stage was set with the egregious acts of the antiwar movement, congressmen and government officials would discuss the merits of Pool's legislation. It was to be a two-act HUAC production, exposing communists along the way.

The power of the HUAC had diminished considerably since the fifties. At that time the HUAC ran the show and even lawyers were reluctant to accompany a witness. In 1966 the radicals saw the hearings as an opportunity to make their case against the HUAC and the War. Rather than defend themselves against attacks by HUAC members, they planned to launch their own offensive, seizing control of the production. Supporting the witnesses were a new breed of lawyers.

Prior to the hearings, there were some legal controversies. That the *Examiner* learned of Smale's subpoena, implied a violation of the House's nondisclosure rules. On another front, the American Civil

Liberties Union filed suit to enjoin the HUAC from holding the hearings. On August 15, a federal judge issued the injunction. The surprise ruling raised its own constitutional questions on separation of power between the legislative and judiciary branches. Pool's plans to defy the court order became moot when an appeals court panel quashed the injunction, just before the hearings began on Tuesday.[53]

The hearing room and halls were filled with antiwar demonstrators. The HUAC was going to have an overtly hostile audience. Just after the hearing commenced, a woman screamed from the entrance door. It was Jerry Rubin's attorney, Beverly Axelrod. Unlike the deferential lawyers of the fifties, Axelrod forewent the normal decorum to seek admission for herself and her client. Rubin entered, dressed in a Revolutionary War uniform, passing out pamphlets explaining the symbolism.

Rubin's wardrobe and demeanor ended any pretense that the hearing was a dignified proceeding. It was going to be a spectacle, in the tradition of the VDC rather than the HUAC. Rubin's willingness to appear as a clown and sacrifice personal dignity was an unusual form of courage, conceded even by his Berkeley detractors. Throughout the first day, the hearings were interrupted by belligerent demonstrations. There were 17 arrests, including two of the witnesses. Spectators painted slogans on the wall. Rubin became a hero of the radical youth culture, inspiring a new left style that would set a standard for outrageous courtroom contempt.

On day two Pool stuck with the HUAC game plan, but it was not working. A witness proudly announced that he was a communist. Lawyers often ignored admonitions from the chair and aggressively pursued their arguments. There were 21 arrests, highlighted by that of attorney Arthur Kinoy who was dragged from the room as he screamed in protest. A picture on the front page of the *New York Times* showed a marshal gripping Kinoy by the neck.[54] The other attorneys and some witnesses departed in protest, leaving the remaining clients without representation.

The HUAC subpoenaed the district attorney of Alameda County as a friendly witness to describe the activities of the VDC. Testifying in his place was the deputy district attorney, Edwin Meese III. Meese was 30 years old, early in a legal and political career that would lead to the attorney generalship under Ronald Reagan. Meese's testimony gave high marks to the VDC's organization and planning. When the committee's counsel requested that he describe the troop train protests, the following discussion occurred.[55]

> ***Mr. Nittle.*** *Mr. Meese, would you describe the incident or incidents concerning the troop trains?*
>
> ***Mr. Meese.*** *The troop train demonstrations, which is probably the best known of the activities, took place on the 5th of August 1965, the 6th of August 1965, and the 12th of August. On the 5th of August, the demonstrations were led by Stephen Smale, who is one of the witnesses that I believe the committee is familiar with. I believe he was to be subpoenaed as a witness, but was not able to be served, if I understand correctly.*
>
> *And a man by the name of Paul Ivory. Stephen Smale is a member of the faculty in mathematics at the University of California, and Paul Ivory, who is an assistant professor, Jerry Rubin–*
>
> ***Mr. Rubin.*** *I object. I would like to make a statement. I am without counsel because of the way my lawyer was treated. I want to represent myself.*
>
> ***Mr. Nittle.*** *I ask that this witness be seated until the end of the presentation of this witness.*
>
> ***Mr. Rubin.*** *I want to make a statement. I want to represent myself.*

Mr. Nittle. *I ask that this man be removed.*

Mr. Ashbrook. *Mr. Chairman, he cannot be removed.*

Mr. Rubin. *My name is Jerry Rubin. It has just been introduced in the record without notice, I want to make a statement about it. I am now representing myself, apparently.*

Mr. Ichord. *Mr. Chairman, I would suggest that the gentleman be seated. He will be called tomorrow and have the opportunity to deny—*

Mr. Rubin. *I am representing myself.*

Mr. Ichord. *—to refute in any way any allegations made against him.*

Mr. Rubin. *I am sorry. My name was just introduced by this gentleman into the record without my knowing about it, and I would like to make a statement.*

Mr. Pool. *What do you want to say?*

Mr. Rubin. *First, I want to introduce myself. My name is Jerry Rubin. I would like to make an explanation as to why I am wearing the uniform of the American Revolution of 1776.*

Mr. Pool. *I don't care to hear that.*

Mr. Rubin. *I am wearing it because America is degrading its 1776 ideals.*

Mr. Pool. *I am giving you the opportunity to make an objection and I am trying to be fair with you, so state your objection.*

Mr. Rubin. *I am making this objection.*

Mr. Pool. *What is your objection?*

Mr. Rubin. *I am making it right now.*

Mr. Pool. *What is your objection?*

Mr. Rubin. *Would you wait one second and let me say it?*

Mr. Pool. *Real fast.*

Mr. Rubin. *I want to do it—*

Mr. Pool. *Your way.*

Mr. Rubin. *This gentleman has just mentioned my name and introduced into evidence.*

Mr. Pool. *What is your objection?*

Mr. Rubin. *And I have not been informed previously that he was going to make statements about myself that may defame my character; previously, so far, what he said about the Vietnam Day Committee has been, I think, complimentary, but he may be going to make statements, and has not yet, and I have not been informed that my name was to be introduced. That's my first objection.*

My second objection is that I do not—that I want the right to cross-examine Mr. District Attorney Meese.

Mr. Pool. *You have been named in the newspapers on many occasions, and I don't think you are surprised by this statement.*

Mr. Rubin. *This is a little different than a newspaper.*

Mr. Pool. *And your second objection—*

Mr. Rubin. *I want the right to cross-examine.*

> ***Mr. Pool.*** *—that you want the right to cross-examination; that is overruled also.*
>
> ***Mr. Rubin.*** *This gentleman is making statements about me.*
>
> ***Mr. Pool.*** *I overrule both objections. You have no others, so just be seated.*
>
> ***Mr. Rubin.*** *This is quite a courtroom.*

Rubin's interruption was in a similar vein to that of Kinoy earlier in the day. This time Pool was more indulgent. Meese then resumed his testimony which he concluded on the following morning. The subcommittee was delighted with Meese's performance, lavishing praise on the witness.[56]

The hearing was a confrontation between the HUAC and the new left. If press coverage determined the score, and it was vital to both sides, then the radicals were winning the battle. The *New York Times* front page coverage focused on the original injunction, demonstrations, and Kinoy's arrest. Meese's testimony on the Oakland Army Terminal was buried in two brief paragraphs at the end of the continuation.[57] Worse yet, the radicals appeared to enjoy the proceedings, even boasting of communist involvement.

It was time to move on to the legislative phase, but several subpoenaed radicals remained to be called. Following Meese's testimony, Pool adopted a new strategy, exploiting the walkout of the attorneys. As much as the HUAC wanted to hear the remaining radicals, it was only fair to defer their testimony until November, permitting them to secure new counsel. However, Pool made the mistake of offering the deferral individually, rather than invoking it unilaterally. A couple of witnesses were called and accepted the deferral, but then Stuart McRae countered the HUAC move. McRae insisted on testifying and resumed the attack from the left.

By the following day, August 19, the radicals had fully regrouped and decided to testify. The phrase "hostile witness" was especially appropriate, as the radicals directed insults and obscene gestures at the committee members. Each of the witnesses responded that they were members of the Progressive Labor Party (Chinese communist). Two of the radicals were dragged out of the room when they refused to leave the stand. Pool had had enough. He ended the investigative phase, excusing the remaining witnesses, much to the chagrin of Jerry Rubin. When Rubin persisted in his attempts to testify, Pool ordered his ejection.[58]

The legislative phase went more smoothly. Congressmen, generals, and Meese expressed enthusiasm for Pool's bill. The final witness was Deputy Attorney General Ramsey Clark, who, like Meese, would later become attorney general. Clark expressed opposition to the bill, stating that existing state and federal statutes provided more than adequate protection.

Smale felt ambivalent over missing the hearings. One year earlier he would have been anxious for a platform to push his antiwar agenda. In 1966 he would have enjoyed the experience, but was content to do mathematics and avoid the aggravation that accompanied a HUAC appearance. Besides, he still had the subpoena credential. It is difficult to speculate on how a Smale appearance might have impacted the chaotic hearings. Certainly he would not have worn a Revolutionary War uniform. Looking back he said "I would have used the situation to express my points of view on the War and troop trains. I would not have clowned around, but I would not have felt bound by the strictures the committee lays down for witnesses."[59]

The HUAC had been in decline since the midfifties, and the August hearings accelerated its demise. Congress had a genteel self-image and the raucous hearings were an embarrassment. Even Senate Republican leader Everett Dirksen observed that Congress had fared poorly.[60] In an editorial, the *New York Times* called for "radical surgery" to abolish the HUAC.[61]

Subsequent years saw an increase in opposition to the HUAC, concurrent with the strengthening of the left by the antiwar movement. In 1969 the HUAC attempted to rehabilitate its image with a change of name to the House Internal Security Committee. The mission was only altered slightly and the makeover did little to deter its opponents. Congressman Robert Drinan, a Jesuit priest from Massachusetts, secured membership on the committee with the express intent of seeking its abolition. Still the House Internal Security committee managed to obtain appropriations for several years. Finally in 1975, following the post-Watergate Democratic congressional election gains, the last vestige of the HUAC was purged from the House.

Smale Versus the National Science Foundation

08

A. Summer fallout

The summer of 1966 had been an extraordinary period for Smale. While the Fields Medal confirmed his place in history as one of the greatest mathematicians of his generation, the press conference and HUAC subpoena validated his credentials as a radical. Steve was well aware of the provocative nature of his Moscow activity, and that it was likely to lead to trouble in the United States. The ramifications were immediate.

As the National Science Foundation (NSF) and the University of California tried desperately to contact Steve, the Smale family traveled through Europe and returned to New York on the SS France, oblivious to the developing crisis. A conflict of scientific politics was forming in which the principal players were Smale, NSF Director Leland Haworth, Indiana Congressman Richard Roudebush, and two Berkeley administrators. The two-stage battle in 1966 and 1967 provided a rare period in which mathematical culture received the scrutiny of both Congress and the public. The roots of this controversy, however, may be found in the history of the National Science Foundation.[1]

B. A brief political history of the NSF

On May 10, 1950, President Harry Truman signed into law the National Science Foundation Act. Out of several missions assigned to the new agency, the early focus was placed on the mandates to support basic research and education in the sciences. The term basic research, as distinguished from applied research, is quite ambiguous. To some extent the distinction is one of motivation rather than substance. Thus the basic researcher seeks fundamental knowledge or understanding rather than practical application.

Prior to World War II there was virtually no United States government support of basic science. Most basic research was conducted by a small number of university scholars. Government support of science was concentrated on the applied side, as exemplified by the founding in 1807 of the Coast Survey (now Coast and Geodetic Survey) and in 1915 of the National Advisory Committee for Aeronautics (now National Aeronautics and Space Administration).

The forces of World War II provided the catalyst for ending the isolation of pure scientists from government support. University professors shifted from basic research to government supported research of the application of their science to warfare. In view of the striking results, one might think that this union of science and government was the inevitable result of scientific patriotism and government pragmatism. However, in 1940 there was neither a mechanism for massive government support of science nor a clear understanding of its potential. That the partnership was implemented was largely the result of the vision and diplomatic skills of a man who would later perform a similar role in the establishment of the NSF.

Vannevar Bush was both a scientist and an adroit administrator. As an electrical engineer he invented the differential analyzer, an early computing device. Bush was the vice president of MIT prior to accepting the presidency of the Carnegie Institution in Washington. In 1939 he was appointed chairman of the National Advisory Committee for Aeronautics. Not surprisingly, a scientist/administrator with a Washington base and the above credentials developed powerful contacts, including President Franklin Roosevelt's influential aide Harry Hopkins. Bush comprehended the potential for scientific contribution to war technology and lobbied effectively for its realization.

The result of Bush's efforts was a 1940 Roosevelt executive order creating a Bush-directed agency that would soon become known as the Office of Scientific Research and Development (OSRD). The architecture of this NSF forerunner established important precedents for scientific autonomy. "By Bush's design, and with the approval of the president, the OSRD was simply not to fill orders for the military; rather it was to be a source of weapons creativity, unencumbered by what technically untutored military men conceived to be useful and possible. For the future of American science, however, the key point was that the work of the OSRD could be contracted to university laboratories, on a flexibly drawn contractual basis, designed to assuage the scientific community's traditional fear of government interference with scientific independence."[2]

The science-government symbiosis thrived in wartime, but would it survive the war? Certainly both parties were willing. From the perspective of the government, basic science had provided the essential foundation and training for the Hiroshima application. The Cold War would provide further incentive. Meanwhile the cost of scientific research and laboratory equipment was increasing beyond the means of university support. Finally, the OSRD had provided the model for a functioning relationship.

Establishment of the NSF required a new law. Navigating the minefields of the Executive Branch, an engaged scientific community, and especially the Congress would challenge even the diplomatic skills of Vannevar Bush. Prior to the end of the war, both Bush and Senator Harley Kilgore of West Virginia had studied possible postwar solutions. Kilgore was a populist who, beginning in 1942, submitted an annual sequence of bills that were among the many casualties of NSF legislation in the forties. Bush disliked the Kilgore plans for a variety of reasons including issues of patent and political control of the Foundation. Moreover Bush had in mind an agency dedicated to the support of basic science. In 1945 Senator Warren Magnuson introduced an unsuccessful

Bush-inspired bill. The various legislative efforts evolved through compromise and a bill actually passed the Congress in 1947, only to be vetoed by President Truman. Further compromise resulted in the 1950 law.

That the NSF was specifically charged with support of basic science (to the exclusion of applied science) can be attributed to Vannevar Bush. Another contentious issue had been the organization and accountability of the leadership. The compromise solution involved a presidentially appointed National Science Board that would advise the president in the choice of a director. Alan Waterman, a physicist with administrative experience from the OSRD and the Office of Naval Research, was selected as the first director. The appointment was briefly delayed due to a specious McCarthy type allegation that Waterman's wife possessed communist affiliations.[3]

The new director proceeded to recruit staff, formulate policy, and prepare for the annual budget ritual. While the Foundation would have substantial freedom in allocating grants for scientific research, the total NSF budget was determined by Congress and the president. For the fiscal year beginning July 1, 1951 (the first year in which grants would be funded) the budget evolved chronologically as follows:[4]

Waterman request	$14,750,000
Truman request	$14,000,000
House approval	$300,000
Senate approval	$6,300,000
Conference approval	$3,500,000

A director could not help but observe the seeming crapshoot nature of the process and its susceptibility to political instabilities in the branches of government. The new agency certainly did not need enemies. The 1950 enabling law directed the Foundation to perform a variety of functions including two that could portend controversy.

Specifically, Congress expected the NSF "to develop a national policy for the promotion of basic research" and to "evaluate scientific research undertaken by federal agencies."[5] Since exercise of this authority could adversely impact the budgets and scope of other federal agencies, Waterman elected to proceed superficially. William Carey, from the Bureau of the Budget, characterized the director as having "ducked out of sight under the table."[6]

As to the Foundation's responsibilities to support basic research and science education, Waterman acted rapidly to establish policy and guidelines. Scholars, who have been on the inside or periphery of the NSF grant culture for their entire academic lives, might be surprised to learn that the funding procedures were formulated in 1951, rather than handed down by Moses. Among the issues facing the staff were whether money should go to universities or individuals, and to what extent would the funding be linked to particular tasks.

The following model for *research grants* was adopted in 1951 and has undergone only minor changes during the life of the Foundation. Individuals submit research proposals through their institution to the NSF. Scientific consultants provide evaluations for the Foundation which makes the final decision. Successful grants are awarded to the institution for use by the scientists. Budgets can include salary support, equipment, supplies, travel, and indirect costs. The latter item consists of a percentage allocated to the institution to defray overhead and administration. Originally there was a 15 percent ceiling on indirect costs.

From its $3.5 million fiscal 1952 budget, the Foundation funded 96 grants totaling just over $1 million. Only one mathematics proposal was successful and resulted in a $19,300 award to Purdue for Lamberto Cesari's project *Asymptotic Behavior and Stability Problems.* Approximately $1.5 million supported graduate and postdoctoral fellowships.[7]

In principle, the grant format provided an ideal separation of science and politics. On one end the scientist/administrator director made the case for budget increases to the Congress and president. At the other end, grant proposals were evaluated by scientists, with awards based primarily on scientific merit. Scholars were given sufficient freedom to pursue their work. For the first decade the process functioned smoothly. As basic science flourished, the NSF sustained impressive budget increases and reached $326 million for fiscal 1963. The director's case was greatly aided by the launch of the Soviet satellite Sputnik in 1957, arousing insecurities about American scientific standing.

C. Project Mohole

Part of the increases in the sixties supported an ambitious, but ill fated, project in the earth sciences. The top layer of the earth is known as the crust. Beneath the crust is the mantle. The Mohorovicic discontinuity (named for a Yugoslavian seismologist) is where the change from crust to mantle occurs. Now the depth of the discontinuity varies, with a thicker crust on land than under the ocean. Project Mohole was to dig a hole through the ocean floor and obtain sample cores of the mantle.

In contrast to other Foundation initiatives, the story of the project eventually received wide press coverage. The tone of the publicity and the story of its birth, at a 1957 NSF panel meeting, is conveyed in the book chapter *Mohole: The Anatomy of a Fiasco.*[8] The requirements of the drilling project substantially exceeded all existing technology, but the NSF's mission was to support ambitious projects in basic science. It all began with a $15,000 feasibility grant in 1958. Other grants followed. An early estimate placed a $5 million price tag on completion.[9]

A two phase plan was conceived with the first part involving a *relatively* shallow hole in shallow water. Phase one was funded for $1.25 million and accomplished in 1961. That same year, phase two was put up for bid, attracting oil companies and other heavyweights of the corporate world. Among the participants was the engineering-construction firm of Brown & Root. The mutually beneficial relationship between this company and the then vice president is documented in the Johnson biography by Robert Caro.[10] Although Brown & Root's proposal was ranked fifth and then third by successive internal NSF review committees, the Texas company had a further asset. George Brown's former college roommate was Texas Representative Albert Thomas who chaired the House Appropriations subcommittee with authority over the NSF.[11] (Zapata Off-Shore placed third then fourth in the evaluations. The president of this unsuccessful bidder was future President George Bush.) Waterman awarded the project to Brown & Root on February 28, 1962. The director insisted that his decision was independent of any political benefit that the Foundation might derive from pleasing Johnson and Thomas. One day after his decision, Waterman appeared before the Thomas subcommittee for the annual budget presentation.

Following one year of unimpressive progress by Brown & Root, the politics of the NSF reached a new audience as *Fortune* published "How NSF Got Lost in Mohole."[12] The article raised the political influence issues as well as detailing questionable management policy by the Foundation. Finally, it alleged that in its administration of Mohole, the NSF had betrayed the legacy of Vannevar Bush. The story appeared as Alan Waterman was retiring after two six year terms. Leland Haworth, a physicist and former director of Brookhaven National Laboratory, assumed the leadership of the NSF. As he inherited the Mohole mess, Haworth was immediately forced to resolve a major dispute in the phase two planning. The issue involved whether to further divide the phase into two more parts. Haworth adopted a 1-1/2 step compromise.

Haworth was stuck with a classic case of cost overrun. In fiscal 1962 Mohole became an explicit budget category in the NSF Annual Report. In the years 1962–65 the respective costs were $1.6, 3.3, 8.0, and 24.7 million. By 1966 cost estimates for completion were exceeding $100 million, and drilling for the second hole remained well into the future. Early that year the political dynamic shifted when Thomas died and was succeeded as subcommittee chair by Tennessee Congressman Joe Evins. Under the new leadership the House excluded Mohole funds in its appropriation bill. President Johnson persuaded the Senate to restore the project. In conference, the House position prevailed. On August 24, 1966, the Senate approved the conference committee report and Mohole was dead. It was a situation that called for Haworth, as director, to focus his attention on damage control and the delicate issues involved in phasing out an expensive project.

D. Round 1

On the day that Mohole died, Haworth initiated a staff inquiry into Smale's European summer activities.[13] The director was reacting to a *New York Times* article about political activity at the International Congress of Mathematicians. He was particularly concerned with the statement that Smale had "been traveling in Europe for the last few months, giving lectures, attending a mathematics meeting in Switzerland and making speeches against United States policy in Vietnam."[14] In the midst of the Mohole debacle, the NSF did not need further harmful publicity. Haworth feared this might occur if Smale's grant were associated to the *Times* comment.

Smale was in charge (principal investigator) of a two year $91,500 NSF research grant to the University of California. It was a typical mathematics grant providing two months of salary, each summer, to Smale and several other faculty members. In addition the grant supported graduate students, travel, publication costs, fringe benefits, and indirect costs. Of particular relevance were Smale's $5556 summer support for 1966 and $1000 budgeted for travel to the International Congress.

Prior to detailing the ensuing controversy, let's first set the main characters. The principals were Haworth and Smale, ironically, both born in Flint, Michigan. In supporting roles were Robert Connick, vice chancellor of the University of California at Berkeley, and Indiana Congressman Richard Roudebush.

It is interesting to observe this example of the influence of the press on government. The August 22 *Times* story also attracted the attention of the State Department that sought further details from the National Academy of Sciences and the American embassies in Moscow and London. Haworth's first questions were directed at establishing the period Steve spent in Europe and whether NSF travel or salary was involved in support. Geoffrey Keller, Division Director for Mathematics and Physical Sciences, contacted Connick who confirmed that Smale was receiving NSF funds for both salary and travel.

Quite a bit of staff activity was now engaged in the investigation, and the Moscow press conference had not yet even occurred. When Haworth learned that Smale had spent the entire summer in Europe, he felt vulnerable to congressional inquiry on two points: (1) Why was the NSF paying Smale's salary while he was in Europe? (2) Was he really doing mathematics? The director was sufficiently concerned that he elected to pursue the matter himself. On August 26 he tried to reach Connick at Berkeley, oblivious that the Smale press conference had occurred earlier that Friday. Connick was unavailable and the director spoke to Budd Cheit, executive vice chancellor. Haworth requested written responses to his two questions along with a statement of relevant university policy.[15]

The call is fascinating both in its timing and substance. With a plate full of Mohole, including a Monday meeting with contractors about the expensive phaseout, did the early Smale issues require personal investigation by the director? Perhaps, in view of Mohole, he was especially sensitive to further bad publicity. Haworth was a hands-on administrator who sensed trouble. His prescience was validated. Not

Early Chronology of Smale-NSF Dispute

August 16	Smale receives Fields Medal
August 22	*Times* article appears
August 24	Haworth initiates staff inquiries, Mohole killed
August 26 Friday	Smale's Moscow press conference, Haworth calls Berkeley
August 27	Front page *Times* article on press conference
August 29	First Haworth–Connick phone call, Meeting to dismantle Mohole
August 30	Connick suspends Smale grant
August 31	Roudebush makes first inquiry to NSF about Smale

only was the Moscow press conference story breaking, but Congressman Roudebush would soon enter the picture.

Underlying Haworth's questions were two factors that were exacerbated by his fears of congressional trouble. The director himself was offended that Smale had attacked the United States government while abroad, especially when his salary and expenses were provided by the government. He would express this sentiment in every conversation with Berkeley, often adding that it did not affect his position in the handling of the case. Moreover, Haworth was a high energy physicist and did not comprehend the culture of mathematical research. He could understand a scientist visiting a powerful particle accelerator such as Brookhaven, but why would anyone travel around Europe and do mathematics. Congress would smell a junket, and he shared that skepticism.

Mathematical research, for the most part, is extremely portable. Smale's primary tool was his brain. He needed to create and work through ideas, occasionally writing things down. Although library work and interactions with other mathematicians have their place, deep thought is the main ingredient. Where and when this might occur depends on the individual and their commitment to the work. Smale would address this issue later, but it was clearly a factor in Haworth's view of the case.

In his Friday conversation with Cheit, the director had moved to contain the issue and prepare for congressional questions. With the Saturday coverage of the Moscow press conference, Haworth's fears were realized. Once again he focused on a particular sentence of a newspaper article. The *Washington Post* stated that Smale "was vacationing with his family in Greece when the HUAC subpoena was issued."[16]

Finally reaching Connick on Monday morning in Berkeley, Haworth repeated the questions and concerns already expressed to Cheit. He specifically wanted to know if Smale had been paid during his vacation in Greece. NSF–Smale case phone records indicate Haworth–Connick communications on 11 days over a 3 week period. As the grant was actually awarded to Berkeley, the University bore responsibility for making appropriate payments.

Connick was in a difficult situation in which he could have stood behind a member of his faculty. The University, however, had developed a large degree of dependency upon the support that it received from the NSF and other government agencies. Smale's grant provided the 20 percent indirect cost share along with other benefits such as graduate student support. A chemist himself, Connick was under pressure from the director of the NSF. Of course Smale would have been able to clarify matters, but he was traveling in Europe and the mathematics department was uncertain of his itinerary.

On Tuesday, Connick learned that Smale had designated July 23 to September 23 for his two months of NSF salary. The July payment had already been made and it was then August 30. Connick took the decision to hold the August check which was to have been deposited in a local bank.

On the next day Congressman Richard Roudebush checked in with a letter to the NSF requesting information about the Smale grant. Roudebush was a conservative Republican from Indiana. He was then at the midpoint of a ten year congressional career, marked by an anticommunist vigor that would soon land him a position on the HUAC. Although the text of the letter was perfunctory, Haworth recognized trouble. He had dealt with Roudebush on a similar case, and knew his style. The congressman would issue press releases and try to influence congressional committees. Haworth pondered his response and contacted Connick.

The conversation began with Connick reading a draft of his official response to the previous calls. Haworth was worried about Congress. If NSF funds supported anti-American lectures and a vacation, there could be budget problems and even restrictive legislation. At this point Connick revealed that he had held up the August paycheck. Connick wanted to avoid any scenario involving Berkeley and NSF funding of a vacation. He was also concerned about the consequences, should the vacation information prove wrong. Recall that the evidence was a phrase in a newspaper article. Connick's contemporaneous notes of Haworth's comments include "Glad withheld check."[17] The director's secretary was transcribing the conversation and the transcript does not contain any reaction by Haworth.[18] While it seems unlikely that the director did not express a position, it is clear that he did not oppose it.

Haworth talked through the problem of dealing with Roudebush. He decided on a prompt response detailing most of the facts, but omitting the check hold. The director was correct in the belief that this would "buy some time," but how much time did he need? Further resolution of the issues depended on obtaining information from Smale. After Moscow, Steve had joined his family in Athens on August 27. From Greece, the Smales traveled to Istanbul, via automobile and ferry. They then traveled through Europe prior to boarding the SS France in Le Havre on September 9. The ship reached New York on September

14, and then the family proceeded to Princeton where Steve would be in residence at the Institute for Advanced Study for the semester. Connick was neither able to reach him nor obtain a firm date of arrival.

During the first week of September the NSF staff was busy investigating a grant that Smale had received from the National Academy of Sciences–National Research Council (NAS-NRC). The NAS-NRC, using funds from the NSF and industry, had awarded $400 grants to individuals for travel and support to the Moscow ICM. As a plenary speaker, Steve was presumably among the strongest applicants. He accepted the grant, and subsequently an additional $1000 from his NSF grant. Was this improper double dipping? The staff sought to identify any mathematician who had accepted ICM travel support from two sources.

September 12 was a congressional nightmare for Haworth. Roudebush issued a press release, Representative William Bray (Indiana) entered remarks in the *Congressional Record*, and Representative Durward Hall (Missouri) sent a letter to the director. Each called for the cancellation of the remainder of Smale's grant.

A troubled Haworth called Connick the next day. The director described the congressional activities and the new information about the NAS-NRC grant. He feared that "the lid will really blow" when the congressmen learn of the "financial irregularities" that were under investigation.[19] At that point there was no official record that the grant payments had been suspended. Presumably they had originally hoped to settle the matter in a few days without anyone noticing a reinstatement, if it occurred. It now was two weeks and there was serious trouble on the horizon. Haworth decided it would be best to put it on the record. Arrangements were made for a sequence of messages. An NSF telegram to Connick stated:[20]

> *This will confirm that costs incurred on behalf of Professor Smale to be charged against Grant GP-5798 will be subject*

to question pending clarification of following points: First, Professor Smale accepted funds from both the National Academy of Sciences and Grant GP-5798 for travel expenses to attend the International Congress of Mathematicians in Moscow in August 1966. Second, press and other reports indicate Professor Smale not performing research during full period specified in your letter of September 1, 1966.

Smale arrived in Princeton as the telegram reached Berkeley. Following phone conversations with Connick and Berkeley mathematics department chair Leon Henkin, Smale composed his now famous response.[21]

Dear Vice Chancellor Connick:

With respect to our telephone conversation, here is a short account of my mathematical activity this summer.

Last week of May, Paris, Institut des Hautes Etudes Scientifiques, $110 honorarium.

June and July, University of Geneva, Paid expenses of about $1,200.

During June I traveled for six days to a conference in Bonn, where I received less than $50 for hotel.

During the first two weeks of August I drove to Athens to leave my family while I spent August 16–27 in Moscow for a meeting. 100 rubles honorarium for talk; $1500 prize money.

After returning from Moscow to Athens, I drove to Le Havre via Istanbul and Paris to meet the S. S. FRANCE, crossing the Atlantic, September 9–14.

I consider that I spent full-time this summer on mathematical activity and research, whether measured from time spent on mathematics or on results produced.

For bookkeeping purposes for the two-month summer National Science Foundation support, one can select a two month subset when I was attending a conference and/or had an office, e.g., June 13–July 30, August 16–27, and September 14–23. However, during the remainder of this time I was also doing mathematics, e.g., in campgrounds, hotel rooms, or on a steamship. On the S. S. FRANCE, for example, I discussed problems with top mathematicians and worked on mathematics in the lounge of the boat. (My best-known work was done on the beaches of Rio de Janeiro, 1960!)

I would like to repeat that I resent your stopping of my NSF support money for superficial technicalities. The reason goes back to my being issued a subpoena by the House Un-American Activities Committee and the subsequent congressional and newspaper attacks on me.

The beaches of Rio de Janeiro phrase would eventually resonate through the mathematics community. At the time Smale was referring to the Higher Dimensional Poincaré Conjecture, as his best known work. Today (in 2000) he suggests that his other Rio result, the horseshoe, is better known.

To appreciate Steve's contention that he worked the entire summer requires an understanding of the *doing* of mathematics. Smale's work relied on the concentrated use of a powerful mind rather than a powerful particle accelerator. It is possible for a mathematician to work on a beach and it is also possible to *not* work in an office. Finally one can work extremely hard, in any environment, and have nothing to show for their efforts. Certification of mathematical work over a short period of time is a difficult business. Connick read the beaches of Rio letter to Haworth over the phone. Given the director's skepticism that Smale had worked in Greece, one can only wonder about his reaction to the Brazil line.

Since NSF allegations about Steve's grant administration play a central role in both this story and its sequel, it is appropriate to examine them in some detail. We separate the issues into the categories of summer salary and travel expenses. The beaches of Rio letter confines itself to the former which we address first. Consider the three questions raised by Haworth.

Haworth summer salary questions:

1. Was it appropriate for Smale to receive NSF salary while in Europe rather than Berkeley?
2. Was he really working?
3. Was it appropriate to receive NSF salary while engaged in anti-American activities abroad?

Discussion:

1. In May Smale had notified the NSF that he would be in Europe for the summer and designated a colleague to handle administration of his grant. The letter was received and acknowledged by the program director with authority over the grant.[22] No objections were raised at the time. It is difficult to see any merit in this allegation which originated in reaction to the August 22 *New York Times* article. Even after Haworth was informed of the notification, he continued to press the issue.
2. Steve was budgeted to receive two months of summer salary. According to a subsequent internal report by Connick, Smale "specified the pay period as the last two months of the summer on the advice of the departmental secretary handling grants, who told him that as long as he worked at least two months the actual period specified did not matter."[23] The beaches of Rio

letter was Steve's response to the demand to certify his work. Connick, with the approval of Haworth, revised the dates of the pay period so as to cover specified days in Geneva, Moscow, and Princeton. Eventually payment was made and everyone seemed to agree that he worked at least two months. The summer of 1965 was another matter, but Haworth's inquiry was confined to 1966.

3. This is essentially a political matter and it was the motivation behind the heavy congressional pressure to terminate the grant. While Haworth felt that Smale's actions were inappropriate in this regard, NSF policy specified the nonscientific criteria under which a grant could be revoked (e.g., being an avowed communist). An FBI check provided several past reports, but the National Science Board found no basis for termination.[24] Haworth subsequently found himself defending the decision to unhappy congressmen.

Smale resolved the questions about his two month $5556 salary, but what mathematics did he do in the summer of 1966? I asked Steve that question in 1993. Despite the 27 year gap, he was able to recall one specific result and the general thrust of his work during the period.

René Thom had produced a map of the torus (doughnut) that, like the horseshoe, was structurally stable and had an infinite number of periodic points. The map had an additional property known as hyperbolicity. The Russian mathematician Anosov proved that assuming hyperbolicity alone, rather than working with the specific map, was sufficient to ensure structural stability. This meant that if one could establish the hyperbolicity of a map, then structural stability was automatic by Anosov's theorem. A difficult question was to determine whether any *Anosov maps* actually existed on spaces other than tori. In Geneva Smale showed that the answer was yes and designed techniques for their construction.

Chronology

October 25, 1965	NAS-NRC application made
November 22, 1965	NAS-NRC award announcement
February 28, 1966	NSF grant approval

Beyond this specific result, during the summer of 1966 Steve was working on the mathematics that would go into his seminal article *Differentiable Dynamical Systems.*[25] Essentially a blueprint for a new field of mathematics, the paper was undoubtedly the most read and cited of Smale's career. Included in the article were deep theorems and profound questions leading to an extraordinary number of papers and doctoral theses. The summer of 1966 was a crucial period in the development of these ideas. Given the NSF's mandate for basic research and science education, history has shown that this paper provided a terrific return on investment.

Once the beaches of Rio letter resolved the summer salary issues, it was time to deal with the travel expense allegations. Here Haworth found two specific improprieties.

A. Acceptance of travel expenses from both NAS-NRC and NSF.

B. Travel on a foreign carrier (SS France).

While both statements are certainly true, the conclusion of impropriety rests on a combination of controversial interpretations. In order to understand the arguments involved, we will quote from the relevant documents.

Smale applied for the NAS-NRC grant on October 25, 1965. The instructions stated "The total grant will not exceed the first class air fare from the applicant's residence to Moscow and return, but it is hoped that it will at least provide air fare plus an allowance for expenses at the meeting." Two questions on the application were pertinent. When asked whether he had "requested or been granted funds which might be used for travel to the 1966 Congress," Steve responded "On NSF contract application."[26] He then agreed to notify them if there were a change in the answer. At that time his NSF grant was still pending.

On November 22, 1965, notification was sent of a $400 grant. The award letter concluded with the following ambiguous sentence: "It is hoped that you will promptly notify us if you will not use this award either because you have other sources of travel funds or because you find it impossible to attend the Congress." While the $400 would not accomplish the original intent, there were no guidelines stated as to its use.

When Smale received notification of the NSF grant in March, what were his obligations to the NAS? Haworth contended that he was bound to inform the NAS. Another interpretation is that once the NAS-NRC decision was made on November 22, the operative policy became the last sentence of the award letter. Since the sum of the two grants did not quite cover his travel expenses, both would be applied and there was no obligation for notification unless his resources exceeded expenses.

NSF regulations prohibited travel on a foreign carrier. Since the $400 NAS-NRC award had no such stipulation, Smale applied it to his $300 tourist fare on the SS France. Haworth's position was that NSF rules were applicable to the NAS-NRC grant because its funds were provided by the Foundation (actually it was also supported by industrial sources).

Altogether there were five allegations of impropriety (1, 2, 3, A, and B). Number 2 was dismissed with the information provided by Smale upon his return, and the National Science Board rejected 3. This author sees no basis for 1 and B. The validity of A rests on the interpretation of ambiguous statements. Despite the extraordinary scrutiny of Steve's summer activities, Haworth's charges of administrative malfeasance could not be established.

Haworth estimated that he devoted 1/3 of his time in September to the Smale controversy. Others observed that the dispute extracted a great toll on the director.[27] Haworth's direct handling of the Smale case, from the original *New York Times* article through his conversations with Connick, was an odd division of labor for the director of a federal agency with a large staff. This overreaction was prompted by a fear of Congress rather than direct pressure. We suggest that the juxtaposition of Mohole exacerbated his concerns and made him feel the necessity to get out in front on this one. As it turned out, he was so far in front that he had little perspective.

It is especially unfortunate that the Foundation failed to celebrate the Fields award to Smale. An American had just received recognition for reaching the highest level of basic science. Not only was he a current principal investigator, but his prize winning work was originated while on an NSF postdoctoral fellowship. Smale's success should have been promoted by the Foundation as validation of its role in basic science.

The actual reimbursement of Smale's travel was complicated. Steve submitted expenses for $1599 against his $1400 advance. At this point Berkeley mathematics department chair Leon Henkin entered the loop between Smale and Connick. Henkin had also received ICM support from both the NSF and NAS-NRC. Steve, now in the background, finally had an ardent advocate.

Henkin was personally offended by Haworth's allegations of impropriety in the two grants. When the NSF ruled that the total allowable expenses could not exceed the $1018.95 cost of round trip tourist air fare to Moscow, Henkin wrote a sequence of letters detailing the fallacies in their reasoning. He even succeeded in enlisting Connick who noted that the NSF interpretations were contrary to Berkeley administrative policy.[28] Despite eight months of letter writing, the Foundation remained adamant and Haworth eventually wrote directly to Henkin closing the episode.

During the fall of 1966 there were a variety of consequences from the dispute. On September 26 the NSF announced a policy change to preclude the funding of international travel through ordinary research grants. The Institute for Advanced Study supported its visiting members through NSF funds and its own endowment. Haworth was sufficiently paranoid about possible congressional attacks that he prompted the Institute to fund Smale's stipend out of the endowment.

The NSF received numerous inquiries from the White House and Congress including 26 senators. For the most part, the officials forwarded letters from constituents, outraged that the government continued to support Smale. Many were reacting, either directly or indirectly, to Roudebush's press campaign. On October 10 he issued a press release that blasted the NSF and divulged information from the confidential HUAC file on Smale. Some of the material was published in a Catholic periodical which erroneously reported that Steve received $13,000 for his trip to Moscow. By the end of the year the mail abated and it appeared that the controversy was over. However, it was fated to arise again the following year.

E. Round 2

On May 29, 1967, the NSF received a $247,886 Berkeley proposal designating Smale as principal investigator. The budget was similar in nature to the previous grant that had been submitted at $152,758 and funded for $91,500. Both were for a two year duration, but the new proposal provided support for a total of eight faculty members, doubling the previous number. Signing off on the submission, as institutional representative for Berkeley, was Sanford Elberg, dean of the graduate school.

Normally a proposal was first sent to external scholars for *peer review*, prior to internal evaluation. The new NSF Director of the Division of Mathematical and Physical Sciences, William Wright, recognized that special handling was required in this case. Instead of soliciting outside reviews, Wright sent a memo to Haworth, routed through the associate and deputy directors.

Round 2 of Smale versus NSF was underway. Once again let's set the main characters. In the forefront of Foundation-University negotiations were their respective representatives, Wright and Elberg. Returning from the previous dispute were Haworth, Roudebush, and Smale. Finally Daniel Greenberg's aggressive reporting for *Science*, engaged the scientific community and made him a high profile player.

To understand how both the NSF and University found themselves in uncomfortable positions, consider the flow of money:

Congress → NSF → University → Scientists

By Vannevar Bush's design, the NSF was meant to insulate scientists from government interference. Since the Foundation's scientific constituency tended to be as ignorant of the Congress-NSF dealings as Congress was of science, the three parties were rarely involved in an issue at the same time. This was the case in 1966, when matters were resolved while the scientific community was largely unaware of the controversy. With Connick's willing cooperation on one side, the pressure on the NSF was mainly from a segment of Congress.

Similarly the University was positioned between its faculty constituency and its funding source. However, when Wright outlined the problem in his memo on May 31, neither Congress nor the mathematical world were yet involved. He was certain that the proposal would receive "outstanding substantive reviews," but he felt that Smale's past behavior made him an unsuitable principal investigator for a quarter of a million dollar grant. The memo outlined two "realistic alternatives," either of which could be invoked with the consent of the Berkeley administration:

a. The University designate a different principal investigator from the group of eight faculty participants.

b. The University separate Smale as a "principal investigator for a much smaller grant confined solely to support of his own research activities."[29]

With the endorsement of Associate Director for Research Randal Robertson, Wright solicited the approval of Elberg. Elberg, appreciating the delicacy of his own position and that of the NSF, requested time to consider his decision. After some reflection, Elberg's response was a polite, but unequivocal, rejection. He then pointed out that Haworth had successfully withstood the political pressure to terminate Smale's grant the previous year. "The Foundation won our case for us last year and did it for Science. I hope you will not renege on this issue today." In a follow up discussion, Elberg argued that Smale deserved to be principal investigator "unless challenged on mathematical grounds or directly charged in writing with malfeasance during the course of administering the grant."[30]

In his original memo to Haworth, Wright had observed that there was plenty of time for staff discussion and deliberation. Surely Roudebush was unaware that a proposal was under consideration. Two weeks after Haworth received the memo, an anonymous NSF staff member tipped off Roudebush. The congressman expressed his outrage in a letter to the director. A few days later, a Roudebush press release threatened trouble in Congress.

As Roudebush increased the pressure in late June, the NSF was at an impasse. Both Wright and Robertson viewed the principal investigator arrangements as unacceptable. No alteration was possible without the consent of Elberg or Smale. Elberg had refused and Haworth ruled out direct negotiation with Smale.

Wright and Robertson recommended a standard rejection letter.[31] Haworth brought the issue before the Executive Committee of the National Science Board on August 12. According to the minutes, "The director said he could not, in good conscience, approve this project covering the work of so many persons only loosely related with Smale as principal investigator, even if found meritorious by a scientific panel, because of the irresponsible behavior exhibited by Smale in the administration of the current grant." He then informed the committee of the failed Berkeley negotiations and the Roudebush activity. The Executive Committee concluded "that the proposed administrative arrangements are unacceptable but that the Foundation will be willing to consider on their own merits two or more proposals covering support for advanced mathematics, one of which might provide some support for Smale and those closely associated with his own research." After an FBI check provided no new information, Wright sent a letter to Elberg on August 31 with the following text:[32]

> *During the interval since our last discussion of the proposal for the renewal and expansion of NSF Grant GP-5798, the Foundation has carefully reassessed the whole matter. We have come to the conclusion that, in the light of Professor Smale's performance in the administration of the present grant, we cannot tender a new grant to the University based on the proposal in its present form.*
>
> *This does not reflect any adverse decision on the part of the Foundation concerning the intrinsic merit of the research proposed. Rather it reflects a decision by the Foundation that the proposed administrative arrangements are unacceptable. It is my suggestion that the University submit for our consideration two or more new proposals which cover approximately the same span of substance. One of the new proposals should confine itself strictly to the needs of Professor Smale in the pursuit of his own research interests without involving NSF support of other faculty members. The other proposal or proposals could then cover the remainder of the substance of the earlier proposal but with different principal investigators.*

Elberg forwarded the letter to Smale who made it available to others. Since the letter provoked a great deal of controversy as it circulated through the mathematics world, it is appropriate to itemize its key points.

1. The NSF declined to fund the proposal and suggested the submission of new proposals in a manner that the University had already refused. Smale and other mathematicians interpreted this as a rejection letter, while Haworth insisted it was not.*
2. The basis was Smale's unsatisfactory administration of his then current grant.
3. The Foundation would consider new proposals, but there was no assurance that they would be funded.
4. Smale must be separated from each of the other faculty members.

Smale replied directly to Wright. He stated that the NSF's action was politically motivated, and that if there were administrative allegations then it was incumbent on the NSF to document them. After concluding that the "NSF has dishonored itself," he declined to submit a new proposal since such a capitulation would make it more difficult for other scientists "to disassociate themselves from Johnson's brutal Vietnam policy."[33]

Daniel Greenberg had copies of both the Wright and Smale letters when he phoned Haworth at his home on Sunday morning, September 10. Greenberg was news editor of *Science*, the weekly magazine of the American Association for the Advancement of Science. He joined the staff in 1961, specializing in the politics of science.

* The author has received both acceptance and rejection letters on his own NSF proposals. He interprets this as a rejection.

Early Chronology of 1967 NSF-Smale Dispute

May 29	NSF receives proposal
May 31	Wright memo informing Haworth of proposal
June 2	Wright phones Elberg who refuses to alter proposal
June 16	Anonymous NSF staff member informs Roudebush
August 12	Executive Committee of NSB considers proposal
August 31	First Wright letter to Elberg
September 8	Smale reply to Wright letter that "NSF has dishonored itself"
September 10	Greenberg interview with Haworth
September 15	Greenberg article appears in *Science*

When Greenberg pressed for the details of NSF dissatisfaction with Smale's administrative performance, the director agreed to speak off the record. Several years later Haworth recalled, "I was pretty annoyed when in the next issue of *Science* Greenberg repeated everything I told him."[34] What Greenberg did report was the previously discussed allegations labeled 1, 2, A, and B. No source was named in the article. Although the NSF would never inform Smale of these allegations, the director divulged them to a reporter. Whether Haworth intended his remarks to appear anonymously in print or simply to influence the article's spin, raises substantial questions about his own notion of propriety.

Haworth was not the only unnamed source for the Greenberg article. Philip Handler, chairman of the National Science Board, also spoke off the record. Following the interview he drafted a statement for publication. Among the seven included points was the Board's finding "that management of this grant has been relatively loose and has not conformed to appropriate standards." Handler closed by "deplor(ing) the actions of those who have sought to conduct in the public press negotiations between the Foundation and the University concerning a purely administrative matter." After the statement appeared in *Science*, Smale wrote to Handler. He requested a "bill of particulars" for the administrative objections, but Handler never replied. Of course the NSB had received its information filtered through Haworth.

Now the NSF was beginning to feel the pressure from its scientific constituency. Serge Lang, a Smale friend and colleague at Columbia, was widely circulating the Wright letter. He was also preparing a long letter of his own for publication in the *Notices of the American Mathematical Society.* Two other distinguished Columbia mathematicians, Lipman Bers and Hyman Bass, proposed that the dispute be submitted to a panel of eminent mathematicians. At the University of Pennsylvania 35 faculty members signed a petition renouncing personal use of NSF funds, unless there was a satisfactory explanation for the grant rejection.

The Greenberg article brought the issue to the attention of a broader audience. A number of factors contributed to the widespread support of Smale from the previously docile scientific community. He was a Fields Medalist, the most prestigious credential in mathematics. Mathematical papers tend to be inaccessibly vertical in nature, relying on a sequence of other papers and results. Most mathematicians are familiar with an extremely narrow portion of the literature. Deep results are only read and fully appreciated by a small number of people. To achieve widespread acclaim among mathematicians is rare, but a Fields Medal does suffice.

There was no doubt that Smale was a great mathematician with controversial politics. It was 1967 and opinions on the Vietnam War were shifting, especially on university campuses. Finally the administrative charges had a bogus sound. With cynicism increasing as to the government's conduct of the War, other scientists quickly embraced Smale's cause.

The August 31 Wright letter had been a disaster. Rather than prompting negotiation, the letter was provoking a negative response that had the potential to escalate out of control. The Foundation needed to make another move. Clearly they would not accept the Bass–Bers proposal which weakened their authority and provided a forum for the administrative issue. Of immediate concern was a Freedom of Information Act request filed by Dan Greenberg. The administrative allegations were under investigation by *Science.* On September 23, the day following Greenberg's formal request, Wright sent a second letter to Elberg:[35]

> *In view of the numerous and widespread misinterpretations which have been placed on my letter to you of August 31 and in view of your recent absence from the country, I would like to reaffirm the position of the Foundation.*
>
> *The Foundation remains convinced that timely negotiations can result in a grant to the University of California with Professor Smale as principal investigator, which would support his research needs and those of his immediate collaborators in a manner completely consistent with our ability to sustain mathematical research generally.*
>
> *I believe our understanding of the suggestions in my letter of August 31 as explored in our telephone conversation early this month is consistent with earlier suggestions made by the Foundation in telephone conversations with university officials in June and July. Now, as then, the Foundation is prepared to make such a grant to the University of California, Berkeley, subject to satisfactory scientific and budgetary review.*

Although the second Wright letter was couched in the format of a clarification, it actually represented a substantial change from the first. There was no longer any suggestion of administrative malfeasance, and the rejection tone was absent. Whereas the previous letter expressly stipulated that Smale must be isolated from other faculty, the new position ambiguously permitted "immediate collaborators." Steve reacted favorably, noting especially that the prospects for future funding had improved from neutral to positive. However, before there could be a resolution, "I would like to have the reported NSF allegations about my administrative performance cleared up once and for all."[36] He suggested implementation of the Bass–Bers proposal.

The new letter provided a fortuitous opening at a time when both Berkeley and the NSF were desperate for a solution. At the University, there was dissension in the mathematics department. Henry Helson, the department chairman, was critical of Smale's position. Some of Steve's colleagues on the grant considered submitting their own proposal. After the first Wright letter, Elberg and Smale had exchanged a sequence of angry memos, concerning the propriety of Steve's release of the letter.

The dean was now concerned about the ramifications of the conflict on the mathematics department. With the second letter Elberg saw an opportunity for mediation. He reconciled with Smale and proceeded to negotiate with the Foundation. With Greenberg's Freedom of Information request pending and as articles about the case appeared in *The New Republic, The Nation*, and *Notices of the American Mathematical Society*, the NSF was anxious to resolve the problem.

The Foundation was now publicly entrenched in the worst possible position, between its scientific constituency and a vocal segment of Congress. Fortunately for the NSF, not all of their advice was polarized. R. H. Bing, an eminent topologist, was vice chairman of the NSF Advisory Committee for Mathematical and Physical Sciences. At the time of the dispute he was in residence at the Institute for Advanced Study. More than likely Smale was the major topic of discussion at the afternoon teas. Bing observed that the administrative charges did not impress the mathematical community. He viewed the dispute as harmful, and sought a solution that would fund Smale within the parameters of the Wright letters.

Among the eight faculty members on the grant proposal, only three worked in the same mathematical area as Smale. Funding a grant to Smale, Kupka, Pugh, and Palis would be consistent with the "immediate collaborators" stipulation. The other mathematicians, Kobayashi, Wolf, Schmid, and Griffiths, could be taken care of in one or two other grants. Bing floated the idea before a few selected members of the mathematical hierarchy. Satisfied with the reaction, he transmitted the suggestion to the NSF.

Meanwhile Smale had requested that Elberg obtain clarification of "immediate collaborators." Wright responded that the NSF would accept a proposal consisting of Smale and three faculty colleagues with Steve as principal investigator (i.e., the Bing plan). All he required from Elberg was a request to split the proposal and specification of the faculty and budgets that would be associated with each grant. After some consultation with Berkeley faculty,* Elberg supplied the names (previously suggested by Bing) for two separate grants of four members each, with Smale and Kobayashi as the principal investigators.

On October 12 Wright informed Smale and Elberg that the two grants would be funded, but that it remained for the staff to finalize the budgets. On the same day Haworth wrote to Elberg reviewing the history of the case. The Haworth letter is notable for an especially creative new spin that the director employed to the administrative issue. "We concluded that in view of the excessive amount of time required on the part of both the University and ourselves in the administration

* Smale's response to Elberg was that details of a split remained premature until the administrative allegations were resolved. In 1993 I interviewed Elberg and asked about the input that he received. He remembered a crucial meeting with Wolf and Kobayashi, but could not recall the details and suggested that I speak to Wolf. Wolf declined to discuss the matter.

Chronology of 1967 *Science* Articles on Smale Case

September 15	Smale and NSF: A New Dispute Erupts
September 22	Handler Statement on Smale Case
September 29	Tracing the Path that Led to NSF's Decision
October 6	Smale: NSF Shifts Position
November 3	Smale: NSF's Records Do Not Support the Charges

of the current grant, it would not be desirable to adopt an arrangement in which Professor Smale would be responsible for the administrative direction of a spectrum of activities, some of which are quite independent of his own work, but that some alternative arrangement should be made. It was, of course, for this reason that Dr. Wright called you early in June."[37]

While the official award of two $87,500 grants occurred on November 17 immediately following approval at a meeting of the National Science Board (NSB), the die had been cast with the letters five weeks earlier. During the intervening month, threads continued to develop in *Science*, the mathematical community, and with Congressman Roudebush. Many scientists received their information from Dan Greenberg's reporting in *Science*. His development of the story is conveyed by the titles.

Among the most avid readers of the *Science* coverage were Haworth and Smale. The director was an unhappy subscriber. When the second Wright letter was reported in the article entitled "Smale: NSF Shifts Position," Haworth went ballistic. He immediately sent pre-emptive explanations to Roudebush and the NSB, categorically denying any change in position.

In the third article, Greenberg provided a detailed history and analysis of the case. Relying on his prior background interview with the director, Greenberg concluded "Privately, NSF explains that Smale is a fine topologist but a bad housekeeper—which is probably a fairly accurate assessment of the realities of the situation." Smale was especially dismayed by this sentence and wrote a letter to *Science* on October 15. Frustrated that neither Haworth nor Handler would provide a "bill of particulars," Steve wrote: "I have been accused in such a way that I can answer only a journalistic account of the accusations themselves."[38] He then responded to each of the published charges. While Greenberg had sanctioned the validity of the allegations in his earlier articles, the reporter had been sufficiently skeptical to file a Freedom of Information inquiry. The Smale letter arrived as the files were made available.

After examining the materials, Greenberg concluded that there was no basis for the NSF charges. Rather than publish Smale's letter, Greenberg presented his own personal detailed analysis in the devastating article "Smale: NSF's Records Do Not Support the Charges." His revised assessment was "(i) NSF is unable, or at least unwilling, to provide documentary evidence to support its allegations of impropriety or substandard performance on Smale's part in the administration of his government grant; but even more important, (ii) at the time NSF made these allegations, it was in possession of documentary evidence which either clearly contradicted the allegations, or showed them to be based on trivial and technical departures from ambiguous regulations." The article concluded with the statement, "At this point, it must be said that there is something putrid about the whole business, and the aroma seems to come out of NSF headquarters."

On the congressional front, Haworth had other problems. Following the Wright letter to Elberg of October 12, Roudebush received reports that the grant had been approved. Twice, he wrote to the director requesting an explanation. When Haworth received the second letter, the budget had already been determined and the decision had been approved by the Executive Committee of the NSB. It only remained for the full NSB to endorse the action at its meeting a few weeks away. The director had a dilemma. His past responses to Roudebush had been prompt. Although he could not delay his reply until after the NSB meeting, a forthright explanation would invite congressional intervention between the director and the NSB.

Haworth's solution was to draft a remarkably disingenuous letter to Roudebush. Negotiations with the University continued but were incomplete, and "no grant has yet been made nor had the Foundation made any announcement regarding the matter."[39] Two and one half weeks later, the congressman was informed that the negotiations had been completed and the grants were approved. Roudebush vented his rage on the House floor. In the Senate similar views were expressed by Karl Mundt.[40]

The bottom line was that the NSF did fund the meritorious research of all eight Berkeley mathematicians. The actual split of grants between global analysis with Smale and differential geometry with Kobayashi was, in itself, a perfectly reasonable scientific division. So why was everyone unhappy?

Roudebush believed that Smale's "disloyalty" made him unworthy of any government support. Whether Steve was a principal investigator of a large grant or received money on a smaller grant, merely affected the text of his press releases. There was no solution that would fund Smale and avoid a Roudebush confrontation.

The skeptical scientific community believed that the NSF had succumbed to political pressure. Haworth pointed out that the Foundation's actions of 1966 and 1967 were initiated prior to Roudebush's intervention. The record clearly substantiates the director's claim in this regard, and it is also the case that both final decisions were contrary to the pressure from the congressman. However, both actions commenced as pre-emptive measures to defend against Roudebush-type attacks, and the actual process was driven by political considerations.

The NSF was unquestionably the biggest loser. In a period of flattening budgets, the Foundation was ridiculed in both houses of

Congress, petitioned in protest by its scientific constituency, and exposed for deceit in a prominent scientific publication. The Haworth administration had walked into an untenable Mohole situation. Ironically Haworth gave birth to the second major NSF embarrassment just as Congress administered last rites to the first.

Smale viewed the results as mixed. He had persevered and received the grant, but there was some ambiguity in the solution and the NSF never retracted the administrative allegations. Steve felt that it was important to withstand what he construed as political intimidation. Smale was fully aware of the power conferred on him by the Fields Medal. If he had yielded to the early NSF pressure, then less powerful mathematicians might be reluctant to oppose the war. Not only do NSF grants provide funding, but their prestige impacts careers as a significant credential in tenure and promotion decisions at many universities.

As to the administrative allegations, Haworth's official history of the case contained the spin of his October 12 letter to Elberg.[41] Rather than a judgement on Smale's management, the problem was the "excessive amount of time by both University and Foundation officials" required in 1966. In an interview many years later, Haworth returned to his original story. "We wanted to limit his supervisory duties as much as possible for the reason that his conduct had convinced us that when it came to running things he was inept." The former director then repeated several of the previously discussed allegations and concluded: "Obviously he was a man who didn't care to play by the rules. In our view it added up to a lousy administrator."[42]

There certainly was a great deal of ineptitude in the two NSF-Smale episodes, but it was not on Smale's side. From Haworth's overreaction in 1966 to the specious administrative charges that only appeared when no informed defense was available, the director's actions were motivated by a fear of Congress. As Elberg reflected, they are "hypersensitive to political currents. They forget their strength."[43]

The Aesthetic Side: Minerals and Photography

09

A. Collecting

As Smale approached midlife he had already exerted his influence on the seemingly disparate worlds of mathematics and politics. Examination of these ventures reveals several common threads, beginning with his capacity for complete immersion in a problem. Whether it was the Poincaré Conjecture or the Vietnam War, Smale focused his considerable energy and intellect on a single endeavor. Among his most vital assets were the self-confidence and audacity to undertake risks in pursuing his objective. The same qualities proved invaluable to Smale when, in the late sixties, he turned his attention to mineral collecting.

The desire to own precious metals and rare gemstones, such as gold and emeralds, dates back to antiquity. For centuries beautiful minerals have exerted a role in shaping aesthetics, culture, and economics. Crown jewels are the trappings of royalty, while diamond and gold rings symbolize marital commitments. Crafted jewelry is the most common expression of society's appreciation for minerals, but there is another manifestation of the attraction. Thousands of hobbyists are allured by the rare acts of nature that yield beautiful crystals. They collect natural mineral crystals in their unaltered state. For some people the hobby evolves into both an obsession and a business.

Two such men are Joel Bartsch and Wayne Thompson who grew up in Texas and Arizona, respectively. Both encountered minerals in their childhood and were overcome by a magical force. Bartsch and

Thompson used the income from paper routes to support their rock collecting habits. As they bought, sold, and traded, increasingly higher quality material was required to satisfy their desires. Thompson recalls an early experience that shaped his philosophy. He visited a local dealer, bringing with him 25 pieces with a total value of about $135. Another collector appeared with a single specimen, considerably more beautiful than any of Thompson's. When Thompson realized that the superior piece was comparable in value to that of his entire basket, he decided "I would never again buy junk. In other words, buy one, whether it's $135 or $10,000 or $50,000, always buy one. Buy the quality, instead of the quantity."[1] Out of their early passion for minerals, Bartsch and Thompson both fashioned careers that would make them forces in the mineral world, and eventually bring them into contact with Smale. Bartsch became a museum curator and Thompson a dealer.

To succeed in the mineral game, as in other aesthetic enterprises, requires *connoisseurship*, the ability to evaluate a specimen. A number of related factors contribute to determine the standing of a mineral specimen. Among the most important are perfection and beauty. Assessing quality, rarity, aesthetic composition, and other features is a complex undertaking. Some people seem to have an instant sense of a piece's quality and economic worth, but most of us are unable to distinguish the ordinary from the extraordinary specimen. One topic of debate among collectors is the extent to which connoisseurship is a congenital or cultivated talent.

Smale's adolescence offered no evidence that he possessed a mineral gene, nor a mathematics one. An intellectually inquisitive young man, he demonstrated some precocity in chess and chemistry. There was little of note on the aesthetic side. Steve enjoyed listening to music; however, his foray at piano lessons was short lived. Unlike other successful mathematicians and mineral collectors, Smale's early years provided no clue as to any talent or inclination in those directions.

Clara was the first to develop an interest in minerals. While in New York she took the children, without Steve, to view the collection at the American Museum of Natural History. On a trip to Washington state in the mid-sixties, Clara and Steve visited a rock shop, purchasing four decorator type minerals at a total cost of about $50. It was a pleasant experience, but the hook was not set in Steve until 1968, when he received a copper specimen as a gift from his father. Shortly thereafter, Steve and Clara explored their local mineral shops, buying a couple of pieces. Different people are moved to varying degrees by music, literature, painting, and other art forms. For Steve, minerals registered a far deeper feeling than his previous aesthetic experiences. His characteristic reaction was to immerse himself in the new interest, learning as much and as quickly as possible.

For any mineral collector, knowledge and appreciation may be gained by reading the literature, talking to other enthusiasts, and viewing specimens as well as buying and selling. However, buying poses considerable risk, particularly to the novice. As the mineral world unfolds to the new collector, there is a compelling urge to gain possession of some of this beauty. Most specimens are available for a price, but appraisal requires expertise. Wayne Thompson advises serious new collectors to abstain from buying for the first year, while they educate themselves and learn their tastes.[2] Steve followed a different approach, seeing buying as an essential adjunct to education.[3]

> *You have to buy first of all. You can't tell without buying. If money means something to you, you learn a lot faster, by mistakes. I bought from the beginning. You learn fast when it is a question of losing or gaining a lot of money.*

Soon Steve was driving by himself to Los Angeles, seeking out California shops listed in a mineral magazine. Without Clara's moderating influence on the trip, he moved up to the $100 per rock range, beginning the exponential price trajectory of his purchases. As with mathematics and politics, Smale was not lacking for aspiration. Why couldn't he build a great private mineral collection, comparable to that of museums? The combination of beauty and competition was a driving force. Steve wanted to own the best and most beautiful crystals, and from early on, had the audacity to believe it was possible.

In the long history of mineral collecting, there have been surges and lulls of activity. Smale's entrance in the late sixties happened to coincide with several events that would boost the hobby to an unprecedented level of popularity. The actions of a few people were especially instrumental in this development. Most notable was Dave Wilber's decision to resign his position as a systems analyst with Prudential and devote himself full time to the mineral business. Wilber had collected since childhood. While working for Prudential, he invested his income in the buying and selling of crystals. Wilber, recognizing that the world's best minerals were undervalued, set out to obtain specimens of the highest quality. Bill Larson was another connoisseur with a lifelong passion for minerals. After studying geological engineering at the Colorado School of Mines, Larson and a partner began a mining company in 1968. That same year Smithsonian curator Paul Desautels published his informative coffee table book, *The Mineral Kingdom*.[4] Desautels' book promoted the aesthetic aspects of the hobby, and made it more accessible to the public.

With these developments collectors began to shift their emphasis from quantity to quality. Smale was a step behind Wilber and others, but he was moving quickly to expand his connoisseurship. At the same time he continued his mathematical development of dynamical systems. There was a cadre of Berkeley graduate students engaged in thesis work under Smale's direction. The mathematical results of the "Smale School" were attracting international attention. When Steve received invitations for long term visits to a British university and a French institute, he planned a 1-1/2 year stay in Europe that included the summers of 1969 and 1970.

Steve arrived at the University of Warwick in the spring of 1969. Continuing his interest in minerals, Steve altered his modus operandi to permit concentration on two endeavors simultaneously. In mathematics he was applying the theory of dynamical systems to various physical models, notably the *n*-body problem of celestial mechanics. There the dynamical system was the evolving state of the universe with each celestial body operating on the others by gravitational attraction. To analyze the asymptotic behavior of the system, Smale created new roles for Morse theory and other mathematical tools.

On the mineral front, Steve continued to seek opportunities to enhance his knowledge. The magnificent collection of the British Museum was perhaps the finest in the world, and there were major museums in Paris as well. Smale frequented these mainstream institutions, but much of the mineral dealing occurs at gatherings known as shows. "I went to a number of mineral shows, meeting a number of the European dealers and collectors, eventually going to their homes or shops."[5] Sotheby's was another important connection where Smale bought and sold. The London auction house had just begun to trade in minerals one year prior to Steve's arrival in Europe.[6] As the buying advanced his collection and education, Smale's price threshold jumped into the hundreds of dollars. Steve returned to the U.S. in the fall of 1970, committed to building his collection. Minerals were not just his hobby, they had become a driving force in his life.

Smale's pursuits of minerals and mathematics were not entirely separate. While the best minerals were located in various locales spread throughout the world, an eminent mathematician received a number of invitations on the international lecture circuit. Smale began to select travel opportunities with an eye to possible side trips for mineral exploration. The most notable example was Brazil with its mineral-rich state of Minas Gerais. Steve's mathematical connection to the country included his legendary proof of the Higher Dimensional Poincaré Conjecture on the beaches of Rio in 1960. Two of the most influential

Brazilian mathematicians were Mauricio Peixoto and Jacob Palis, both of whom were followers of Smale's approach to dynamical systems. Peixoto and Palis hoped to upgrade the profile of the IMPA institute. A visit by Smale to Brazil offered a mutual benefit. It boosted the emerging Brazilian mathematics initiative as well as providing Smale with both an enthusiastic environment to work on dynamical systems and a base to explore the country's remote mineral sources.

When Peixoto and his associates hosted a major conference on dynamical systems during the summer of 1971, Smale was among the headliners. The conference was to take place in Salvador on the extreme northeast Brazilian coast. Smale and a few others were invited to Rio de Janeiro for residencies at IMPA prior to the conference. The latest developments in the field were explicated on Brazilian soil, but, given the long lag typical in mathematical publication, did not appear in print for two years. Attendance at the meeting was an imperative for any scholar in the field. The proceedings of the meeting[7] included Smale's first paper on mathematical economics. It marked the beginning of his long foray into the subject.

The Minas Gerais region has a long history of mineral production, dating back to the discovery of diamonds in the eighteenth century.[8] It continues to be a fertile area for diverse specimens, particularly tourmaline. Since Minas Gerais is a long difficult drive north from Rio de Janeiro, the 1970s Brazilian mineral market was beyond the immediate reach of virtually every international collector. The few dealers able to venture to the remote region controlled the trade on both ends.

Accompanied by Clara and Jacob Palis, Steve drove from Rio to Salvador via a route through Minas Gerais. In the remote town of Teofilo Otoni, Steve bought tourmaline directly from the important Brazilian mineral dealer Jacinto.[9] Buying from the local source, rather than an international dealer, was to become a distinguishing feature of Steve as a collector.

Earlier in the year, the annual meeting of the National Academy of Sciences provided another professional travel opportunity. This organization, begun under Abraham Lincoln in 1863, is the most prestigious fraternity of American scientists. Whenever a roster of new inductees appears, university presidents anxiously scan the list, hoping to identify members of their faculty. Smale's election in 1970, ten years after the solution of the Higher Dimensional Poincaré Conjecture, was overdue. It seems likely that Smale's political activity caused the conservative body to overlook his accomplishments. Attending the 1971 meeting in Washington enabled Steve to view the minerals at the Smithsonian, then the repository of the most outstanding collection of the Western Hemisphere.

For several years Steve had been politically dormant, marshaling his energy for mathematics and minerals. Out of the Washington and Brazil trips, vestiges of the Lone Ranger of the antiwar movement emerged. Following the National Academy of Sciences meeting, Steve wrote a scathing commentary for the *New York Review of Books.*[10]

> *I had my first contact with the business processes of the NAS that day. It was in the "math section." The dozen members at that meeting averaged in age at least sixty-five, it seemed to me. The "math section" devoted its time to two themes that permeated the whole NAS meeting: selection of new members and commemorations of deceased members. It was really fantastic as these three days passed to see how this group of America's most celebrated scientists meeting together could be so dominated by the question of just how to increase their membership and ways to remember their dead.*

Smale was especially harsh on Philip Handler, the new president of the National Academy of Sciences. Recall that Handler, as chair of the National Science Board, had both openly and covertly criticized Smale's administration of his National Science Foundation grant. Recent comments attributed to Handler had dismissed the environmental impact of automobile exhaust emissions. Throughout the *New York Review of Books* article Steve referred to Handler as "Mr. Smog."

Political controversy preceded the Brazilian conference. Brazil was in the midst of a succession of military regimes, with accompanying terrorist actions. A European mathematician proposed a boycott of the meeting, arguing that attendance gave tacit support to the government.[11] Smale's position was to attend the meeting and confront whatever issues arose. Soon after his arrival in Rio de Janeiro, the philosophy was tested. An IMPA graduate student, Magalhães, and his wife, Pandolfi, were in serious trouble. There were government allegations that the woman was involved in a bank robbery and murder. She remained in jail, awaiting trial, while her husband lost his fellowship. The foreign mathematicians learned that the government had obtained its information by torturing the couple and others.[12]

Together with mathematical colleagues in Rio de Janeiro, Smale decided to draft a letter petitioning the judge on behalf of Pandolfi. By the start of the Salvador conference, the letter had aroused strong feelings. Many foreigners were eager to lend their support to protest the treatment of Pandolfi, but the Brazilian mathematicians found themselves in an uncomfortable position. Twenty-eight years later the organizers recalled that the protest action "made us quite nervous, perhaps even more than [it] should, since 'large' meetings and demonstrations of a political nature were not allowed at that time ... More important was the fate of Magalhães' wife and thus of himself, as well as our students and even guests, not to say about our dream to forge a new era for mathematics in the region."[13] The organizers viewed their subsequent reactions as in the interest of these constituencies. "We did talk to the participants about this during the first week of the meeting. We did not and could not censor and much less suppress a letter, but we did express our hope that it would be written in the best possible way to achieve its main objective."

Smale and Shub were among the foreigners who interpreted the actions of their Brazilian colleagues differently. Shub was incredulous at an announcement from one of the organizers that "certain unspecified types of meetings between Brazilians and foreigners would not be tolerated."[14] He saw the letter as a mild form of protest that the Brazilians were attempting to suppress. The Berkeley dynamical systems subculture was shaken. In the sixties they had been close friends, thriving in a free environment on common mathematical and political passions. Now the friendship was challenged by conflicting issues of human rights and Latin American mathematical development. No solution was apparent when Magalhães came to Salvador and met privately with Shub and Smale.[15] He discouraged the letter, arguing that it would not help his wife and might hurt him. The project was dropped.

The Pandolfi protest and reaction was a time of intense emotion for many of the principals. There was much discussion of the contemplated actions by the foreigners. Were they being impudent? One of the overriding contentions was the claim that if the foreigners sent the letter, then the government would retaliate by declining to accommodate the development of IMPA. Since the letter was not sent, even hindsight cannot shed further light on the implication. There is no doubt, however, that IMPA did flourish, becoming a major center for Latin American mathematics.

The political episodes did not alter Steve's focus on minerals and mathematics. He was learning mathematical economics and developing his ideas on the theory of general equilibrium. Meanwhile he kept abreast of the developments that were upgrading the status of mineral collecting. Among the growing number of shows, the most important is one held in Tucson, dating back to 1955. Prior to the 1972 show Bill Larson and his partner Ed Swoboda hosted a party at the Tourmaline Queen Mine in San Diego County. Ostensibly to celebrate their recent discovery of blue cap tourmalines, the event attracted major museum

Steve and Clara at a mineral show with some of Steve's mineral photographs in the background.

curators and collectors.[16] Everyone then proceeded to Tucson. The "First Mine Bash" highlighted the important blue cap pocket hit while simultaneously boosting the Tucson Mineral Show to the pre-eminent standing that it continues to enjoy. Steve's teaching obligations prevented him from attending the Mine Bashes, but he became a fixture at the Tucson Mineral Show.

Tucson annually offers the compelling temptation of a full range of specimens, available for sale or barter. It is there that many collectors extend their financial limits. Mineral purchases tend to advance in jumps, as the buyer becomes comfortable with the $100 plateau before moving on to $500 or perhaps $1000. After the first deal at a certain level, it is easier to make the next. For Smale a great leap occurred around 1973. As part of the activity surrounding the Tucson Mineral Show, a major dealer passed through Berkeley. Steve was struck by a beautiful azurite malachite from the Tsumeb mine in Southwest Africa. He and Clara made the purchase for $4750.[17]

It was a lot of money to spend on a single specimen, but the community of collectors was about to be rocked by a much bigger deal. In late April, 1974 Dave Wilber purchased 52 minerals from Peter Bancroft for the unheard of sum of $400,000.[18] Wilber had raised the bar to a new level and it remained to be seen whether the market would respond. To succeed required a small critical mass of followers, supporting the new price standard. Smale was a part of this group, investing large sums of money in the highest quality minerals. Once again Steve's self-confidence and audacity were important ingredients in assuming the risk. In subsequent years the market affirmed his decisions and rewarded him generously.

Among the Bancroft specimens was a superb kunzite crystal with an added historical feature. It was the same piece that American gem expert George Kunz held for a famous photograph taken in 1904 when this gem species was named in his honor.[19] The combination of its superb quality and historical significance was an overwhelming attraction to a serious collector. Bill Larson traded a magnificent blue cap tourmaline crystal to Wilber for the kunzite.

To make a great mineral deal requires an opportunity. Sometimes a collector is fortunate enough to be at the right place at the time when terrific material becomes available. This may occur as the mineral comes out of the ground, but there are other possibilities. Smale's mathematical travel increased his opportunities. A visit to a British university in the mid-seventies placed Steve in London just as a dealer gained access to an early nineteenth century collection. Up until that time, the material had been known only to the heirs of the original English collector.[20] Smale seized the opportunity to obtain several fine pieces before his competitors were even aware of their existence. Foremost was a pyargyrite, an ore of silver. At the time neither Smale nor the dealer fully appreciated the prominent role the pyargyrite would play in Steve's collection. Their later discovery of an 1812 lithograph depicting the specimen, greatly enhanced its status.

As a mineral collector, Smale had come a long way since 1968 when he received a mineral as a gift from his father. Shortly afterward Smale began pursuing his objective of amassing a world class collection. A mind and drive that had previously cracked intractable mathematics problems was adding mineral connoisseurship to its conquests. By 1976 Steve had moved to the front rank of collectors. It was a remarkably brief time to make such strides. How had he done it? As in mathematics there was the single minded focus on the goal, supremely confident that it was within his grasp. To advance his collection Steve had exploited his mathematical travel opportunities, making key deals in Brazil and Europe. Most importantly, Steve had the audacity and confidence to assume the risk of quickly closing a major deal that others might have lost to the delay of a long deliberation or solicitation of a second opinion. Just how good was Steve's collection? The answer came at the 1976 Tucson Mineral Show. *The Mineralogical Record* covered the competition for the McDole Trophy.[21]

The awarding of the McDole Trophy has been a standard feature on Saturday night, and this year was no different. The trophy, for those who don't know, is awarded for "Best Rocks in the Show," as McDole would have put it. Ed McDole was a beloved, rough-hewn old character, a fantastic yarn-spinner, and also a dealer in amazingly good mineral specimens. He travelled around the Rocky Mountain states in a dusty Lincoln Continental and consistently surprised people with the fine, mostly self-collected, specimens he stashed in the back. He's dead now, but some of his friends felt he deserved to be immortalized by a trophy which is now eagerly competed for every year. There is a little ceremony the winner must go through Saturday night before he can receive his trophy. It seems that dark rum (black, actually) was McDole's favorite drink, and years ago when the first trophy was awarded a bottle of rum that had been Ed's was brought out from which the winner was poured a good-sized slug. The requirement has been continued to this day; only after downing this initiatory drink is the winner officially declared "an Old Bounder" (and also probably a fire hazard) and given his trophy. This year John Barlow, who won last year but was called out of town before the ceremony, was first to take his overdue drink. In an impressive display of a cast-iron esophagus, John waved away water for dilution, drained the rum in one gulp, and jumped off the stage to a wild ovation from the audience! This year's winner, Steven Smale, (a professor of mathematics at the University of California, Berkeley) took a little longer but finally got his down. Anyone who plans on winning next year had better start practicing this feat! Incidentally, after every show the hallowed bottle of rum is filled back up to the top with more rum so that "there will always be some of Ed's

> *original rum in this bottle." Ah, the wonderful traditions we have!*

The McDole Trophy certified that Steve's collection was the state of the art in 1976, but there were others at that level, all seeking the same high quality material. To remain on top, a collection must be dynamic rather than static. Smale fully appreciated the importance of obtaining new specimens, but with fewer financial resources than some of his competitors, Steve needed to make the most of his buying, selling, and trading.

An extremely risky opportunity arose with the rubellite tourmaline strike at a Minas Gerais mine in April 1978. Steve was in Rio de Janeiro when he heard of the discovery and maneuvered a speaking engagement at Belo Horizonte, the capital of Minas Gerais.[22] From there he could drive to Governador Valadares and possibly make a deal for the tourmaline. In retrospect it might appear simple, but the venture was fraught with uncertainty. Wayne Thompson recalls the circumstances.[23]

> *When that pocket came out, it confused everybody because the tourmalines were so much better than anything that had come out previously—the matrix, the color, the crystal size. The prices were phenomenally high. I think everybody was standoffish thinking, wow these are great, but they hit a huge pocket, undoubtedly they will hit more pockets. When they hit more, there will be more material, maybe the quality will even be better and the prices will come down because there's more material available. There was a lot of confusion. Steve was smart enough to go down. He had a contact.*

At the time there were two gigantic obstructions to making a deal: reaching the dealer and pulling the trigger to make the transaction. Smale's mathematical connections placed him in Rio at a key moment and then moved him to Belo Horizonte. Carrying large sums of money to Governador Valadares and returning with gems offered its own danger. Belo Horizonte mathematicians enlisted one of their graduate students to accompany Steve and act as his bodyguard. Another bit of luck produced the entree to the mine manager Jonas. The wife of a Brazilian mathematician was a friend of Jonas' wife. She facilitated an arrangement for Steve to receive a respectful meeting with Jonas, enabling the opportunity for a deal.[24]

In Governador Valadares, Jonas' agent showed Steve a great deal of material. Finally, a flat appeared with specimens that Steve recognized as outstanding. He quickly decided to make a purchase. To take the decision required confidence in his connoisseurship and assumption of the risk that no better specimens would become available.

Steve met with Jonas who had a gun on the table. There was little negotiation as Steve found the asking prices to be reasonable. However, he lacked sufficient resources to purchase all the material he desired, and Jonas did not have a layaway plan. Consequently Steve later engineered a second part to the transaction. Another dealer acted as a middleman, purchasing more tourmalines from Jonas, and allowing Smale time to make the payment. For the moment, the two part deal cornered the rubellite tourmaline market. Wayne Thompson gives his perspective.[25]

> *Steve was smart enough to stick his neck way out and spend for that particular period of time what seemed to be an incredible amount of money for a batch of tourmalines that had just come out of the ground. I remember thinking, my gosh, what a risk. I would never do that because I expected, like everybody, there would be more. Well there were never any more, and, as time proved, he made a brilliant maneuver. It was a risky maneuver, but it was a brilliant move on his part. Yes, there had been other people around. Probably some of them may have seen some of what he had seen. Most of them were probably afraid to pay the money. They were*

probably confused. Steve was smart enough and time proved him to be correct. He spent the money and got the pieces. In the end the price didn't matter what he paid. The fact is he owned the pieces. That would be like squabbling over buying Rembrandts or Monets in the 40s for $30,000 each or whatever the price may have been. At the time he paid a lot of money, but time proved him to be right. I'd say he's had a knack for that. Consistently lucky and smart.

The next step for Smale was to recoup his investment. To accomplish this he marketed most of the new specimens, saving the very best for his own collection. In the mineral business, the boundaries are sometimes blurred between dealers and collectors. Bill Larson wore both hats, but Smale's reserved demeanor had already made him an enigma to his mineral colleagues. All knew that Steve had an outstanding collection, and some realized that he was a mathematician, but few were aware of his accomplishments. It seemed peculiar for the mathematician/collector to move into the treacherous dealing territory. Joel Bartsch, curator of gems and minerals at the Houston Museum of Natural Science, recalls seeing Steve marketing the rubellite tourmalines at the Tucson Mineral Show.[26]

All of a sudden I turned the corner and walked into this mineral room and there was this mathematician guy selling minerals which seemed totally off the wall to me. He had fabulous stuff and he was selling it quite readily because it was superb, beautiful crystals with great color and in my eyes he seemed to be sort of unaware. In his own heart he knew what he had, but in my eyes I don't think he understood the impact of what he had.

The story of the 1978 Smale purchase of the Jonas Mine tourmalines is practically a legend among mineral collectors. A mathematician uses his connections to reach a remote locality, and has the guts to make a highly speculative gamble which paid off financially as well as for his collection. Smale's deal was a huge success, but a view persists that Steve overpaid for the minerals. According to Bill Larson, Steve was "an inexperienced retail collector who didn't understand how to buy from foreigners."[27] Smale was accustomed to bargaining with dealers and collectors. In this context the lower asking prices at the mine were attractive. Larson, as well as others, maintains that Steve accepted prices that were several times the going rate, launching an inflationary spiral. Costs increased at the source while middlepeople and dealers continued to mark up by their previous factors.

Steve's response to these allegations is that they are a combination of truth and legend.[28] He points out that others do not know what he actually paid and they may have a false impression that overestimates the price. That said, he concedes that he is not a hard bargainer with third world people, adding that he wants them to seek deals with him in the future.

The value of a mineral is an ambiguous notion. At the mine in a remote location, a specimen might sell for a fraction of what it fetches at the Tucson Mineral Show a month later. Once an outstanding piece finds its way into a collection, what is its value? By its nature the specimen is unique and can only be appraised by recent sale prices of comparable pieces. On this basis connoisseurs might estimate a market value, but the specimen is unlikely to be available at that price. The priority of the owner is usually to obtain the finest collection. There is little incentive to diminish that collection by giving up a key piece at the price a comparable specimen may have recently sold. Compensation is required for the value of the specimen *and* the damage to the collection.

For this reason trades are not uncommon. Sometimes the trade is one for one, but a standard technique for the emerging collector is to offer several lesser pieces whose *aggregate* value is significantly greater than the desired piece. So what is the value of the outstanding specimen? Is it the hypothetical market value for which it may not be available? Or is it the price that must be paid to pry it out of the hands of the owner, a price which perhaps no one would pay? When these gaps are bridged and a deal is consummated, there is considerable interest among high end collectors, particularly if there is a big jump above the market standard. With the new threshold there are reappraisals of peer specimens.

Consider the blue cap tourmaline that Bill Larson traded to Dave Wilber. The tourmaline had been Larson's second selection in the blue cap draft with his partner Swoboda. While the rock was a runner-up based on size and value, its stunning perfection ranked aesthetically with any specimen in the world. Smale was struck by the piece at first sight. The singular blue cap epitomized the quality that Smale wanted in his collection. Eight years later the specimen belonged to Tom Morris who had purchased it from Wilber. At that time Bill Larson estimated the value at $20,000–$25,000, but an offer in that range was likely to fail. Morris had control. Smale realized that only a spectacular *beyond market* bid could gain him possession. He wanted the blue cap and his $40,000 proffer shocked the 1980 mineral world. Steve got the piece. The actual deal expanded to include other minerals and people, but the new market value of $40,000 for the blue cap led to a fresh round of appraisals.

As with the rubellite tourmalines in Brazil, Steve's deal had broad ramifications for the top end mineral market. Was he foolish to give up $40,000 for the blue cap? In 1980 many experts questioned the wisdom of the purchase. However, investments of this nature cannot be evaluated in the short term. Fifteen years later, Smale looked like a genius. Connoisseurs agreed that it was the finest piece in his collection, appraising its value in the hundreds of thousands.

While distinguishing himself in his career as professor of mathematics at Berkeley, Smale's parallel pursuit had made him a force in the mineral world. To refine his collection he made repeated trips to Brazil (normally in conjunction with a mathematical invitation) as well as to Morocco, Pakistan, and other cosmopolitan locales. Through these

Photograph by Smale of Blue Cap Tourmaline out of the Tourmaline Queen Mine in San Diego County. The height of the specimen is about five and one half inches. Steve purchased it from Tom Morris in 1980 and traded it to Wayne Thompson in 1997.

ventures he successfully competed with the full time dealers for the scarce new specimens. Staying abreast of current mineral developments is a demanding task.

Smale cultivates contacts with other collectors, keeping track of their acquisitions and remaining alert to trade possibilities. Smale's profile as a trader contrasts sharply with that of his buying at the source. Rather than accept reasonable offers for his holdings, he tenaciously negotiates for the best possible deal. In this setting Smale's competitive nature emerges. After the deal is completed, he anxiously waits to learn whether other collectors perceive him as the winner.

From the beginning there were several scholarly aspects of Smale's climb to the top of the mineral world. He read the literature and asked questions, assimilating the ideas of others. Once he had grasped the big picture, Steve shaped his own approach. While buying at the source was his most distinctive imprint, Steve has made other innovations in the aesthetic. Two terms in the connoisseur vernacular are sometimes attributed to Smale: *economy* and *horizon*. The notion of economy of a specimen is that "every part should have a reason for being there."[29] This prerequisite for perfection does not, however, confer equal status on all the parts. Since the eye catches the top of a piece, the *horizon* has greater importance. For Smale the horizon should be crystallized and beautiful. Another of Steve's priorities is that the display of the mineral possess an essential three dimensional character. Unlike other collectors he wants specimens that maintain their appearance from perspectives all the way around, rather than the 300 degrees that might be seen in a museum case. These subtle aspects of Smale's taste do not substantially alter the bottom line. Other collectors may weight certain factors differently, but arrive at comparable evaluations. The significant observation here is that, as in mathematics, Smale learns the conventional wisdom but is not bound by it.

In the drive for beauty and perfection, damage is the evil force. Many of these natural crystals are easily broken. While repairs may cosmetically disguise the flaw, the existence of the imperfection operates with varying impact on collectors. For example, Steve is much more deterred by repairs than is Wayne Thompson. Ironically, the azurite which moved Steve to the $5000 purchase plateau was a repaired specimen. For many years it was a vital piece in Steve's collection, as Thompson saw potential for gaining possession.[30] Finally, in 1996, a deal was made. Steve likes to point out that, unlike some other collectors, he always declares the repair on the label of his displays. Smale's diligence in recording the provenance or history of his pieces has prompted a higher standard throughout the hobby.

When Steve received the McDole Trophy at the 1976 Tucson Mineral Show, he was certified in the highest echelon of private collectors. Since that time he has aggressively worked to improve his position as the standards ratcheted upward. At the 1996 Tucson Mineral Show several connoisseurs were asked to rank Smale's collection. Most replies began with disclaimers concerning issues of individuals versus museums and quality versus quantity. The British Museum and Smithsonian possess millions of specimens, filling the continuum from sensational to junk. Smale holds 500–1000 pieces, all choice. Thus the average quality of the specimens has little comparative significance. However, selecting the best several hundred pieces from any of the leading museums yields a portfolio beyond the reach of any private collector. Narrowing the comparison pool to individuals raises issues of scope. Smale specializes in gemstones, especially tourmaline, while others seek broader coverage. Finally some collectors are secretive about their holdings. Given these constraints, the consensus was that the Smale collection was among the top five. Only one collection was deemed superior to that of Smale. This belongs to a wealthy individual who is known as "the black hole" because his acquisitions are never displayed or marketed.

Most collectors have a few favorite pieces that are deemed unavailable for trade. For Bill Larson the provenance of his kunzite gave it a special place. Another person might be attached to a particular specimen because of the intriguing circumstances that led to its acquisition. Sentimentality has less of a role for Smale. His collection is dynamic and he is always willing to consider a deal that improves its overall standing. For years Steve rejected offers for his signature blue cap tourmaline; nevertheless, in principle, it was available. Then, late in 1997, Wayne Thompson put together the right combination of a dozen outstanding pieces that reached Steve's threshold for a trade. After 17 years, Smale finally exchanged the blue cap, secure in the belief that he was advancing his collection.[31]

How did the same person achieve standing at the top of both the mathematics and mineral worlds? Powerful intellect and focus were necessary but not sufficient. In both endeavors Smale was aided by a combination of personal qualities. He had the audacity and self-confidence to take enormous risks. Steve attacked the seemingly intractable Higher Dimensional Poincaré Conjecture and dealt one-on-one with Jonas in Brazil. Many other mathematicians and collectors lack the confidence to invest themselves in these situations. Smale sought out challenges and followed through, all the while driven by a competitive urge to reach greatness. An intriguing question is the extent to which Smale's personal and intellectual endowments might transfer to other endeavors or whether he was simply fortunate to find his own natural niches. The serendipity of the mineral gift from his father seems to suggest an individual with the capability to follow almost any path that sparks his imagination.

B. Photography

Mineral collection had proved gratifying to Smale in a variety of ways. He enjoyed both the accomplishment of building a great collection and the accompanying financial dividends. Then there was the beauty. Smale is not an emotionally demonstrative person, but he held a deep aesthetic appreciation for fine crystals. Ownership amplified his sense of pleasure and satisfaction.

Smale's initial engagement in minerals was enhanced by the publication of *The Mineral Kingdom*. Desautels' text provided an expert overview of the subject while the color photography of Lee Boltin captured the beauty. As Steve devoted his energy to collecting minerals he contemplated taking up the photographic aspect. That he lacked any significant camera experience was a surmountable obstacle for Smale. He could learn.

The first stage was apprenticeship. In the 70s Steve made an arrangement with an experienced local photographer. Pat Craig came to the Smale house for a few hours each week, photographing minerals as Steve assisted and learned.[32] There are a number of delicate issues including lighting, background, and positioning. Steve studied the technique and began to experiment on his own with a 35 mm single lens reflex camera. Meanwhile he read the literature and conceived a more ambitious project.[33]

The next stage began in 1979 when Steve purchased a 4x5 Sinar camera. The large-format view camera was well suited for Steve's aspiration to produce high quality studio images. The output of the Sinar is a 4x5 inch transparency. When enlarged to a standard 8x10 inch print size, the product is a more faithful image than that from a 35 mm camera. Using the view camera, Steve posed his minerals for individual color shots. The immediate project was a calendar featuring a Smale crystal opposite each month. To produce the pictures Steve focused his energy on photography.

At that time theme calendars were not the pervasive industry that they are today. The 1981 crystal calendar was a Smale self-publishing venture. Steve and Clara arranged for the printing. Clara handled the marketing, primarily to mineral enthusiasts. Approximately 7000 copies were sold, completing Steve's second photography stage.[34]

Mineral photography offers formidable technical challenges. To capture the three dimensional crystalline features requires complex lighting arrangements. Reflection from background material further complicates the problem. By photographing his own specimens Steve had the freedom to handle and position the fragile objects. Moreover, he already had an intimate knowledge of each crystal, facilitating the choice of an aesthetic perspective.

Following the 1981 calendar Steve withdrew from intense camera activity, but he remained intrigued by the potential for mineral photography. The most successful practitioners were Erica and Harold Van Pelt. The Van Pelts were commercial photographers who contracted their services to individual collectors, producing beautiful color images. Despite the high level of appreciation for their work among the mineral community, this genre of photography remained outside the art world.

After a decade away from the camera, Smale's mix of vision and self-confidence was activated. He determined that he could transform mineral photography into an art form. The still life images of Edward Weston's pepper and Robert Mapplethorpe's flowers were widely regarded as masterpieces of photography. Why couldn't Smale produce pictures of his own minerals which achieved comparable acclaim? It was an audacious idea. Weston and Mapplethorpe were two of the most prominent photographers in the history of the craft. Approaching the age of 60, Smale was not content just to maintain his status at the pinnacle of mathematics and mineral collecting. Now he was driven to make his mark on photography. In addition to his desire to become known as a great photographer, Steve was motivated to produce a perpetual record of his own collection.

In May 1990 he purchased a Sinar 8x10 camera and immersed himself in photography, devoting 40–50 hours per week. Smale painstakingly set up mineral shoots in his home studio, idealizing the beauty that he saw in his rocks. He arranged for a Berkeley gallery owner to produce the actual prints, employing the vivid cibachrome

Photograph by Smale of Nevada Sulphur out of the Washoe Quicksilver Mine. The specimen is about four inches in width and was obtained in 1977 from Bill and Linda Leach.

color process. A plan was taking shape. Each image was to be marketed in a limited edition of 12. The initial price was in the $100 range increasing by 60 percent after every third sale. To promote the venture, Smale planned to produce a calendar and coffee table book while galleries displayed his prints in one person shows.

The next stage of photography engagement was underway and would last a couple of years with about six months of intense activity.[35] Steve restricted his photography practice to specimens from his collection. He experimented with a number of artistic approaches. Inspired by Mapplethorpe, Steve introduced shelf settings for the minerals. Smale's philosophy for mineral art photography was evolving. Shadows were important and to be included and controlled rather than eliminated. He made images of specimens together or with other objects, but decided it was best to focus each picture on a single specimen. Smale believes that "the specimen itself has all the components necessary for the composition."[36] His task was to devise lighting, backgrounds, and orientations that best displayed these aesthetic qualities. Since each specimen was already selected for Steve's standards of perfection and economy, the output was a visual idealization of a magnificent object. Thus Smale was choreographing two orders of perfection.

Things moved quickly from the May camera purchase. In September a show of Steve's prints opened in the Berkeley gallery of the printer. There was an arrangement for a 1992 calendar, this time with a professional publisher. Steve was preparing the text to accompany the calendar, selling prints, and producing new images. Considering the brevity of his photography experience Smale had achieved remarkable acclaim, but just how good were his pictures? Two photography experts viewed a print and the calendars prior to commenting on their technical and artistic merit.

Jane Jackson, director of Jackson Fine Art, rated the technical quality as "superb."[37] She praised Smale's handling of lighting and shadows as well as the color backgrounds. Smale "is doing a wonderful job in terms of blending in the colors to the pieces of the mineral and the way they are being lit." Stephen Scheer, a photographer and the chairperson of photography in the Lamar Dodd School of Art at the University of Georgia, was also impressed by the lighting and use of color, noting the gradation of light in the background. "Technically they are very much of a high professional quality … As a practitioner of photography, whatever motivated him, he figured out how to do it. They are very sharp and they are very crisp."[38]

Both Jackson and Scheer observe that the resulting look is that of commercial, rather than artistic, photography. Commercial (or illustrative) photographers normally shoot still lifes of table top products for advertisement. Just as the marketing of fine glassware and other goods demands high quality photography, Smale's pictures promote the beauty of his crystals as pristine acts of nature. In the art of Weston and Mapplethorpe the pictures possess a sensuality that transcends the utilitarian value of the subject. Scheer concludes of Smale:

> *He has a very straight, grounded approach. You can look at that in two ways. 1. He is deliberately doing that so as not to destroy the integrity of the mineral. or 2. He doesn't have any other ambition for the mineral other than documenting it in the most beautiful way that he can, because these pictures are very beautiful …*
>
> *He is not trying to fictionalize here to create things that may not be truthful to the actual mineral, just for the sake of photography. The artist, who may not know enough about the minerals, might start to do that.*
>
> *Art museums are interested in photography that is less perfect than this, that has more accidents, more exploration.*

The yield of Smale's third phase was an inventory of quality images that documented his collection. *Beautiful Crystals* calendars appeared for the years 1992 and 1993. By 1997 nearly 100 cibachrome prints had sold. Smale had taken mineral photography to a high level, but it had not become an art form. Acknowledgment of his work was primarily confined to the mineral world.

Smale intended to return to photography, but a mid-90's extended residency at a Hong Kong university introduced new complications. Steve was separated from his home studio and involved in new enterprises. He remained optimistic that he was on track to make mineral photography into an art form. Next on the agenda was the coffee table book for which he had great expectations. Although Scheer and Jackson could see a successful book venture, they remain dubious that the photographs will be collected by art museums.

It is too early to make a definitive evaluation of Smale as a photographer. If he ever does resume serious engagement, further creative growth is likely. Moreover, the standards for artistic photography evolve. Vintage prints of certain images that seemed mundane fifty years ago are now eagerly sought by museums at high prices. Whether or not Smale's mineral photographs attain this status, the drive to become known as a great photographer is yet another example of his audacity and self-confidence.

Adventure and Physical Risks

10

A. First ascent in the Grand Tetons

Smale's risk taking went beyond his mathematical, political, and mineral enterprises. Throughout life he enjoyed the exhilaration derived from a variety of physically challenging adventures. An early clue to this predilection came on the East Berlin trip following his junior year in college. That summer, together with Vince Giuliano, Steve attempted to experience as much of Europe for as little money as possible. When they reached the Alps, Steve wanted to try his hand at mountain climbing.

In Zermatt, they purchased an ice axe and clothesline, but a guide was beyond the budget. The two novices then climbed some of the smaller mountains near the Matterhorn. Vince recalls that he was fortunate to survive. "We didn't know what we were doing. That was a very reckless thing."[1] Steve was less daunted by the danger, rather it whetted his appetite to become a genuine mountaineer.

An opportunity arose two years later. A member of Steve's University of Michigan cooperative, Jack Hilberry, had a connection to a guide at the Grand Teton National Park. It was a pivotal time in Smale's mathematical development. In June 1953 the department chair informed Steve that his academic progress was unsatisfactory. The summer provided time for Steve to reflect on his future. He began the break by earning some money as a welder at a Flint automotive plant. Then he left the $2 an hour job to join Hilberry and hitchhiked to the Grand Tetons, spending a month in the Park.

The Grand Teton National Park is located in the vicinity of Jackson Hole, Wyoming. Steep rises of rock with ice and glaciers lead to the peaks of the Teton range, 7000 feet above the elevated valley floor. The tallest of the mountains, Grand Teton at 13,770 feet, was first climbed in the late nineteenth century. There are a variety of approaches to the top, some of which withstood attack for decades. Whether it is to pioneer a new route or follow an established one, throughout the twentieth century the Tetons have remained a primary objective for climbing enthusiasts in the United States.

Steve's experience with Giuliano in the Alps counted for little in the Tetons. He was essentially starting from scratch. Characteristically, Steve learned by study and doing. "We went to climbing school for a day. The first day we were there I think we climbed Mount Jenny and got lost after dark coming down. We were doing a lot of risky things."[2] As they developed their skills, Smale and Hilberry became acquainted with the network of climbers in the park. There was Willi Unsoeld who, with a partner one decade later, would make the legendary first ascent of Mount Everest along the treacherous West Ridge.[3] Another young climbing star was William Buckingham who shared Smale's interest in mathematics.

By the end of the month Steve considered himself a mountaineer. He was ready for a challenge. When Buckingham formed a group to attempt a first ascent, Steve enlisted. He and Hilberry would share the second rope while Buckingham led the first. The objective was Grand Teton along a modification of a route pioneered by Robert Underhill in 1931. Steve recalls that his day began on a note of uncertainty.[4]

> *I remember on the morning of the big climb, the first ascent, I woke up sick. I was ready to cancel it, but after breakfast I felt better and we did it. That was kind of a little dangerous because our route took us through where stone was falling down from the mountain.*

The "Direct Underhill Ridge" climb is described in *Teton Classics*:[5]

> *This difficult variation stays more on the crest and avoids the often wet chimneys of the regular route. From the Lower Saddle, a conspicuous tower of white rock can be seen on the skyline of the Underhill Ridge. Instead of traversing around onto the west side, the route climbs the chimney between the white tower and the main buttress, then joins the original line. It was first climbed by William Buckingham, Steve Smale, Ann Blackenburg, Charles Browning, and Jack Hilberry on 30 August 1953.*
>
> *Steve led his rope over the chimney. At the summit they met Unsoeld's group who had successfully completed a first ascent along a different route. Unsoeld then aided Steve in a long free hanging rappel from the top.*[6]

For the return trip to Michigan, Steve relied on hitchhiking, his standard mode of travel. Smale had routinely hitchhiked to New York and other parts of the country. However, he and Hilberry were unable to obtain a ride out of Laramie. Frustrated, they decided to "jump a freight." It was the first time for Steve and he sought advice.[7] "We discussed how to do it with some people around the freight yard. They said to wait until the freight is leaving the town so the bulls* will not be present." Smale and Hilberry managed to climb into a freight car that was unoccupied except for meat hooks and other equipment. Then the train stopped,

> *and they closed the door on the outside so we were locked in the freight car. I was a little worried we would get shunted aside. A couple of hours later the car stopped and we used*

* railroad police

these meat hooks to start whacking on the side of the car. And then somebody from the engineer's car came along and let us out and threatened us with jail. I think at that point we were just outside of the next big town and Hilberry ran and hopped on the freight again as it was leaving. I couldn't quite make it with my backpack so I saw him in town a few hours later.

When Steve reached Ann Arbor for the fall term, he enrolled in Raoul Bott's topology course and began his serious engagement with mathematics. Steve maintained his interest in climbing, returning to Grand Teton National Park a few years later with Clara and his sister Judy. Steve and Judy climbed Symmetry Spire on that trip. Periodically he took on other mountains, often accompanied by mathematicians. The last major effort was a rope climb of the North Maroon Bell near Aspen in 1976.[8] Skiing in the Alps and scuba diving in French Polynesia offered a similar thrill. In pursuing each of these endeavors Steve selected an ambitious itinerary. Still they remained isolated diversions rather than foci at the level of mathematics and minerals.

B. Sailing to the Marquesas

Smale's interest in sailing evolved differently. As a student at Michigan, he occasionally rented a boat to sail local lakes. The outings were enjoyable and evoked, in Steve, a desire for further participation in the sport. Mathematical travel offered additional opportunities. If a conference were scheduled in an exotic port, some participants might be interested in joining a sailing excursion, but organizing the venture was another matter. Assuming the responsibility of renting a boat and enlisting a crew required substantial initiative. For most people the associated work and aggravation were too much of a burden. On several occasions Steve led boat trips at conferences. The most ambitious effort took place in the British Virgin Islands where he arranged for a conference to take place on a sailboat.[9]

An inevitable issue for a sailing enthusiast becomes whether to purchase a boat. Ownership is a big step. Aside from the considerable expense (a boat is a hole in the sea where you throw all your money) there is the time factor. Could Steve maintain a boat while doing mathematics and staying on top of mineral collecting? An opportunity to experiment arose one day in the early eighties.

Steve's dynamical systems colleague Charles Pugh owned a small sailboat that he kept at the Berkeley Marina. Pugh was about to begin a one year sabbatical, and the boat was excess baggage. First he tried to sell it, but was unsuccessful. Weeks passed and Pugh was becoming desperate. Finally he went to the Berkeley mathematics building and posted a solicitation for a boat sitter. That same day Smale, an office neighbor in the large building, saw the notice and volunteered to look after the boat. Pugh was elated, never having realized that a solution was so nearby.[10]

Steve sailed the boat in the Bay, gaining experience and confidence. When Pugh returned, Steve was ready to purchase a boat of his own. His ultimate objective involved ocean travel for which Pugh's boat was not suited. Smale analyzed the problem of boat selection. There were the dangers of getting too much and too little. Smale elected to proceed in two steps. He would first buy a provisional boat on which to hone his sailing skills and gauge his interest. If this were successful, then he would buy the ultimate boat "that would be safe and comfortable, something I could take into the ocean, so that I would never feel that I wanted a third boat … The second boat was to be the final boat."[11]

For the temporary position Smale selected a 30 foot Catalina, a popular boat with a ready resale market. Following a suggestion from his children, Steve renamed the boat *Red Emma.* The name commemorated the turn of the century anarchist Emma Goldman. Over the next two years Steve captained *Red Emma* and read books about sailing. Meanwhile he thought about the final boat and its mission.

The goal became a major ocean voyage from Berkeley. The constraint that it take place over a summer disqualified an around the world itinerary, but left plenty of room for creativity. What would Steve Smale choose as his destination? To the suggestion of Hawaii, he responded that it was "not interesting enough to just go to Hawaii. Everybody goes to Hawaii. Sailors go from San Francisco to Hawaii and back. That's the main popular route."[12] Going beyond the conventional choice Smale selected a more southern Pacific island group, the Marquesas. "If you look at the charts and maps and winds, its the obvious choice. It's exotic, exciting, challenging, and serious and doable. All the right things."[13]

Red Emma had served her purpose and would be sold. For the trip to the Marquesas Steve needed a strong, high quality cruiser that could safely withstand big waves and rough weather. He wanted comfortable accommodations for himself and his crew as well as adequate room to store all the provisions. There were tradeoffs to consider. A heavy boat is slow but more stable. Then there were the variety of sail rigs. Following careful deliberation Smale purchased a previously owned 43 foot Hans Christian ketch, retaining the name *Stardust.* The teak furnished 16 ton cruiser was equipped with three staterooms and an auxiliary diesel engine. The ketch rig features two masts, distributing the sails in smaller, more manageable components than other configurations. Now he was ready to focus on the trip to the Marquesas.

To ensure a safe voyage Steve went to work on the logistics. He is an inveterate list maker whose *laissez faire* manner belies his attention to detail. Recall it was Steve, rather than Jerry Rubin, who oversaw the mammoth organization of the Vietnam Day Committee. For the month long trip to the Marquesas, Steve and his crew were totally dependent on themselves and *Stardust.* A lot could go wrong and, as captain, Steve assumed the responsibility of being prepared to cope with any contingency. For a couple of years he devoted considerable energy

Steve and Charles Pugh on Steve's sailboat.

to the project, culminating in the months prior to the summer 1987 voyage. Preparations included the following areas:

Gaining an intimate understanding of *Stardust.*

Recruiting a crew that was both competent and good company.

Bringing *Stardust* to the highest level of fitness.

Learning whatever he needed to know.

Obtaining all necessary provisions.

Finding companions for the trip was a gradual process. There was flexibility in the number of passengers. Considerations were the accommodations provided by the three staterooms and dividing the 24 hours of daily watch time into reasonably sized chunks. Steve's son Nat and his wife Hermine were expecting to participate until they learned that they were expecting a baby. Another woman expressed interest, but then withdrew. To Steve's regret it was going to be an all male crew.[14]

At some point in 1986 Steve invited Charles Pugh to join the voyage. Pugh asked to put off his decision until the first of the year. On January 1, 1987, Smale phoned Pugh to get an answer. Pugh was excited by the opportunity for a great adventure. He saw the trip as "the chance of a lifetime" and accepted.[15] The other crew member was Brazilian mathematician Welington de Melo. De Melo received his Ph.D. under the direction of Steve's student Jacob Palis. He was an athletic individual who had logged quite a few sailing hours and had valuable experience with bad weather.

To prepare *Stardust*, Steve hired consultants and engineers. A professional diver inspected the hull while other experts tuned the mast, rigging, and engine. Steve arranged for the installation of special equipment including a life raft and a short wave radio receiver. He demurred on a transmitter, noting the expense and unreliability. The boat did have a VHF radio which was capable of two way communication with ships in close proximity.

Navigation is a vital element of any voyage, particularly when crossing the ocean and out of land sight. Steve needed to follow an accurate course to reach the Marquesas. With a small error he might miss the islands entirely. The position of a boat can be calculated from its distance or angle from known objects. The classical measurement device is a sextant that finds the angle between the horizon and a celestial body. These readings are distorted by the rocking of the boat, leading to an unreliable determination of position. When Steve was planning his trip, satellite navigation technology had recently become available. These instruments calculate a boat's position by honing in on signals from satellites above the earth. Smale installed the satellite navigation equipment and became confident of its operation.

Steve was less certain of his mechanical skills. In preparing *Stardust* he relied on experts to perform certain tasks. At sea no contractor would be available. Pugh filled a void by assuming the role of mechanic. He took a course on diesel motors and "read several books about the mechanics of sailboats and what to worry about, what to prepare for. Especially I was interested in the electrical starting system for the motor, because I figured if we couldn't get the motor turned on it would be a disaster; that we wouldn't have any electricity and that we therefore wouldn't have any communication with the satellites."[16]

As Pugh prepared to deal with mechanical problems, Smale took on the responsibility of ship's doctor. Steve "read a lot about emergency medicine at sea" and discussed the issues with his own doctor.[17] The weather posed the greatest threat to safety, and Smale examined the history of storms in the Pacific. The itinerary for the voyage was Berkeley to the Marquesas to Hawaii and back to Berkeley. What was the probability of an encounter with a hurricane and how did storm paths evolve? Effective evasive action relied on careful monitoring and an understanding of hurricane tracks.

Smale, Pugh, and de Melo would need to work as a team. They practiced together with some short trips, but Smale ruled out a substantial shake down cruise. Where would they go he asked? Out and back to some point in the ocean was not interesting. A trip to the islands off Los Angeles required at least a week which was too long. Prior to his departure for the Marquesas, Steve's overnight sailing experience in *Stardust* was a three day coastal trip in which he anchored overnight. Self-confidence, audacity, and risk taking emerged again. Steve was confident that his own thorough preparation would lower the risk to an acceptable level. To him, careful planning mitigated the need for a major ocean test run. The logic escaped outside observers.

A crucial element of planning was deciding what to bring. Think about a projected several month voyage with just two resupply stops. A thousand miles at sea, there are neither hardware stores nor pharmacies.

If a tool or antibiotic were not stowed before leaving Berkeley, then it would not be available for use on the way to the Marquesas. The scope of necessities was immense and any omission jeopardized the success of the trip. The problem of planning the inventory had a vastly different intellectual nature than that of the Generalized Poincaré Conjecture; nevertheless Steve attacked it with a similar intensity. It would be an adventure, but he gave a "huge amount of thought to make the voyage fun and safe."[18]

Steve and the crew met several times to discuss these issues. Food received considerable attention, particularly in view of the refrigeration limitations. It was unclear how long the ice would last and when they would become dependent upon canned food. Pugh planned to bake bread. To wash it down there was wine and champagne. They stocked jugs of water to supplement the boat's reservoir.

The first destination was Hiva Oa, the island in the Marquesas where Paul Gauguin died. Hiva Oa is about 3000 nautical miles from Berkeley. (A nautical mile is slightly longer than a mile. This unit of distance is derived from the length along a latitude meridian covering an angle of one minute.) The conservative route was to proceed south along the coast to Mexico and then sail westward. If there were a problem, then land was within easy reach. However, Smale was concerned with time. He was committed to spend the fall at the IBM Watson Research Center in New York. Following the direct path to Hiva Oa was quickest. With the return via Hawaii it was likely to require about three months, leaving just a couple week cushion. Smale elected to take the shortest route.

Sailing across the ocean, there was no opportunity for anchorage. To keep the boat on course, respond to weather changes and be alert for approaching ships it was imperative that someone be on watch at all times. This meant that each sailor was responsible for eight hours per day. Before leaving they adopted a schedule with the day divided into nine shifts. Smale handled the periods: 3:00–7:00, 11:00–13:30, and 18:30–20:00.[19]

Stardust departed from Berkeley at 12:21 on Friday May 8, 1987. Just over an hour later they passed under the Golden Gate Bridge. To document the voyage Smale maintained a log in which he recorded navigation data, impressions of the trip, and other information. On each day around noon he computed the "miles made good" for the past 24 hours. This was the difference in the number of nautical miles from their destination at the beginning and end of the period. Over the first three days they made good 434 miles. Early on the morning of May 12 Steve wrote:

> *Barometer seems pretty consistent at 1020 millibars.*
>
> *4:36 Voltage meter at 11.2 (Battery #1). Gentle winds still wind off starboard quarter about 130° for our course of 180° magnetic. Need to put up more sails. Yesterday Welington set a record 11.3 boat speed under spinnaker & main.*
>
> *4:51 Beautiful full moon, mostly clear sky. North star in the right place.*
>
> *6:01 It has been a very fine watch this 3–7 one with the sky finally clearing. With Bach Cantatas (Strictly Art) on the tape watched the moon set as the sun, Venus and Mars are rising. Last night saw a sail boat a schooner I think.*

Later that day Pugh noted a passing sea turtle and Smale recorded the position of a gale from the short wave radio.

Eight hours a day was quite a bit of watch time, but there was little else to do. Pugh adopted a routine of sleeping twice a day, before and after his midnight watch. His quarters were located in the bow and he recalls falling asleep quickly, conscious of the water flowing past the boat.[20] Reading was the primary diversion for everyone. As captain,

Smale assumed the responsibility of "chief worrier." He had a "list of things to monitor" and shifted into worry mode whenever he heard "funny noises."[21]

Steve noted potential problems in the log. On May 9 he wrote "Need navigation light fuse fixed!" Then on May 13 it was "Short wave radio needs attention. Can't turn light off." On the same watch Smale began to worry about a more serious problem. "Elec bilge pump & its trigger are working well. But WATCH the bilge." The bilge is the lower inside portion of the hull. By design, water collects in this region. When the water reaches a certain magnitude, a trigger switch activates an electric bilge pump. Alternatively the water can be removed by a manual pump. Smale's log entry was prompted by the observation that the bilge pump was effectively performing its function, but was being activated too often. Subsequent monitoring confirmed the inference of a leak. Finding and repairing the leak offered significant challenges. Eventually it was associated with engine use and then located in the *aqua-lift*, a muffler type device. The mechanic Pugh slowed the hemorrhaging with a "towel scrap, form-fitted by [a] piece of plastic from a 2" plastic pipe, pressed by a couple of bungees."[22]

The leak incident increased attention on engine use. The engine served two important functions. It charged the batteries to provide electric power and was a backup source of propulsion. When the wind was calm, as was typical in certain regions, it was tempting to turn on the motor. However, engine use was constrained by the 120 gallon tank capacity for diesel fuel. If the supply were exhausted prior to reaching Hiva Oa, the consequences could be serious. After one week of travel the numbers were alarming. The distance covered was 919 nautical miles (out of 3000) and the engine had run for 55 hours. A rough rule of thumb for gallons per hour indicated they were consuming fuel too rapidly. Pugh set out to calculate the remaining supply. He took a dipstick measurement and did a volume computation based on the geometry of the tanks. Pugh estimated that only 1/6 of the fuel remained to cover 2/3 of the distance. The situation called for a far more conservative engine use policy. Smale's log entry of May 15 indicates that they were coping with the new austerity program.

> *16:01 We all just went swimming off the boat for first time. The water is nice. Since we have measured the fuel, our worries about it have made us shift to the spinnaker. Have made 2 miles since about noon! We have been in an incredibly long calm. It certainly would be good to find the trade winds! Also have started to use no electricity except compass light, sat nav. & wind log-meter.*
>
> *Barometer has been rising up to 1022-1/2 & has just fallen a bit.*
>
> *Use hand pump for the bilge! For lunch had roast potatoes and filet mignon cooked over the Bar B-Q with the BV Cabernet.*

Despite all of Smale's advance thought and preparation, it appeared that he had made a critical blunder by failing to conserve fuel. To compensate it was necessary to shut down the engine for an extended period. This in turn required minimizing the power drain on the batteries. There were choices to be made. They switched from the electric to the manual bilge pump but continued to run the satellite navigation off the battery rather than rely on a sextant. As for propulsion, they could only hope that the trade winds were nearby. Then on the following day:

> *07:48 This morning at 05:00 it seems as if the trade winds have begun to move us along decently. Up to then the weather for a few days has been very calm. From noon yesterday till 03:00 today we moved 33 miles closer to Hiva Oa. The*

spinnaker has been in use continuously since 11:00 yesterday, but now boat speed is 6 knots.

Last night a good sized whale passed us going north spouting as she went. The night watches have been tough. We could hardly leave the whale.

They had indeed reached the trade winds. For the next week sail power alone could sustain *Stardust* at a speed of about 6 knots (nautical miles per hour). The fuel problem had been deferred, but not eliminated. Beyond the trade winds was the calm region near the equator known as the doldrums. Moreover, the satellite navigation continued to draw on the battery necessitating periodic engine use for recharging. Steve's preparation did offer a partial solution to the energy problem. Among the special equipment he stocked for the voyage was a solar energy panel. He deployed the panel which produced sufficient energy to run the satellite navigation. On May 19 after 11 days at sea they were nearly half way to Hiva Oa. Things were going well and Smale recorded the following log entries:

13:40 Just finished my watch. Pretty hard, under spinnaker and main 15–20 knots of tradewinds. Moderately heavy seas Boat Speed about 8 knots.

Temperature 82° It is getting hot!

Saw many, many flying fish today on watch. Also several birds! I am surprised to see the birds so far from land.

During watch, Charles relieved me for lunch cooked by him. Filet over charcoal-boiled potatoes, carrots with lemon & garlic Bacigalupi pinot noir.

We are conserving on electricity since our main battery is down, being nourished by solar power. One problem is we don't use auto pilot and the wind vane isn't functioning very well.

Here is the days run for day #11

1722-1588=134 miles for 11th day.

At 13:50 It is 1577 miles to H.O.

Fix at 13:30 is 15° 32.80N 131 26.33W

16:10 At 6:30 this evening we celebrate a bit early the half way mark with French champagne and smoked salmon

16:10 Have not used engine since about 11:00 on May 15

The wind vane is a steering mechanism. When engaged, it is intended to allow the watch person to remove his hands from the steering wheel. Smale had tested it in the Bay along with the satellite navigation and solar panel. Unlike the latter two devices, the wind vane did not perform under the rigors of the trade winds in the ocean. To permit some freedom from the wheel, Smale and de Melo rigged up a shock cord restraint. Another annoyance was the noise from the propeller shaft which persisted despite Pugh's makeshift remedies. However, the most important topic was the wind. As it maintained a solid double digit velocity, *Stardust* covered 155 miles on Day 12.

The afternoon of May 20 was hot and sunny. Smale cooked his "typical chicken lunch with Italian sun dried tomatoes olive oil garlic & oregano," augmented by Pugh's freshly baked bread.[23] The seas were big and there was a strong steady wind. *Stardust* was moving rapidly under the large spinnaker sail as the engine continued to rest. Conditions were ideal. Pugh recalls that he was below when "I heard this noise. It was like a tremendous thud. I thought we'd hit something. I guess it was the sail hitting the water, or the rope snapping, or something like that."[24] The halyard, a rope used to hoist the spinnaker, had broken.

The first task was to retrieve the sail from the ocean. The strong wind became the enemy as the three men successfully struggled to haul in the big spinnaker.

The captain and crew faced a crisis. To restore the spinnaker, a repair was needed at the top of the mast. Going up the mast was both frightening and dangerous. The method of ascent was to sit in a canvas chair while the others hoisted it up the remaining halyard. It was essential to maintain a grip on the mast as it swayed in the wind and the boat rocked in the waves. If contact were lost, the mast would batter the dangling climber.

Smale decided that the repair was essential and urgent. Who would climb the mast? When neither Pugh nor de Melo volunteered, Smale took on the responsibility. The former mountaineer was approaching the age of 57. He hugged the mast with his arms as he carried a new halyard to the top. The swaying increased with altitude. Working in the wind about 50 feet above the water, Smale managed to install the new halyard. The crisis was averted. Steve had responded courageously and constructively to a sudden disaster that threatened the safety of the voyage. Still they were a long way from the Marquesas, and other unknown dangers might be lurking ahead.

With the support of the trade winds *Stardust* traveled 944 miles in the second week for a total of 1863. Sustaining that pace they would reach the Marquesas in another nine days. However, the doldrums were somewhere on the horizon. During Smale's early morning watch of Saturday May 23 the boat speed slipped below 2 knots. Steve turned on the engine for the first time in over a week. A few hours later the wind picked up and Steve shut down the motor. The diesel boost allowed them to make 131 miles on day 15.

They were now in the vicinity of the doldrums. Somewhere in the remaining 1000 miles they could expect to pass from the doldrums to the favorable southeast trade winds. The uncertainty of the borders was an additional complication to a strategic problem. When to use the meager, but unknown, quantity of remaining fuel? Smale's position was unambiguous: Do not drift when there is diesel in the tanks. On his next watch, Steve cranked up the engine again.

A difference in philosophy had emerged. Pugh and de Melo sought to conserve fuel while Smale insisted on progressing. Each time that the engine started, it was during Smale's watch. Some of Smale's May 24 log entries indicate his impatience with the circumstances:

> *04:07 … In eleven hours have gone maybe eleven miles! The doldrums!*
>
> *… Our worst night! Welington woke up Charles (& me) to take spinnaker down. At the same time it started raining with variable very light winds. Then we drifted aimlessly for several hours, mostly north.*
>
> *Now Charles has us on a course, very slow; and I am on watch …*
>
> *In 4 hours have gone 5 miles mostly west.*

A couple of hours later Smale started the engine. It remained on until 9:03. Even so they only managed 58 miles on day 16. Then at 11:44 he wrote:

> *The doldrums: at 10:00 took the spinnaker down. Since we have been drifting, mainly toward San Francisco. It is ironic that the Moet & Chandon bottle we threw overboard after our doldrums celebration start*[ed] *drifting away from us to the Marquesas.*
>
> *In any case we decided to use our quickly dwindling diesel fuel to try to find the south east trades. So we just turned on the engine.*

Smale was the captain and his view on diesel use had prevailed. At 16:50 the engine stopped. They were out of fuel with over 900 miles to Hiva Oa. Pugh found a positive spin—finally he was freed from worrying about how much fuel was left.[25] The next morning a breeze appeared. Was it the onset of the trade winds? Day 17 was 91 miles.

There was a new problem. The ice had finally melted. Considering that they lacked refrigeration and were approaching the equator, it was remarkable that the ice survived as long as it did. The event was expected and the menu shifted to canned food. The beverage selection was especially impacted. No more cold beer and there would be no champagne to toast the forthcoming crossing of the equator.

The trade winds were the key. Every gust offered hope, but it remained premature. On the next two days *Stardust* covered 111 and 68 miles. The wind appeared infrequently, sometimes accompanying squalls. Smale reviewed the weather in a log entry on May 27:

> *18:00 Have had two more substantial squalls this afternoon. Familiar pattern by now. Winds rise to close to 20 knots, very heavy rainfall for 10–15 minutes. Then back to an almost calm. Since noon our progress has been hardly measurable. Our worst day (also, since ice is gone, not eating as well). We started fishing today.*

Half an hour later another squall appeared, but this one was different. After it passed, the wind continued. Within the first degree of the northern hemisphere latitude, they had found the trade winds. On the morning of May 28, *Stardust* crossed the equator, covering 76 miles on day 20. For each of the next four days, she took a 125 mile chunk out of the remaining distance to Hiva Oa. If the satellite navigation was correct, they were almost there. On June 1 Smale declared victory.

> *06:18 Well this great passage is nearing its end. Almost 24 days at sea without sight of land. By tonight, we may see land again and we expect to anchor Atuona early tomorrow (Tues.) A.M.*

Atuona is a harbor town on Hiva Oa. Although the target was just a small volcanic island in the vast Pacific Ocean, Smale was completely confident that *Stardust* was on course. He was right. On the morning of June 2, Hiva Oa appeared in the distance. To reach the harbor required navigating between two rocky islands. Smale described the end of the 3000 mile voyage in the log:

> *It turned out that the last few miles into Atuona was a much greater ordeal than expected. Instead of 3 hours it took all day. We had no fuel & the winds were very light and variable. The last couple of miles were the worst. We would hold the sails to catch a puff of air, always worried about being washed onto the rocks with the big swells. Just before the final breakwater, I couldn't keep our direction, so forced to throw out anchor to keep us off the rocks.*

While drifting aimlessly in the doldrums was frustrating, here it was treacherous. If only they had saved a little fuel. Now they would have to remain at anchor until a suitable wind arose. Then two men appeared in a dingy powered by a small motor. Their offer of a tow was gratefully accepted. The tiny boat pulled *Stardust* the final few hundred yards into the Atuona harbor.

Smale had sustained two injuries on the 25 day voyage, both associated with lurching of the boat. His head was bloodied by a blow from the boom and he was scalded while making coffee. The ocean had exacted her toll on *Stardust*. The pounding surf and steady diet of salt water were a challenge to the hull and engine. Leaks and broken parts needed attention. The propeller shaft remained a problem.

The anchorage at Atuona provided an opportunity to service the boat. Climbing the mast to deal with rigging remained a difficult task, but at least there was shelter from the big waves and wind. For some repairs it was necessary to dive under the boat.[26] In their three days on Hiva Oa Smale, Pugh, and de Melo worked on repairs and restocked provisions. The Marquesas were an exotic place to visit but lacked some of the more modern port conveniences. To refuel the diesel tanks required many greasy trips in a bumpy life raft.

With the fuel restored it was easy to travel to other islands in the chain. They remained on the Marquesas for three weeks. What did they do when not occupied with repairs? "We dined out! Usually the only diners. We explored uninhabited areas of the islands, long hikes, socialized with various round the world sailors, visited little settlements, swam and snorkeled, climbed mountains, etc. Pretty nice, pretty exotic."[27]

The next stop was Hawaii, 2000 miles to the northwest. The route included another equator crossing and a rematch with the doldrums. This time, however, they expected to have plenty of fuel when the winds died down. *Stardust* departed from the Marquesas early in the morning of June 24. An initial run of engine power took her out of the harbor and into the ocean. The first omens were mixed.

There was a problem with the engine's water pump that Pugh repaired by tightening a belt. On the positive side were the brisk trade winds. Following an early shift to sail power, *Stardust* covered 135 and 152 miles on the first two days. With the winds came heavy squalls that battered the boat with rain and waves. Difficulties from the first leg began to recur. The propeller shaft and bilge were acting up again. Still they were making great progress, 157 and 145 miles under sail for the next two days.

After crossing the equator the winds began to weaken. Day five was 121 miles. Meanwhile the large accumulation of bilge water implied that there was a significant leak. It was June 29 and they were in the doldrums. That night an engine start up was followed by a dramatic increase in the bilge problem. There was no choice but to shut down the engine, and unclear whether it was of further use on the leg.

With a nearly full fuel tank Smale endured a 64 mile day. The slow pace continued into Smale's early watch on July 1, as his impatience grew. He turned on the engine and brought them to 77 miles for day 7, but the price was dear. The water pump deteriorated beyond function. Serious repairs were required and could not be performed until reaching port in Hawaii. Totally dependent on sail power, the next two days were 69 and 53 miles.

The trade winds were somewhere to the north. Not all winds were friendly. Tropical storm Beatrice was forming far off to the northeast, but moving west. A collision was possible and Smale closely monitored weather reports on the short wave radio. Sitting in the doldrums without a motor, there were no evasive options. Then on the morning of July 3 the trade winds suddenly appeared. *Stardust* covered 152 and 168 miles over the next two days to maintain her distance from Beatrice. Smale noted the 25 knot wind and 12 foot waves in the log and compared the experience to mountain climbing. "The sailing is pretty spectacular."

Day 12 was a record 175 miles. They were within 500 miles of Hawaii and moving well. Beatrice and Calvin posed less of a threat. The wind held up for the next few days. On the morning of July 9, just 10 miles from Hilo, the wind died. Steve described the approach:[28]

> *We could not use the engine at all arriving in Hawaii and the wind was light and sporadic. Thus we had a spinnaker up as we tried to maneuver a narrow passage between underwater reefs as we were approaching very close to the port. I was more aware and worried about this than either Charles or Welington since I could lose the uninsured boat!*

We had to zigzag back and forth tacking, getting very close to the reefs. But at one point according to the chart we were extremely close to some underwater rocks even a few feet, and we turned just in time. I recall that as we eventually docked there was a little crowd watching to see us arrive in this very unconventional way.

Repair of the engine was essential. The task was performed in Hilo by Pugh and a local shop. The motor became operational again. In restocking provisions a new problem arose. For safety considerations Steve had selected compressed natural gas, over propane, as the fuel for the stove. The reason was that propane is heavier than air and sinks into the bilge. Unfortunately, compressed natural gas was unavailable in Hawaii.[29]

The week and a half in Hawaii was divided between Hilo and the Kona Coast. By prior arrangement Clara joined Steve for this portion of the trip. In Hilo they viewed the mineral collection at the Lyman House museum. The time for rest and relaxation was a pleasant diversion. Remaining was the 2000 mile sailing leg from Hawaii to California, and the variety of challenging obstacles that it posed.

The first difficulty was that they were headed against the wind. The trade winds blow in the direction from California to Hawaii. This precluded Smale from following the geodesic course home. The desirable winds were the prevailing westerlies, lying farther to the north. However, to reach the westerlies it was necessary to traverse an intermediate region known as the horse latitudes where the movement of air is light and variable. Compounding these difficulties, a large elusive high pressure system heavily influences the dynamics of the wind in the north Pacific. At the center of the high it is calm, as in the doldrums. Adding another element of adventure was the possibility of a tropical storm. Smale monitored weather reports, carefully noting the locations of the high and any tropical depressions.

Boats cannot sail directly into the wind. At best, most cruising boats can sail at about a 45° angle into the wind. This point of sail, called a beat, is physically demanding because it forces the boat to angle into the waves that are formed and driven by the wind. Traveling in this way is a jarring experience for both the boat and crew. However rugged, it was the only viable option for the initial portion of the trip. When *Stardust* departed from Hawaii on the morning of July 18, the basic plan was to head north and eventually turn right, dodging the high pressure system.

For each of the first six days *Stardust* beat about 100 miles. Tropical storms posed no immediate threat, and the location of the high was dancing around to the northwest. On the seventh day, *Stardust* reached the horse latitudes and progress slowed to 55 miles. They made the turn to California and began a northeastern course. With little wind and a large supply of diesel, Smale relied heavily on the engine. The high was positioned to the west and he expected the wind to pick up as the separation increased. On the way to a 144 mile day on July 28, Smale was confident that the plan was working.[30]

> *03:15 Under engine since 20:30 last night. But now the winds look very good and will put sails when someone else is awake. It looks like we are well past the high (we went under the high) and should expect good winds from now on.*

It was a reasonable expectation, but weather is not bound by any rules of engagement. On the next day the wind was weaker and the barometer was higher. The shock came with the weather report. Overnight the high had leaped toward them and landed just to the southeast. Since the air was calm and there was fuel in the tank, Smale leaned on the motor. As the diesel burned, the high seemed to stalk *Stardust*. On the afternoon of July 30 Smale knew that the fuel was practically gone. At 15:26 he wrote: "The high still has us in its grip. Wind speed all day about zero. The engine still serves us well." Two hours later it stopped. Out of fuel, they had failed to shake the high and were still 767 miles from San Francisco. They were not much closer on the morning of August 1 and Smale seemed discouraged:

> *On this day 14, it looks like our progress to S.F. will be a bit more than 60 miles, which is terrible. On the other hand it all came from beating into gentle winds or calms (thus no spinnaker). Thus it doesn't look so bad. One can't really estimate an arrival in S.F. because of the unpredictability of the high.*

The weather report introduced a new worry. There was a gale to the northeast, the direction of their course. Early in the morning of August 2, Smale observed that the winds were increasing and the barometer falling. Had they finally escaped the high only to get clobbered by the gale? Attention shifted to following the position of the storm. As the gale remained a threat, *Stardust* picked up the pace, making good 125 miles on day 16. It was tough going, beating into a strong wind with substantial seas. The next three days averaged 150 miles each and carried them to the Farallon Islands. Within 30 miles of the Golden Gate, Smale made his last log entry at 07:44 on August 6: "Certainly expect to arrive Berkeley this afternoon. Have been almost becalmed for a couple of hours outside Farallons." It was a bit optimistic. No wind materialized and as night approached they were contending with heavy fog and big ships. If only they had saved some fuel. The radio transmitter had a range of about 20 miles. It was of little use in the open sea, but was now sufficiently strong to reach the Coast Guard. Steve ordered a few gallons of diesel and they delivered. At 21:40 *Stardust* passed under the Golden Gate Bridge. The trip from Hawaii to Berkeley was just under 20 days.

Looking back Pugh said "it was a heck of an adventure. I am so grateful to have been able to do it … As well as conceiving the trip, what I admire Steve remarkably for is the organization of it. His seeing how it should work, bring the right stuff. I think we very rarely had a time we said well I wish we had brought this and it was in some sense predictable that we should have brought it."[31] Steve had successfully designed, planned, and executed the most ambitious and exotic sailing mission that could be packed into a summer. It was another stunning accomplishment. With the goal attained Smale lost interest in sailing. Going across the Bay became anticlimactic. Within a year he sold *Stardust*. There were parallels to solving the Higher Dimensional Poincaré Conjecture and leaving topology.

Smale accepted risk as an investment in accomplishment. The nature of the risk varied with the endeavor. It was physical on the sailing voyage, financial with minerals, and intellectual energy in mathematics. The opportunities were not chosen recklessly, but always with great care and measured against the potential and magnitude of the payoff.

With Steve's approach to mineral collecting, the risks could be physical as well as financial. Reaching the most promising sources required traveling to inaccessible parts of the world. Pakistan provides a good example.[32] Superb mineral specimens have been found in the north, a remote region of mountainous wilderness. The material somehow makes its way through a network of local dealers to the city of Peshawar in western Pakistan near the Khyber Pass. Smale was interested in buying minerals in Peshawar, but first he wanted to deal directly with the agents near the mines. Getting there was a nontrivial problem. In 1988 a mathematical engagement placed Steve in India. From there he made his way to Islamabad where he hired a car and driver for the trip north. Carrying a great deal of cash, Steve set out along the Karakoram Highway. The conditions were primitive and treacherous. The narrow road wound around mountains with washed out portions exposing travelers to steep drops. By the time Steve reached the city of Gilgit, the strain of the trip had introduced second thoughts. He tried to arrange for air passage back to Islamabad, but was unable to book a flight.

Pushing further north toward the Chinese border Steve saw some spectacular scenery, but the mineral venture was failing. The road to one of the best mines was closed by an avalanche. Steve never did succeed in connecting with a broker. Returning to Gilgit, the plan was to take the Karakoram Highway back toward Islamabad and then detour to Peshawar for some serious mineral dealing.

Just south of Gilgit an unanticipated danger arose. A Muslim religious war was underway between the Shiite and Suni sects. The dispute was over the timing of the transition from the fast holiday of Ramadan to the feast of Ede. People were dying and it was a bad time to travel the Karakoram Highway. Men with guns blocked the road. Further ahead of Steve a car was in flames. Smale's driver managed to edge his way through the obstruction and eventually delivered Steve to his hotel in Peshawar.

The trip to northern Pakistan was an exception to Smale's characteristic success at risk taking. The danger was greater than he anticipated and the mineral objective was unrealized. Still there were great specimens to be had in Pakistan. In his subsequent trips Steve went directly to Peshawar where he purchased aquamarines and other fine minerals.

Other People

11

A. Family

Through his mathematics and political leadership Smale touched the lives and careers of many people. With his restrained demeanor, however, it is more difficult to evaluate the effect of others on Smale. Those close to him say that Smale can become highly emotional citing, for example, his deeply felt grief at the death of his father. This chapter takes a closer look at some of his more significant personal interactions.

For most of his childhood Steve lived in the isolation of rural Michigan. In his sparsely populated farm neighborhood, cultural influences largely defaulted to family life. While other children were imbued with religious and patriotic values, Steve and Judy saw their father reject such traditions. Lawrence Smale was a complete iconoclast. The activities that he promoted were intellectual: science, music, reading, etc. Steve acknowledges that his father was a "big influence" on his life,[1] but part of this was in developing the independence to have opinions of his own.

The next generation of Smale children also consisted of a son and younger daughter. The influences on Nat and Laura included a strong dose of Berkeley culture. There were other unusual elements to their upbringing as well. They participated in family trips abroad, seeing a great deal of the world at an early age. Their parents were gracious hosts for individual guests and groups. The Smale home was the setting for mathematics, mineral, VDC, and Black Panther gatherings, intermingling serious intellectual and political matters with some of the more free-spirited parties of the times. The diverse guest list included Jerry Rubin and Eldridge Cleaver as well as Raoul Bott and Bill Larson.

Parental roles followed traditional patterns of the time. Clara took care of the children and ran the household while Steve engaged in his work. The demarcation enabled Steve to focus all of his energy on his current interest as it varied between mathematics, politics, and minerals. Steve's primary presence at home was to work in his office or entertain associates. He refrained from fatherly activities such as playing ball with his son or tinkering with the family car. From Nat and Laura's perspective, Steve's interest in his children increased with their age (and his).[2]

Steve and Clara were permissive parents, with Steve taking the lead in matters of discipline. There were few rules, however, and little pressure on the children to perform at school. Their initial reactions were different. Laura was a conscientious student, but Nat was less serious. He viewed high school as the terminal phase of his education. Rock climbing and scientology captured his energy. Nat's involvement in scientology led to a major conflict with Steve. Eventually Nat drifted away from the movement and Steve came to compare it to his own youthful involvement with communism.

Nat's high school grades reflected his underachievement. When he failed algebra, Clara urged Steve to intervene. Steve refused to help his son with mathematics. This *laissez faire* approach to his children's educational progress should not be confused with indifference. Steve enjoyed vicarious pride in Laura's academic success, but he let Nat follow his own course. After all, Steve had no special loyalty to academia. He could well imagine himself in other careers. Nat seemed bright enough. If his interest and ability led outside academe, Steve could be supportive. As long as it wasn't scientology.

Following his graduation from high school, Nat took a job in a cafe. Steve began to consider investing in a restaurant for Nat to manage. However, Nat's perspective was changing. A friend from the coffee house business returned to college and reported a positive experience. After 1-1/2 years of work, Nat was psychologically ready for college. His academic preparation was another matter. Nat enrolled in a community college and began to address his high school deficiencies. A course in biology captured Nat's interest and he became a serious student, transferring to Berkeley to complete his undergraduate education.

The courses at Berkeley offered some surprises. Molecular biology was disappointing, but a challenging mathematics course opened a new direction. Just as Bob Thrall and Raoul Bott had inspired Steve at Michigan, Moe Hirsch's small honors section influenced Nat to pursue mathematics. He completed the major and remained at Berkeley for graduate school. Nat's mathematical field was in the overlap of geometry and analysis. For a thesis advisor he selected Rick Schoen, a rising star at Berkeley who was soon to lured to the University of California at San Diego. Nat completed his doctoral research in San Diego. After a few years of prestigious postdoctoral appointments, he settled in at the University of Utah and achieved tenure. Nat had taken an unusual path, just as his father, and became a research mathematician, just as his father.

Like her brother, Laura did not proceed directly to college after high school. She spent a year in Europe, but there was little doubt that college was in her future. In 1978 she enrolled as an undergraduate at Berkeley. Laura became interested in hormones and animal behavior. She remained at Berkeley to earn a Ph.D. in biological psychology. While both her parents were enthusiastic about her educational success, she recalls the differing nature of their interest. "My father always wanted to know what kind of grades I was getting, and my mother wanted to know what I was learning about and she would like to talk to me about it."[3]

Laura's career continued to advance, even as uncertainty in her personal life unfolded. Following a long term relationship and subsequent marriage to a man, Laura moved to a lesbian partnership. News of the change required some adjustment by the Smale family, in part because her husband had been well liked and accepted by Laura's parents and brother. Nevertheless the personal changes did not interfere with continuing professional success. Laura completed two years of postdoctoral research in New York, and then, with her partner Kay Holekamp, began a four year field study of hyenas in Kenya. Upon their return both joined the faculty of Michigan State University. Laura received tenure in 1996.

Steve loved to see exotic places. He made the most of his opportunities and enjoyed sharing them with his family. In 1960, when Nat and Laura were infants, they lived in Brazil. Over the next decade there were several residencies in Europe as well as trips to other parts of the world. When they were away from home, Steve seemed more present to Nat and Laura. As the children reached adulthood, Steve continued his ambitious travel regimen, often accompanied by Clara. In the summer of 1989 he put together an itinerary that integrated mathematics, minerals, adventure, and Laura's career: China, Hong Kong, Bombay, Kenya, and Pakistan.

Laura and Kay's home was a collection of tents in a Kenyan national park. It was the world of the Masai people and an extraordinary collection of animals. Laura and Kay learned Swahili to communicate with the people, but it was the social hierarchy of the animals that they came to study. Moving about in jeeps they observed the interactions of the hyenas, especially in the early morning and at night. Although the jeeps and Masai provided some protection from the lions, hippopotamuses, elephants, and other wild animals, it was a dangerous environment that would deter most Americans.

After obtaining provisions in Nairobi, Steve and Clara, moved into the guest tent. They were in the field for nearly two weeks. Steve thrived on the adventure, made more special by the role of his daughter in enabling the opportunity. Some days he drove a jeep by himself, counting the different mammal species encountered along the way. Other times he walked with Laura along the river, armed respectively with spear and machete.

It was the third consecutive summer with a dangerous endeavor: sailing in '87, northern Pakistan in '88, and lions in '89. After Kenya he went to Peshawar, opting against further risks along the Karakoram Highway. However he had inadvertently stepped into some excitement earlier in the summer. The Chinese leg coincided with the protests in Tiananmen Square. For several days Steve witnessed the building drama, recalling the emotion of the Free Speech Movement in Berkeley. The crackdown occurred on the day after his departure for Hong Kong.

Steve's desire for physical challenges continued beyond the age of sixty. During the summer of 1992 he joined an ambitious family outing in the Zion National Park. While Clara looked after the grandchildren Steve, Nat, and Hermine (Nat's wife) undertook the 15 mile hike through the Narrows of Zion Canyon. Distance was just one of the obstructions to enjoying the magnificent scenery. "This wading route follows the Virgin River through the spectacular Narrows of Zion Canyon. Here, fluted walls rise for a thousand feet, crowding the waters into a narrow channel that twists and turns far beneath a thin slot of sky. Although the gradient of the hike is moderate, the near-constant wading over slippery boulders makes the hike a wearying undertaking."[4]

B. The baby in Peru

Steve had an uncanny knack for appearing on the scene as events transformed other people's lives. Coincidence placed him at the Free Speech Movement, Tiananmen Square democracy demonstrations, and Muslim war near Gilgit. In each case he was a fascinated spectator who walked away with a great story to tell. During the summer of 1981 a tragic circumstance connected the lives of Steve and Nat with an unlikely Peruvian family. It all began innocuously enough.

Steve was invited to participate in a mathematics conference in Brazil. Of course he would use the opportunity to reach some South American mineral sources, but there was an additional aspect to the trip. Nat had just completed college. To celebrate his graduation, Nat accompanied Steve for six weeks of travel through Peru and Brazil. It was to be just father and son. First they flew to Lima.

The main attractions in Peru were the mountains, jungle, and minerals. Following a week in the capital city they were ready to see the Andes. Both Steve and Nat were climbers, but they planned to restrict their activity to hikes rather than technical ascents. Peru's highest peak, Mount Huascarán at 22,205 feet, is located near the mountain city of Huaraz. Just a few years earlier a road was completed over the 200 mile distance from Lima to Huaraz. Steve and Nat made the drive north to Huaraz in their rented Volkswagon.

In Huaraz they obtained a hotel room from which they would base their excursions. It is a magnificent locale for mountain lovers. A road from Huaraz follows along a high valley with Mount Huascarán and other snow capped peaks on the east side. To the west are smaller mountains. On the morning of June 27 Steve and Nat set out with Steve at the wheel. The trip would take them through the town of Yungay.[5]

Among the residents of Yungay were Mercedes Girhaldo Chimchay and her six-month-old baby Sabino. Mercedes was a 36-year-old illiterate homemaker who spoke the Quechua Indian language. While Steve drove along the road toward Yungay, Mercedes carried her son to the bus stop. There they waited on the west side for the bus to Huaraz. As the sun rose and the temperature increased, Sabino began to cry. Seeing shade on the opposite side of the street, Mercedes took her baby to a more comfortable spot. A few minutes later the bus appeared to her right. Mercedes reacted by running across the street to the bus stop. Unfortunately, she neglected to check for traffic to her left. Steve arrived at that very instant. He swerved, but was unable to avoid hitting her. She fell to the pavement with Sabino in her arms.[6]

Steve stopped the car and returned to the scene. The mother and baby were seriously injured and in need of medical attention. Steve and Nat helped them into the car and drove to a hospital in Huaraz. Gradually they learned that Mercedes had sustained a broken arm. Sabino's status was more grave. His skull was fractured, and later that afternoon he died.

On top of the horrific experience of killing a baby, albeit accidentally, there were legal issues to settle. What were Steve's criminal and civil liabilities? The Huaraz police appeared at the hospital to begin their investigation. They took possession of Steve's passport and car keys, but did not incarcerate him. To improve communication a young Mormon missionary was enlisted to serve as Steve's Spanish-English interpreter. It was then determined that jurisdiction belonged in a town closer to the accident site. The case was transferred over to Sergeant Dionicio Sotomayor, commander of the Anta civil police.

Steve was uneasy with the prospect of placing his fate in the hands of the local authorities. Aside from the cultural differences, language difficulties, and normal ambiguities associated with an accident, there was ample reason for concern. The political situation in Peru was unstable. In 1980 democracy returned following 12 years of military dictatorship. Under the new freedom, several political parties were struggling for power. Working outside the political system was a revolutionary movement known as Shining Path.[7] From its base in the south Peruvian Andes, Shining Path was forcibly spreading a Maoist ideology throughout the country. According to their dogma a wealthy American intellectual was a class enemy. The problem for Steve was that there was no way to be certain of the justice system he was approaching.

The Mormon missionary had become a supporter who would remain with Steve throughout the ordeal. There were few other resources available to Smale. He decided to rely on the American embassy only in his contingency plan. First he would defer to Sergeant Sotomayor. If things went badly then Nat was instructed to contact the embassy.

Sotomayor proceeded expeditiously, taking depositions from Steve, Nat, and Mercedes. The stories were consistent and there was no evidence that Steve was speeding. Sotomayor concluded that Mercedes had caused the accident by carelessly bolting into the street. Steve was absolved of criminal responsibility, however he had a civil obligation to make reparations to the family of the victims. It only remained to reach a mutually agreeable sum. Everything proceeded amicably until the father of the baby returned to town. Steve recalled that he was "venal" and wanted as much money as he could get.[8] They settled on a few hundred dollars. On June 30, Steve and the father appeared in court to formalize the agreement. Steve and Nat were then free to resume their tourist activities.

Despite the legal exoneration, Smale was left with the trauma of having ended a young life. During a 1997 interview about the incident, Steve assumed a quieter tone and a rare hint of emotion crept into his voice and face.[9] He denied, however, that the accident has had much of a long term impact on him, noting that he does not dwell on the past.

C. Students

Mike Shub was born in Brooklyn in 1943. Undergraduate education at Columbia in the early sixties brought Shub into contact with Smale and his mathematics. It was the beginning of a significant mathematical, political, and personal relationship. Both left Columbia for Berkeley in 1964 where Shub began graduate school and Steve rejoined the faculty. Shub and another student drove the Smale car across the country. At

Berkeley Shub participated in Steve's seminar on dynamical systems. Despite these interactions, there was no expectation for either that the relationship would evolve into thesis supervision.

At this stage of his career Smale was yet to officially direct a thesis. Recall that Spanier received the credit as Hirsch's advisor. Following Chicago, Smale moved from IAS to IMPA to Berkeley to Columbia (with time in Lausanne) and back to Berkeley. He was hardly at any university long enough to attract a student. During the first year back at Berkeley, social and political issues drew Shub and Smale together. Shub was a foot soldier (his words) for the Free Speech Movement and Steve was a faculty supporter. When Smale assumed a leadership role in the Vietnam Day Committee, Shub was a willing recruit. All the while Shub and his wife socialized with Steve and Clara.

Shub needed a thesis project and, like Smale a decade earlier, had some hopes of proving the Poincaré Conjecture. The problem remained open in dimension three and (at that time) in dimension four. Meanwhile, another graduate student solicited Steve for a topic. Smale offered a dynamics question that the student related to Shub. When the student decided to pursue a different direction, Shub asked Steve if he might take up the problem. It was a reasonable question. Good problems are a precious commodity that some mathematicians are reluctant to share. Smale may have wished to work on it himself or perhaps he had already given it to another student. Smale's response was revealing. "Nobody owns any problem."[10] Shub became Smale's student during the 1965–66 academic year.

There are no rules to specify the level of interaction between a thesis student and his or her advisor. Some might schedule weekly meetings while others might get together as the need arose. In one extreme case a Ph.D. student spoke twice to his advisor, first to agree to the relationship and then to hand in the thesis. Smale adopted his typical *laissez faire* approach. Students were welcome, but it was up to them to initiate the contact. Given Steve's mathematical prestige and his lack of facility at small talk, some students could be intimidated by the prospect of going to his office. Shub, however, had enough confidence in his mathematical and social skills to be a frequent visitor. Even so, there were some boundaries.[11]

> *I used to stop by at Steve's office almost daily to say hello and tell him anything new or ask questions. He was always happy to see me and to hear anything new. But he wasn't too interested in technical details or vague ideas. When he started biting his lower lip, I knew it was time to go.*

Without realizing it, Shub had gotten in on the ground floor of a rapidly developing enterprise. Nancy Kopell received a problem from Smale at about the same time as Shub. Jacob Palis came to Berkeley from Brazil and began to work on proving Smale's conjecture that Morse–Smale diffeomorphisms are structurally stable. All of a sudden Smale had three excellent graduate students working on pieces of his program in dynamical systems. Others would soon follow. When Steve left for Europe in the summer of 1966, there was sufficient momentum for the students to organize a dynamical systems seminar.

Both Kopell and Shub received their Ph.D.s in 1967. They were Steve's first official students and he had a responsibility to authenticate the theses. In that period, prior to word processors, just producing the final copy of the thesis was an ordeal. Corrections often necessitated retyping entire pages and insertion of mathematical symbols required special attention. Shub's wife typed the copy that he presented to Steve for his approval. When Smale returned it, his comments merely consisted of some minor grammatical corrections. This left Shub with two concerns. He was reluctant to request that his wife retype the pages. Smale quickly allayed that fear by stipulating that the corrections could wait until the manuscript was typeset for publication. Shub's other concern was with the validity of his results. It appeared that Smale had skipped over the mathematical substance. When pressed, Steve conceded that he had neither read nor did he intend to examine the paper in detail. The conversation proceeded along the following lines:[12]

> ***Shub:*** *I hope it's correct.*
>
> ***Smale:*** *You gave a seminar talk about this right?*
>
> ***Shub:*** *Yes.*
>
> ***Smale:*** *People listened to it?*
>
> ***Shub:*** *Yes.*
>
> ***Smale:*** *Well time will tell. If it's important then people will read it. The errors will come out.*

Smale's cavalier attitude toward certain formal aspects of his job belied his intense desire to discover the theory underlying dynamical systems. Ever since meeting Peixoto in the summer of 1958, Smale had steadily refined his own vision of the subject. He incorporated these ideas into the paper *Differentiable Dynamical Systems* which appeared in 1967, but was available to his followers the previous year. That one remarkable paper contained all the background, direction, and problems necessary to initiate a talented mathematician into the new subspecialty, and Berkeley then was awash with bright young mathematicians.

A powerful group was forming around Smale. Joining from the faculty were Charles Pugh and Moe Hirsch. Pugh had arrived in 1964 as an assistant professor trained on the analysis side of a related area. Not only did he and Hirsch prove major theorems, but they were highly accessible and helpful to the graduate students. Berkeley graduate students responded by reading Smale's paper and choosing to work in dynamical systems. It was an exciting opportunity for a graduate student to engage in the creation of a new field. John Franks, who was initially turned off by the complexity of a Smale lecture on the horseshoe, became Steve's student. At the Smalefest in 1990 Franks reflected on the Berkeley atmosphere.[13] "Anyway, with this large group of people, mostly graduate students, it was a heady time, but we didn't really realize it, I think. Being naive graduate students, as he said, we assumed this is the way it always was for students in graduate school."

Palis and Franks were among the staggering number of seven students who received Ph.D.s under Smale's direction in 1968. Three more finished the following year and then Rufus Bowen and John Guckenheimer in 1970. Smale had 14 Ph.D. students graduate over a four year period. Fourteen is an extraordinary number of students, and it was amazing that Smale managed to oversee so many projects while proving his own theorems and starting to collect minerals. That all these people selected Steve was a tribute to his knack for formulating interesting problems, especially since some other advisors spent more time with their students and offered more help in finding them positions. At that time, however, the academic job market in mathematics was still good, prior to its collapse in the early seventies, and students were not overly concerned about finding employment. Moreover, it was fun to be part of Steve's group.

If there were anything more striking than the quantity of Smale's mathematical offspring, it was their quality. Shub and Kopell received positions at Brandeis and MIT, respectively, and then went on to outstanding careers as research mathematicians. In 1990 Kopell was selected for one of the Macarthur Foundation's so-called genius awards, providing five years of salary without any strings. Shub settled at the IBM Watson Research Center. Smale's subsequent students from that period enjoyed comparable success. Palis returned to Brazil and became director of IMPA and president of the International Mathematical Union. Bowen was a professor at Berkeley and entering mathematical stardom when his life was tragically ended by a cerebral hemorrhage at the age of 31. At Northwestern, Franks continued the Smale tradition of discerning beautiful structure in dynamical systems, and Guckenheimer, at Cornell, was innovative on the applied side.

Maintaining Smale's 1967–70 pace of thesis supervision would have been inconceivable, but he continued to be one of the most prolific advisors among mathematicians. As of this writing (in 2000), the number of his doctoral students was in the midforties. Especially noteworthy in the later generations were Bob Devaney, David Fried, and Jim Renegar. By a curious coincidence, Moe Hirsch's son Michael solved a Smale problem for his doctoral thesis while officially being directed by Berkeley topologist Rob Kirby.

Out of the Smale school of dynamical systems came a large number of collaborations. Fundamental papers were coauthored by Palis–Smale, Hirsch–Pugh–Shub, and Bowen–Franks. One of the most enduring partnerships commenced in the eighties when Shub and Smale joined together to work on problems in the theory of computation. With Smale living in Berkeley and Shub in New York, the bicoastal collaboration depended on periodic visits and sustained independent activity. As the team produced a steady stream of deep papers, the former mentor and student became close friends and colleagues.

D. The Jenny Harrison Case

In 1978 Smale strongly supported the appointment of mathematician Jenny Harrison to a tenure track assistant professorship at Berkeley. When the department voted to deny her tenure eight years later, Smale was among the abstentions. Harrison lost her appeal in 1989 and filed a civil lawsuit against the University alleging gender discrimination. It was an issue ripe for an explosion. No topic is more sensitive to academics than tenure. Employment data showed a striking lack of women in tenured positions at Berkeley and the other leading mathematics departments. As Harrison's case dragged out over the next several years it generated considerable interest among mathematicians and became one of those rare mathematical stories that receives press attention. The following paragraph is taken from a lengthy article about the case that appeared in the *Los Angeles Times Sunday Magazine* in 1993.[14]

> *According to Harrison, her "official mentor" was Smale, at least early on. But Smale had distanced himself from her after a personal relationship between them crumbled. Someone wiser might have weighed the win-factor of getting close to one's married sponsor and decided against following her heart. When it came time for her tenure hearing, Smale abstained from voting, saying that he "didn't feel strongly one way or the other" about her qualifications. "I don't think my opinion of her was based on personal things at all," he says. And he didn't "support her later on," he says, "because she didn't seem as exciting a mathematician as she seemed originally."*

Both Smale and Harrison refuse to confirm or deny that they had an affair.* However, speculation about their relationship and its pertinence to Harrison's tenure denial added a new dimension to Smale's notoriety. The case itself is difficult to evaluate from afar. To gain a fully informed opinion requires an assessment of Harrison's research and an audit of the procedures employed in her evaluation. With the obstructions of the vertical nature of mathematics and the confidentiality of the proceedings, even an outside mathematician is challenged to determine the merits of the case.

Harrison received her Ph.D. from the University of Warwick in 1975. Her thesis was a strong piece of work that was published in *Annals of Mathematics*, an outstanding journal.[15] As a highly regarded new Ph.D., Harrison spent the next three years in a sequence of

* Smale has a reputation among mathematicians for engaging in extramarital relations. In a 5/19/96 interview he stated "I have had relations outside my marriage with other women," but adds that "some sort of exaggerated picture" may have resulted from his prominence.

prestigious postdoctoral positions at the Institute for Advanced Study, at Princeton University, and then at Berkeley on a Miller Fellowship.

In the seventies Smale and other Berkeley mathematicians campaigned to appoint women to the faculty. By 1977 they had little to show for their intentions. Only two women, Marina Ratner and Julia Robinson, held positions, and aspects of each appointment raised questions about Berkeley's commitment to gender equity.[16] When Harrison arrived on campus to begin her Miller Fellowship, she was an obvious person to consider for a faculty position. She had recently obtained another interesting result (on ergodic diffeomorphisms) and her research program looked promising. Then Harrison announced that, in collaboration with Colin Rourke, she had a counterexample that resolved what remained of the Seifert Conjecture, a major problem in the interface of topology and dynamics. Although both of her projects awaited authentication, it appeared that Berkeley had the opportunity to hire a gifted young scholar who both met their standards and was a woman. A difficulty was soon discovered in the Seifert solution, but Harrison then claimed that the work could be rehabilitated to a still significant result. Recall that Smale had to close a gap in his original proof of the Higher Dimensional Poincaré Conjecture. Steve continued to push for Harrison's appointment. The offer was extended and accepted.

Both of Harrison's recent results collapsed.[17] After a year at Berkeley she was given a leave of absence to accept a position at Oxford University. Harrison remained in England for three years and then returned to Berkeley in 1982, announcing another Seifert counterexample. It took a long time for the result to pass the scrutiny of the mathematics community. In late 1985 Harrison submitted a revised manuscript that was accepted for publication. During this period she married Berkeley dynamicist Charles Pugh.

With the Seifert work accredited, the tenure process began. A departmental review committee was appointed and letters of evaluation were solicited. Harrison was in her eighth year on the Berkeley faculty and in the eleventh year since her Ph.D., significantly further along in her career than other mathematicians who had been considered for tenure at Berkeley. There were at least two other unusual aspects of the case, her gender and the failure of her earlier projects.

Harrison's research portfolio featured her thesis and the Seifert counterexample. These were two significant results and worthy of tenure at a very good place. The question was whether it was enough for Berkeley, one of the top few departments in the country. Opponents could say that after eight years she had no more results than were the basis for her original hiring. Supporters could point to the new elements in the substance of the successful Seifert, and give her additional credit for her perseverance.

Smale was not a member of the review committee. For its recommendation to the department on Harrison's tenure, the committee vote was one in favor, four opposed, and one abstention. Moe Hirsch voted affirmatively. The tenured faculty met on March 27, 1986, to discuss the case, and each member was faced with making a decision. One could supplement direct knowledge of Harrison's work by considering letters of reference, the report of the review committee, and opinions of colleagues. Smale "didn't think that the case was clear enough one way or another" and decided to abstain.[18] Steve adds that he abstains in many matters. For example he does not vote in presidential elections. When asked whether his decision to abstain was based in part on whatever personal involvement he may have had with Harrison, Steve responded:

> *I think it's fair to say that just on the objective merits of the case, if such be. You know I thought a lot about it at the time. I spent a lot of time trying to think about what I should do. I was focusing on the case as it stood. Did it justify supporting her in that situation. I looked at her record. I*

think the decision you could really say was mixed on what I thought of the case. It's hard to estimate all the friendships and conflicts and how they bias one's decision. One can make lots of rationales for these things, I'm conscious of that. I don't pretend that there isn't some kind of effect on me. I think it's fair to say that I did try hard to look at what was justified.

Given Smale's stature in dynamical systems and his role in Harrison's appointment, it would have been natural for others to have looked to him for guidance. While Smale was satisfied that personal involvement was not consciously a part of his abstention, department members only knew that Harrison lacked his endorsement. Steve made no disclosure to his colleagues of any personal factors that may have subconsciously impacted his analysis. When the faculty voted on Harrison's tenure, there were twelve in favor, nineteen opposed, and seven abstentions.

The departmental vote is a crucial determinant in the outcome of the tenure process, but at Berkeley there are several subsequent stages of faculty and administrative review. Harrison still had some hope of overturning the department's recommendation. To succeed she needed all the support she could muster. Pugh requested that Smale write a letter for Harrison. Smale complied, providing Harrison and Pugh with a copy. Harrison then requested that Steve withdraw the letter.[19] The tenure review continued into 1987 with Harrison losing at every stage. The final step of the formal process occurred in April when Chancellor Ira Heyman wrote Harrison that the decision on her tenure was negative.[20] That summer Smale and Pugh tabled their discussion as they sailed to the Marquesas.

In 1988 Harrison filed a complaint with the Committee on Privilege and Tenure (P&T), a university faculty committee that hears grievances on tenure and other matters. A description of the committee's mission is excerpted below. Richard Abrams is a history professor who served as chair during a period following the Harrison case.[21]

> *P&T is not charged with examining the substance of departmental tenure reviews, but only with finding out whether or not procedures were properly followed (this includes examining whether nonprofessional or nonacademic criteria entered into a decision about a case). Abrams says that because the procedures for deciding tenure are complex and have many steps, it often happens that faculty unwittingly break the rules. Such errors may or may not affect the outcome of a tenure decision. Abrams says P&T tries to "get to the heart of the matter" and judge whether mishandling occurred, and if so, whether it had any substantive effect on the case. By contrast, in a court of law, a technical infraction, such as when a police officer does not read the Miranda rights to a suspect, often means that a case gets thrown out of court.*

Harrison's complaint alleged that procedural errors and gender discrimination were factors that contributed to her tenure denial by the Department of Mathematics. The P&T inquiry was frustrating for both sides. Many aspects of the tenure review were confidential and Harrison could not be certain of all that had taken place. She hired an attorney and attempted to make her case, but the rules were different than in a court of law. Members of the department were trained as mathematicians rather than administrators. Nevertheless, these faculty members carried out activities such as soliciting outside letters, writing summaries, and serving on committees that were services to the University, requiring time and energy that might otherwise have been devoted to research. Now those contributions were being examined under a microscope and impugned as bigoted. Moreover Harrison singled out several members of the department and accused each of making specific statements that were derogatory of women.

P&T held hearings and devoted considerable time to the complaint. Out of an examination of Harrison's procedural challenges, the committee found one technical violation and some ambiguities, but concluded that there was no substantive effect on the case and its outcome. To consider the allegations of gender discrimination, P&T questioned faculty who Harrison alleged to have made sexist statements. It was an exercise that was certain to arouse bad feeling. The committee determined that "Those accused of the statements either denied making them or provided a context or plausible explanation for the statements from which this committee deduces that gender discrimination neither existed nor played a role in the case. Many of the statements of alleged bias were taken out of context, from discussions when an individual was explaining his or her observations or views of societal events or was freely explaining and expounding on views of affirmative action."[22] The committee of three men and two women went even further, criticizing Harrison's motives and credibility. "We are convinced that there was no connection between the decision to deny Dr. Harrison tenure and gender discrimination. We find it unfortunate that Dr. Harrison has concluded that any one who does not believe that she warrants tenure is biased, and that he or she could not have a legitimate reason for a negative view and position."

The 1989 P&T report was confidential and would not become public until it emerged later in a court proceeding. The mathematics department and its supporters point to the findings as a complete vindication. Harrison characterized the P&T hearing as a "kangaroo court."[23] Between these polarized views there is ample room for more modest conclusions. P&T's thorough report was largely limited to the objections raised by Harrison at that time. It was not a blessing of the entire tenure proceeding.

Harrison was convinced that a male mathematician with her record would have received tenure. Shortly after losing the P&T verdict in 1989, she filed a civil lawsuit alleging sex discrimination. Harrison faced long odds. The University of California was a formidable opponent with attorneys on staff and the resources to hire additional legal muscle. Harrison herself had already run up a substantial tab by retaining counsel through the P&T proceedings. To make her case, Harrison intended to demonstrate that her credentials were comparable to men who had recently received tenure in the mathematics department. First she needed to overcome the University's claim of confidentiality over the relevant case files. There was every reason to expect that it would be a protracted, expensive process.

A breakthrough came in early 1991 when her attorney Anne Weills obtained access to the records of the eight male mathematicians who had been promoted to tenure during Harrison's time at Berkeley. Shortly thereafter Weills and Dan Siegel eased Harrison's financial burden by agreeing to take her case on a contingency basis. More backing was provided by the recently formed Support Committee for Jenny Harrison headed by Charity Hirsch, whose husband Moe continued his own efforts. A campaign was underway to mobilize public support. Harrison made her case in a variety of interviews. When Smale battled the NSF in the sixties, his Fields Medal convinced other mathematicians that his scientific credentials were beyond question. It was impossible for most mathematicians to evaluate the technical quality of Harrison's work. One of her strongest weapons became impeaching the Berkeley mathematics department's attitude toward women. *Science* ran an article in June 1991 with statistics showing that women were barely represented at Berkeley and the other leading mathematics departments. Amidst a variety of quotes about Harrison from her more zealous supporters and opponents was the following paragraph.[24]

> *Somewhere between the extreme positions is Stephen Smale, a mathematician at Berkeley and a winner of the Fields prize (math's equivalent of the Nobel Prize). Smale thinks Harrison was a "borderline" candidate for tenure. Although*

> *he says her case was "not very strong," he thinks she was better than at least one man who was tenured and that her relative merits with others were "arguable." Smale adds that he isn't sure whether improper procedures occurred in the tenure decision, but "the fact that she is a woman brings out all sorts of emotions and partisanships. Weaknesses get magnified—that's where sexism plays a role. Other people support her because she is a woman."*

Overall, Steve felt that the paragraph conveyed a reasonably accurate depiction of his views. With the last sentence, however, he did not mean to imply that Harrison's gender was the only consideration in some of the affirmative votes, but rather that it was a factor. Smale's comments undercut the portrayal of Harrison as a victim of sex discrimination. Smale said that the *Science* article "affected my life a lot." It was a "big bone of contention with Jenny and Moe" and he came "under attack" by them for what he had said.[25]

Later in 1991 Harrison was diagnosed with throat cancer and began to receive treatment. The following year her lawsuit reached the deposition stage. Smale was among the long list of Berkeley mathematicians selected by Harrison and her attorneys for this pretrial questioning. Was the overlap of Smale's personal and professional life to be a target of the Harrison game plan? The prospect of a hostile deposition was a serious concern for both Steve and the University. Smale spent a great deal of time with an outside lawyer who had been retained by the University to work on the Harrison case. In these discussions Steve learned that Harrison lawyer Dan Siegel had threatened that there would be "blood on the wall" at the Smale deposition.[26]

Prior to the date scheduled for Smale's deposition, the University and Harrison appeared to make progress toward a settlement. Steve's testimony was postponed, but then rescheduled when the negotiations stalled. These events repeated, without Smale being deposed, until a settlement was finally reached in March 1993. Harrison was to receive a fresh evaluation. The entire architecture of her customized review procedure was supposed to remain confidential. On July 1, 1993, Harrison was appointed as a tenured full professor to the Berkeley Department of Mathematics.

Harrison issued a press release declaring a victory over sex discrimination. Many of her colleagues were outraged by both the process and the outcome. They felt betrayed by the University. Having been cut out of the loop, the mathematicians were left to conjecture at both the rationale behind the University's decision to settle and the review procedures that had taken place. Some of the details of the process were leaked and appeared in *Science.*[27] As to their motivation, university administrators have been cryptic. Among the possible factors was a desire to avoid the publicity that might have resulted from testimony by Smale.

Smale the Mathematician

12

A. Depth and breadth of contributions

The ultimate measure of a mathematician is the depth and intrinsic interest of his or her research. By that standard Stephen Smale ranks among the foremost scholars of the past half century. Time and again he produced profoundly original work that set a course for others to follow. It is remarkable that his successes spanned such a broad range of areas. Substantive surveys of Smale's forays into differential topology, economic theory, dynamical systems, computation, nonlinear analysis, and mechanics are available in the proceedings of the 1990 Smalefest, a conference celebrating Steve's sixtieth birthday.[1] This section provides a brief summary of Smale's mathematical *oeuvre*.

Smale became serious about mathematics in 1953. At that time he was a graduate student in Raoul Bott's algebraic topology course course at Michigan. Bott himself had just learned topology as a postdoctoral scholar at the Institute for Advanced Study. He was eager to share the sophisticated concepts of fiber spaces and Morse theory. Steve became Bott's first thesis student and constructed a fiber space argument that led to the classification of regular curves on manifolds. The mathematical substance of Smale's thesis was an impressive piece of work, but the entire project went unnoticed. The results were mistaken as a straightforward generalization of the Whitney–Graustein theorem. Only Thom, and not even Bott, appreciated Smale's achievement in his thesis.

Smale's first academic position was an instructorship at the University of Chicago. There he taught elementary mathematics to undergraduates in the Hutchins College. Shortly after arriving at Chicago in 1956, Smale obtained his first famous result. Steve pushed the fiber space argument of his thesis to classify immersions of the sphere into Euclidean spaces of arbitrary dimension. Later he generalized even further, allowing the sphere to assume any dimension. However, it was a special case of the first result that attracted all the attention. Any two immersions of the sphere into three-space were equivalent. In a mathematical sense it was possible to turn the sphere inside out. The counterintuitive nature of the result fascinated the mathematical community. Finally mathematicians were reading Steve's proofs. Embedded in the paper was another surprise. Rather than giving a straightforward recipe for an eversion, Smale's fiber space development was so abstract as to provide little insight into a comprehensible geometric realization of the process. The combination of the amazing result and unlikely proof multiplied the accomplishment to launch Steve's career.

In his two years at the University of Chicago Steve continued to explore topology. He also had an opportunity to interact with René Thom and absorb the valuable tool of transversality. Next Smale went to the Institute for Advanced Study to begin a National Science Foundation Postdoctoral Fellowship. His residency at the Institute briefly overlapped with that of Brazilian mathematician Mauricio Peixoto. At the time Peixoto was stymied by a problem in dynamical systems. He understood the situation in two dimensions, but was unable to move the theory to dimension three and beyond. A mutual friend had suggested that Smale might have some ideas and Peixoto sought him out for a consultation. Steve quickly saw that transversality and Morse theory were key to Peixoto's problem. A mathematical theory for dynamical systems began to unfold in Smale's mind. If correct, he could give a reasonably complete description of the behavior of solutions to the differential equations that pervaded mathematical application. Smale was moving into an area of analysis, but restructuring its framework so as to admit his topological tools.

Steve was in Rio de Janeiro for the first six months of 1960, working at IMPA with Peixoto and others. During this period he discovered two extraordinary results. The first was in dynamical systems. Steve's initial paper on the subject had just appeared. Along with the theorems were his visions for future development. In Brazil Steve learned that his model failed to account for certain phenomena already established in differential equations. This type of setback is inevitable in mathematical research and can be discouraging. One of Smale's strengths was his capacity to respond constructively to data that contradicted his ideas. Steve set out to reconcile the literature from differential equations with his own concept for dynamical systems. He realized that things were more complicated than he first envisioned. Out of this analysis came his horseshoe example and a vital new element to the theory. The horseshoe is an innocuously constructed function that exhibits the essence of what would later become known as chaos.

The significance of the horseshoe was linked to the subsequent growth in popularity of dynamical systems and chaos. It would be years until its importance was appreciated by mathematicians. In contrast to the horseshoe, Smale's second Brazilian result sent immediate shock waves through the mathematical world. He proved the Higher Dimensional Poincaré Conjecture for all dimensions five and greater. Conventional wisdom in topology had dictated that the solution of this problem must be preceded by the three dimensional version which itself was regarded as the most important in topology. Not only had Smale solved a huge problem, but in successfully bypassing dimensions three and four, he completely altered the mathematical perception of dimension. In some sense dimension three was more complicated than five and higher.

Smale followed up in 1961 with his greatest result, the *h*-cobordism theorem. It was a tool of awesome topological power that made it possible to investigate the structure of higher dimensional objects. With the Poincaré Conjecture and *h*-cobordism theorem Smale had broken the dimension barrier in topology. Higher dimensions suddenly appeared tractable. At this point, rather than lead the assault along the mathematical trail that he had blazed, Smale shifted his focus back to dynamical systems.

Smale's theory for dynamical systems evolved in a sequence of increasingly accurate approximations. The initial vision captured several crucial elements, but precluded the existence of the horseshoe construction. As Steve came to understand the horseshoe, he realized that it represented an essential trait rather than some pathological anomaly. His next iteration of the theory expanded the original picture so as to feature important aspects of the horseshoe. Intellectually he was demonstrating a scientific objectivity, embracing phenomena that disturbed his earlier beliefs. Smale's second approximation to the theory of dynamical systems was temporarily set aside in 1962–64 as he explored the mathematical world of infinite dimensions.

Smale made two important contributions to infinite dimensional analysis. In both cases he was able to identify a significant part of the infinite world for which there was a valid analogy to finite dimensions. First it was with Morse theory and then the mathematically appealing Morse–Sard theorem. Smale saw how to generalize these deep concepts to an even more abstract setting. Having made his mark on infinite dimensional theory, Smale returned his attention to dynamical systems.

In 1965 Smale constructed an example that forced him again to revise his beliefs about dynamical systems. However, other aspects of the theory were becoming increasingly solid. He laid out the third iteration of his vision in an influential survey article that appeared in 1967. It was the blueprint for a new mathematical subspecialty. The centerpiece of the article was the Ω-stability theorem, resolving an enormous

Steve and Nobel Economics Laureate Gerard Debreu.

piece of the theory. Smale's proof was a *tour de force* that showcased his new techniques. Even with the Ω-stability theorem, much work remained. To highlight pieces of his program in need of development, he annotated the survey article with a wide array of research problems. The exposition together with the problems served as a primer for mathematicians entering the field, particularly the flock of Berkeley graduate students who became Steve's advisees.

By the late sixties Smale had attracted a cadre of mathematicians to implement his theory for dynamical systems. He then turned to applications of the theory. Unlike others who viewed pure and applied mathematics as separate disciplines, Steve sought unity. For him the application of dynamical systems was a natural progression rather than a shift of interest. He saw that dynamical systems offered a model for the time dependent systems of mechanics. In particular, the n-body problem (understanding the movement of celestial bodies operating under mutual gravitational forces) was ripe for Morse theory. Electric circuit theory was another physics problem that Steve was able to reframe in the language of dynamical systems.

Smale considered a variety of mathematical applications, but economics eventually captured his full attention. It all began in the best tradition of academics. Gerard Debreu, a Berkeley colleague in economics who would receive the 1983 Nobel Prize, approached Steve in his office with some questions. Out of their discussions Smale began to see roles for Morse theory and dynamical systems in the study of general economic equilibria. To apply mathematical methods to an economic model often requires the stipulation of hypotheses that are motivated by mathematical, rather than economic, considerations. Conceivably, these hypotheses can be mitigated with alternative mathematical techniques. Smale became a student of economic theory. Then he recast the framework, retaining certain features and modifying others. By injecting mathematical elements never before applied to economics, he managed to remove the traditional hypothesis of convexity but still succeed in obtaining powerful conclusions. Smale's uncanny capacity to select and customize appropriate mathematical tools is one of the distinguishing features of his brilliance.

Smale's economics phase endured throughout most of the seventies. General economic equilibria was the thread that ran throughout Smale's economic research. A general economic equilibrium occurs when supply meets demand and can be viewed as a price vector for which the excess demand function (demand minus supply) assumes the value of zero. After gaining a deep understanding of the nature and structure of economic equilibria, Steve was led to seek constructive methods for their computation. Unlike the sphere eversion where he was content with having established existence, for economic equilibria Steve wanted an effective mechanism for locating the solution. What good were equilibria if the market were unable to reach one via price adjustments?

The mathematical problem of determining the *zeroes* of a function has long been studied within the field called numerical analysis. Newton's Method is a classical numerical algorithm for addressing the problem. The traditional Newton theory is local in nature, requiring some advance knowledge of the general proximity of a solution. In a paper that appeared in 1976,[2] Smale introduced his *Global Newton* approach. With Smale's globalization, almost any starting price vector initiates a mathematical economic process that converges to an equilibrium.

To make his economic analysis work, Smale was combining a variety of mathematical tools. Along with the Morse–Sard theorem and other techniques that he had already honed in dynamical systems, Steve began to mix concepts from numerical analysis. Numerical analysis involves the study of algorithms for the efficient solution of mathematics problems. It was Smale's first substantive encounter with the subject and, as usual, he sought to understand it in his own way. When he looked for criteria to compare competing algorithms, Steve decided that the subject lacked an adequate foundation for his investigation. By the late seventies a tug of war had begun for Smale's interest. Would he continue to focus on the motivating questions from economics or was he shifting to a self-contained analysis of algorithms?

In 1981 Smale published the state of his thinking on algorithms.[3] He placed the issues in an "environment" that included the diverse areas of algebra, economics, numerical analysis, and theoretical computer science. Using Newton's Method as a paradigm, he argued that the thrust of the existing theory failed to address the central problem. To gain some understanding of Steve's objections, consider the nature of Newton's Method. It is iterative in that it begins with a guess and proceeds through a sequence of approximations, each derived from the previous, hopefully converging toward a solution. The process terminates when it reaches an approximation that is within some negligible error tolerance of the actual solution. Calculus methods have shown that the asymptotic rate of convergence is fast, that is if the sequence ever gets close to the actual solution then it is effectively just a few iterations from high accuracy. Reaching the vicinity of fast convergence may or may not require many iterations. To Smale the genre of asymptotic results failed to address the fundamental problem of determining the *total cost of convergence.* How was the expected number of iterations related to the size of the problem? The answer would need to account for the fact that the algorithm might have vastly different levels of success with two problems of the same size. One feature of Smale's framework was to consider the potential problems as comprising a large space. The nature of his solution was to be statistical in the sense of determining that, outside some stipulated probability of failure, all problems can be solved at a reasonable cost.

Smale examined a variant of Newton's Method to illustrate the flavor of theorem he advocated. The space of problems consisted of complex polynomials with the degree determining the size of the problem. He proved that if zero were the initial guess, then, outside of a small probability of failure, the Newton iteration would lead to rapid convergence after a number of steps related to the degree and stipulated probability. The technical relationship was another polynomial, indicating that the cost of computation grew in a reasonable manner as a function of the problem size. It was a creative new analysis of an old algorithm. The proof was long and formidable, employing an eclectic mix of powerful tools.

While Smale's interest had shifted to the analysis of algorithms, the motivation from economics continued to exert its influence. A genre of mathematical problems that frequently arise in economics is known as linear programming. A typical example involves the selection of a market basket from a variety of different foods. Assume that the choice includes *n* commodities (rice, chicken, spinach, etc.). Each food carries a cost per unit as well as contributing nutritional benefits toward daily requirements such as vitamins and iron. The problem is to select a vector quantity of foods so as to minimize the total cost of the purchase while simultaneously satisfying all the nutritional needs. Linear programming problems with n commodities and m constraints suggest the use of matrices. Many years earlier George Dantzig conceived the simplex algorithm, which became the method of choice for solving linear programming problems. Theoretically it was known that the iterative simplex algorithm was reliable in that it would determine the solution after a finite number of steps. What was unknown was how the number of iterations related to the size of m and n. While substantial experience suggested that the number of steps was reasonable, it was possible to construct problems in which too many iterations were required.

It remained to reconcile the existence of worst case examples with the prevailing body of positive empirical data. Was the simplex algorithm truly fast or had it been lucky in just happening to confront problems it could handle? Smale sought to understand the conflict by determining the *average* number of iterations required to solve a problem. It was an ambitious task. Normally one thinks of an average of a finite number of possibilities. Calculus permits the notion of average to extend to values of functions over infinite domains. For Smale each linear programming problem corresponded to a point in an $mn + m + n$ dimensional space. Associated to the problem was a function whose value was the number of simplex algorithm iterations required for a solution. Even with Smale's clever framing of the problem, the complexity of the function together with the high dimension of the space would have deterred many mathematicians from proceeding much further. Smale succeeded in proving that, for any fixed number of constraints, the average number of simplex iterations grows reasonably with n (more technically it is sublinear). Steve's creative theorem provided a theoretical foundation that validated years of successful experience with the simplex algorithm.

In 1985 Smale returned to the *Bulletin of the American Mathematical Society* to set forth his ideas.[4] He examined several classical algorithms of analysis from his new perspective. Global efficiency rather than local asymptotics or worst case was to be the criterion for evaluation. He then asked, "What is the fastest way of finding a zero of a polynomial?" Smale used the question to highlight what he saw as a gap between theoretical computer science and numerical analysis.

Numerical analysis is the bridge between mathematics and scientific computation. With the extraordinary development of computational power and its proliferation, there was an increasing reliance on numerical algorithms. Smale felt that these problems eluded the grasp of theoretical computer science, leaving numerical analysis without a strong foundation. The algorithms of numerical analysis involve computation of real and complex numbers. The fundamental model of theoretical computer science was the Turing machine which operated on discrete quantities. Smale stated his "belief […] that the Turing approach to algorithms is inadequate for" dealing with root-finding algorithms. As he had done with dynamical systems, Smale boldly presented his vision as an agenda for development.

> *Although the definitions of such algorithms are not available at this time, my guess is that some kind of continuous or differentiable machine would be involved. In so much of the use of the digital computer, inputs are treated as real numbers and the output is a continuous function of the input. Of course a continuous machine would be an idealization of an actual machine, as is a Turing machine.*

> *The definition of an algorithm should relate well to an actual program or flow-chart of a numerical analyst. Perhaps one could use a Random Access Machine (RAM, see Aho–Hopcraft–Ullman) and suppose that the registers could hold real numbers. Then one might with some care expand the list of permissible operations. There are pitfalls along the way and much thought is needed to do this right.*

> *To be able to discuss the fastest algorithm, one has to have a definition of algorithm. I have used the word algorithm throughout this paper, yet I have not said what an algorithm is. Certainly the algorithms discussed here are not Turing machines; and to force them into the Turing machine framework would be detrimental to the analysis. It must be added that the idealizations I have suggested do not eliminate the study of round-off error. Dealing with such loss of precision is a necessary part of the program.*

Unlike some other mathematicians who were becoming interested in aspects of computer science, Smale did not become a hacker. His primary use of computing machines was for email. However he was fully engaged in a new program of designing a model for real computation. As always Steve found creative new uses for topology. In 1987 he began to answer his own questions as "On the topology of algorithms I" appeared.[5] That same year Lenore Blum and Mike Shub joined Steve to collaborate on refining the model. Their approach was to work over the algebraic structure of a *ring,* thus encompassing the bit world of Turing machines and the real-complex number fields of numerical analysis. In 1989 the three authors published "On a theory of computation and complexity over the real numbers: *NP*-completeness, recursive functions and universal machines."[6] In the general ring setting the fundamental structure (Turing machines) and problem ($P = NP$?) of theoretical computer science became special cases that assumed their places along with the real and complex number variations of numerical analysis. With Felipe Cucker on the team the ideas were more fully expounded in the 1998 book *Complexity and Real Computation.*[7]

As of this writing (in 2000), Smale is completing his second decade on algorithms and real computation, twice the length of any of his other mathematical phases. The recent work includes the signature Smale formulation of problems and creative transfer of mathematical tools to new areas. Despite his considerable effort, complexity and real computation has not captured the attention of the mathematical community as did Steve's earlier developments of topology and dynamical systems. Perhaps it is just too soon to evaluate.

B. What made Smale a great mathematician

What made Smale a great mathematician? Certainly he was endowed with a powerful intellect, but that alone does not account for the extraordinary results enumerated in the previous section. Any mathematician can name brilliant colleagues who have failed to make much of a mark with their research. Smale possessed a combination of personal and intellectual qualities that somehow propelled him to the distinguished rank of the top few mathematicians of his generation. No doubt each member of this class had a unique collection of attributes that accounted for his or her success.

The first twenty-five years of Smale's life provided no suggestion of singular mathematical talent. High school and college contemporaries remember Steve, merely, as "a bright guy." Among the graduate students working in topology at Michigan, Smale was overshadowed by Munkres. Even in retrospect it is difficult to detect an indication of vast intellectual potential. The most unusual attribute of Smale's early adult life was his ambition for travel. In the five successive summers following his sophomore year of college, Steve planned and executed trips to New York, East Berlin, San Francisco, Grand Teton National Park, and Argentina. Only the last excursion was incomplete, and then it took a civil war in Guatemala to prevent Steve from reaching his destination. The penchant for exotic travel revealed qualities of risk taking, audacity, drive, and confidence, all of which emerge in Smale's research.

One of the most astonishing features of Smale's work is the combination of depth and breadth. His modus operandi was to focus on a specialty for several years, solve some major problems, formulate others, and then move on. How did Smale manage to surmount the vertical barriers that obstruct others? A key ingredient was Steve's approach to learning. For most mathematicians learning confers a conventional wisdom. Certain techniques and tools are part of an established orthodoxy that the scholar implicitly adopts in following the literature. An overwhelming amount of mathematical research fits into a pattern of understanding previous work and pushing it a bit further. At the other end of the learning spectrum is the skeptic who accepts nothing from others and must create everything. This reinventing-the-wheel approach is extremely time consuming and rarely successful. Smale possessed an independent insight that permitted him to benefit from the work and ideas of others, but rise above the blind acceptance of conventional wisdom. He understood things in his own way, sometimes following the established path and other times striking out on his own.

Unlike his father who was an iconoclast, Steve respected conventional wisdom. He also knew that it was dynamic and required evaluation. Making such judgements in the abstract world of mathematics is extremely difficult and involves risk. Steve possessed the requisite audacity and confidence as well as the intellectual power to develop his own approach and make it work. Once convinced that he was moving in the appropriate direction, Smale worked relentlessly to turn ideas into conjectures and conjectures into theorems. Energy, focus, and intelligence were among his weapons. Finally, although Smale had enormous confidence in his chosen perspective, he understood that it might still be just an approximation to truth. When confronted with evidence that he was wrong, Smale embraced the new information and used it to adapt his approach.

Smale's development of dynamical systems provides insight into his genius. Steve did not just take Peixoto's work and push it a little further. Rather, he understood its essence and reconfigured the definitions so as to apply to all dimensions. Then he proposed how the pieces might fit into a theory. Some of his ideas were correct and some were not, but the intellectual leap was of staggering proportions. When later faced with contradictory data, Smale neither wasted time in denial nor abandoned his original vision. He had the confidence to believe that he was on the right track, but understood that he might not have the entire picture. The new information was used as an asset to rehabilitate his theory. Most importantly, Steve gained a deeper understanding of the forces at work and identified profound new ingredients to incorporate into the model. Further corrections would be needed, but Smale's prescient insights at each stage were truly remarkable.

Steve was driven to succeed and certain that he would. A common thread that runs through Smale's mathematics, mineral collecting, political activism, photography, and sailing is his ambition for greatness. Most people will not reach the top class in any of their endeavors, but Smale was there in practically everything he undertook. Part of his motivation was the exquisite feeling of personal satisfaction derived from proving a difficult theorem or acquiring a beautiful mineral. Other ingredients were the desire for money and prestige. Smale's reticent manner belied his intense competitive spirit. He was interested in his standing and enjoyed his many awards. While Steve would say that prizes were not his objective, he was well aware of their existence, particularly ones for which he felt he was long overdue. The National Medal of Science is annually awarded to about ten scholars. It is the United States' highest honor in science. Smale was disturbed as he was passed over each year. Finally in 1996, thirty years after the Fields Medal, Steve received the National Medal of Science.

There are many smart people who work hard and aspire for success. What really distinguished Smale as a mathematician was the combination of his independent insight and his courage to focus on high risk ventures. On his ascents through the vertical world of mathematics, Smale somehow saw where to depart from the established path. Whether he constructed an entirely new route or forged a

Clara watches President Clinton award the 1996 National Medal of Science to Steve.

Clara, Steve, and daughter Laura at banquet honoring National Medal of Science winners.

fruitful connection to a previously distant theory, Smale succeeded in implementing abstract concepts in ways that nobody had ever thought of them before. Once his arguments were understood, they became part of the conventional wisdom, even seeming natural to some mathematicians. Each, however, was a huge departure from the existing theory and it was Smale who saw how to execute it.

C. Berkeley to Hong Kong

Smale was rewarded for his theorems with a high salary and low teaching load at Berkeley, one of the world's finest mathematics departments. His office provided a spectacular view of the Golden Gate. With outstanding graduate students under his direction, Steve had attained the benefits commensurate with his lofty academic status. His influence over the direction of the department was mixed. He was unsuccessful as a candidate in departmental elections for the position of chairperson. In the vital area of hiring, some of Steve's nominees were offered positions at Berkeley while others were not.

Overall both Smale and Berkeley benefited from their long association. The University contributed a stimulating environment and Steve brought scholarly prestige. For many years Steve's Fields Medal was a unique distinction in the mathematics department, although there were other departments that could boast of having Nobel Prize winners. To reward Nobel laureates Berkeley created a perquisite that was a striking blend of utility and irony, special reserved parking places. Given the keen competition for spaces on the burgeoning campus, it was unclear whether the prize or the parking place was more coveted. In 1990 Vaughn Jones became the second Berkeley mathematician to receive the Fields Medal. Following the award the University made a policy decision to elevate Fields Medalists to the parking status of Nobel laureates. Steve then received a reserved parking space, labeled with his name, located across the street from the mathematics building.

Reaching the age of 60 in 1990 did not diminish Smale's ambition. He remained driven to gain respect for his work on real computation, to continue elevating his mineral collection, and to become known as a great photographer. As many of his contemporaries looked toward retirement, Steve pursued his projects with the same focus and vigor that accompanied his earlier triumphs. Meanwhile budgetary constraints at Berkeley were forcing the University to make major adjustments. One objective of a new austerity program was to induce early retirements among the faculty. For Smale the golden parachute offered the opportunity to maintain his income and retain his office without any future teaching obligations. In 1994 Steve retired from the Berkeley faculty, accepting research residencies over the following year at IBM, Barcelona, and Rio de Janeiro.

At this point Harvard geometer S.-T. Yau entered the picture. This 1982 Fields Medalist was an influential force throughout the mathematical world. Yau phoned Smale to inquire whether Steve were interested in a high salaried position in Hong Kong. Steve pondered the overture. Nothing prohibited him from accepting a new position and the money was certainly attractive. Though formally retired, Steve was continuing his research and his work was portable. Clara was reluctant to move to Hong Kong, but Steve was intrigued by the new opportunity. Had it been known that he were movable, domestic offers may have arisen. They did not. He responded positively to Yau, leading to an offer from the City University of Hong Kong as Distinguished University Professor of Mathematics. That the financial terms were outstanding and the status of the University was low fit in with Smale's values. Once on a trip to Paris he had taken pride in spending more money for a fine meal than on his cheap hotel room.[8] Steve accepted a one year contract beginning in 1995 with the understanding that, if he were interested, an extension would likely follow.

Many American mathematicians were surprised when Steve went to Hong Kong. For them the prospect of moving to a new culture on the other side of the world was a deterrent, but not for Smale. What mattered was that it was a nice offer and an interesting place. One of his biggest adjustments involved the mineral collection. Steve kept abreast of developments by maintaining frequent contact with the other leading collectors, most of whom resided in the United States. Living in Hong Kong would limit these interactions, but increase Asian opportunities at the source. Steve was well placed to obtain his pick of the fine specimens coming out of India, Pakistan, and China. Photography moved to the back burner with Steve's studio remaining at his home near Berkeley.

Steve became an instant academic star in Hong Kong, attaining a status far beyond what he had in Berkeley. There was plenty of support for him to bring in visitors and collaborators for short or extended stays. His teaching duties consisted of only one course per year. Moreover Smale became well connected with the presidents of all the Hong Kong universities and other influential leaders. They listened to his ideas for scientific development and regarded them with respect. Steve saw that "Asia, China, and Hong Kong have a huge amount of power that is not shown in the development of scientific research."[9] For years Asian students had gone to the West for graduate study. To make progress required a diminution in this flow of intellectual talent. Steve believed that Hong Kong was well positioned to start a new graduate oriented institution that could reach the level of Harvard and Berkeley.

Steve's year in Hong Kong came at a time when the financial prospects for both Berkeley and the National Science Foundation were bleak. In Hong Kong, investment was abundant. While Americans were concerned about the coming conversion from British to Chinese rule, Steve saw advantages to consolidating the relationship with the mainland. Shortly after his arrival he renegotiated his contract to a three year agreement, ensuring that Steve would be in Hong Kong during the historic changeover. Steve held no formal administrative position, but he was becoming active in promoting science and mathematics in Hong Kong. He saw himself as an instigator and found it to be a comfortable role. Smale recognized that the changeover offered a golden opportunity to boost consciousness in scientific development. Leading the Vietnam Day Committee had given him insight into the casting and staging of successful outreach events. Steve proposed that the city host a major conference featuring Nobel luminaries and influential scientific administrators from around the world. The concept was warmly received by university officials who took over the logistics of implementing Smale's concept.

The conference *Great Science under One Country, Two Systems* took place on July 4, 1997, a few days after the changeover. Under the theme "Developing a High Level Research and Education Infrastructure for Hong Kong and China in the Context of One Country, Two Systems," there were public lectures by John Polyanyi (Nobel Prize in Chemistry), Richard Karp (Turing Award in Computer Science), David Mumford (Fields Medal in Mathematics), Sheldon Glashow (Nobel Prize in Physics), and Henry Taube (Nobel Prize in Chemistry). To conclude the day, Smale participated in an international panel whose topics included "The Role of Hong Kong in Bridging the Scientific Communities in China and the West." The person who in 1966 had condemned the world's communist and capitalist superpowers was now championing institutional advancement on behalf of a modern hybrid. His new status brought him invitations to participate in some of the high level celebration associated with the changeover. At one banquet Steve was seated with the presidents of the leading mainland universities, adjacent to the table of Tung Chee-hua, the new head of the Hong Kong government.[10]

During the fifties and sixties Smale had fought major battles against university, scientific, and government establishments. Indeed part of his success was based on a skepticism of conventional wisdom. After his political activism ended in the early seventies, Smale remained branded as a radical. Some in positions of scientific leadership were fearful of what Smale might do if he were admitted to their club. Despite all of his great theorems Steve was an outsider in his own country. It took a move to the emerging capitalist-communist environment of Hong Kong for Steve to finally reach a position of partnership with the establishment. Smale had become an insider. In 1997 he signed a new contract continuing his Hong Kong association to 2001.

Six Years at City University

13

A. New possibilities in Hong Kong

When this book was originally published nearly 25 years ago, Stephen Smale was 70 years old. At that point in life, most people start thinking about slowing down and embracing retirement, if they have not already. But Steve is not like other people.

Instead of hitting the brakes, he sped ahead, moving around the world to Hong Kong, where he would find new momentum and energy in both his personal and professional life. As in the past, his quiet self-assurance, independent insights, and willingness to take risks served him well in Asia. It was clear that Steve was no fading star. Indeed, two of his three most-cited papers were published in the 21st century.

Steve thrived at CityU, where he was an academic giant with the freedom and financial backing to pursue his professional interests. With his prestigious contract extended from 1997 to 2001, Steve and Clara made themselves at home in Kowloon Tong, a leafy neighborhood on the Kowloon Peninsula. Compared with other urban areas, the affluent district felt quiet and suburban. Not quite in the American sense—you would be hard-pressed to find cul-de-sacs or sprawling grassy backyards—but in hyper-dense Hong Kong, the courtyard gardens, palm-lined streets, and low-lying buildings created a charming, laid-back atmosphere.

Remarkably, the neighborhood was void of skyscrapers, and not just by chance. In the 1920s, Kowloon Tong was planned as a "garden city" with wide streets and detached single-family homes inspired by European villas.[1] In addition, the government had imposed building height restrictions in areas, including Kowloon Tong, under the flight path of Kai Tak International Airport. Known for its hair-raising landings, the airport operated on the shores of Kowloon Bay until 1998, when it was replaced by the much larger Hong Kong International Airport on Chek Lap Kok island to the west.[2]

Just behind Kowloon Tong, Beacon Hill and Lion Rock rose above the rooftops, creating a natural border between the city and the New Territories—a hinterland full of wetlands, country parks, farms, and alluring coastlines.

The Smales lived comfortably in CityU's Tak Chee Yuen faculty housing estate on Tat Chee Avenue, right by campus. It was extremely convenient. Steve could walk to work in minutes, while Clara could easily stroll to parks, restaurants, and grocery stores. For adventures farther afield, the couple had their pick of public transportation with easy access to buses, an MTR station on the Kwun Tong line, and what was then the Kowloon Canton Railway East Rail line (now the MTR East Rail).[3]

They also lived near a new mall called Festival Walk, which opened in 1998. As the largest shopping mall in Hong Kong back then, it housed several upscale restaurants, shops, a cinema complex, and even an ice-skating rink. Recalling those early years in Hong Kong, Smale remembers taking visitors to

Clara at home in the apartment near CityU.

Festival Walk to dine at a high-end Japanese restaurant, where he usually ordered tuna sashimi.[4]

While living in CityU faculty housing, the couple made friends with other academics and their spouses. "It was easy to manage socially. Clara managed pretty well because there was a community of wives of mathematicians," says Steve. "They had social events and clubs, and they all lived in faculty housing … Clara could meet people in the apartment, which was pretty big just for two of us."[5]

When it came to work, Steve kept a disciplined schedule. On a typical day, he would rise between 5 and 6 a.m., read the news, eat breakfast, head into the office, and often return to work after dinner. As was his preference, Smale had a light teaching load, with just one course per semester, enabling him to spend more time on mathematical problems.

His first substantial project in Hong Kong was *Complexity and Real Computation*,[6] a book stemming from Blum, Shub, and Smale's 1989 paper, "On a theory of computation and complexity over the real numbers: *NP*-completeness, recursive functions and universal machines," which became known as the "BSS Paper." Referring to the first letter of the surnames Blum, Shub, and Smale, the BSS paper provided a model for scientific computation over real numbers. It was a novel way to study calculus-based algorithms, opening up new possibilities for numerical computation and analysis.

Blum explains:

> [Our model] *is very different from the classical theory of computation ... which comes from* [Alan] *Turing's model and is looking at discrete computation. In the 1930s, Turing wanted to capture the notion of computation. He* [created] *a model using a series of zeros and ones. What was really neat about Turing's model was, that whatever you can compute in the cloud or on a supercomputer, you can actually compute in this very simple model. That model worked well. It was really the genesis of all computing today. But it didn't always capture the kind of computations or algorithms that people were doing in more numerical analysis, scientific computation, and other* [areas].[7]

Just as Turing's work formed the foundation of computer science, the BSS model played a pivotal role in the mathematics of numerical computation. Journeying back in time may help us appreciate how the paper led to the book, which was influential in its own right.

In July 1989, the BSS paper had just been published. Soon after, a young Uruguayan algebraic geometrist and complexity theorist named Felipe Cucker at the Technical University of Catalonia (often called BarcelonaTech) began using the model. With encouragement from his former colleague, Tomás Recio, at the University of Cantabria, Cucker mailed one of his papers to Smale with a letter mentioning the possibility of visiting Berkeley. To his excitement, Cucker heard back a few weeks later.[8]

"I received a handwritten letter from Steve, saying that the paper looked interesting, and he would be delighted to see me anytime in May," recalls Cucker. Thanks to Recio, who had a research grant to fund the trip, Cucker visited Berkeley in May 1991. His first encounter with Smale came sooner than expected. Most likely due to jet lag, Cucker woke before dawn his first morning in Berkeley and found his way to Evans Hall, the mathematics building.

"I decided to use the time to get familiar with the building and locate Steve's office. To my enormous amazement, not only did I find Steve's office, but I found Steve in it," recalls Cucker. "He was already there working at 6 a.m."

During the visit, Cucker quickly established a rapport with Smale and Blum, who offered him a warm reception and excellent hospitality. Blum found him an affordable place to stay and arranged for him to give a talk at the International Computer Science Institute. She also introduced him to Leo Harrington, an expert in descriptive set theory, who pointed

Cucker to the information needed to finish an article he had been working on.

Smale and Cucker met again a month later at the Symposium on the Current State and Prospects of Mathematics, a conference at the Institut d'Estudis Catalans (IEC) in Barcelona, where several Fields Medalists, including Smale, gave presentations[9] on recent developments in their respective fields. "In his talk on the theory of computation, Steve mentioned I had obtained some results from the Blum–Shub–Smale model," says Cucker.[10] "I remember some people in the audience started looking around because he was the only speaker who mentioned someone in Barcelona."

Shortly after the event, Manuel Castellet, then the Secretary General of the IEC, asked Cucker to put together a semester-long conference on the foundations of numerical computations during the second half of 1993. Naturally, they hoped Smale would participate and spend three months in Barcelona. Smale accepted the invitation, as did Blum, Shub, and two of Steve's Ph.D. students, Gregorio Malajovich, a Brazilian, and Maurice Rojas, a Colombian-American from L.A., among other prominent mathematicians and computer scientists.

"It was the beginning of a very fruitful collaboration—the beginning of many things. In particular, Lenore, Mike, and Steve proposed the idea of writing a book on real computation," says Cucker. At the time, real computations were being done somewhat experimentally, and Smale saw value in establishing a more foundational view.

By the end of the spring of 1993, the four mathematicians agreed to write a book together to develop the fundamentals.

They met again in Berkeley at the beginning of 1995 and collaborated for the next eight months. At the same time, Steve was organizing a large conference on the mathematics of numerical analysis to drum up more interest in the foundations and cultivate greater unity between mathematics and computational science.

In July of that year, just as Steve turned 65, the four-week AMS-SIAM Summer Seminar in Applied Mathematics took place in Park City, a picturesque mountain resort town outside Salt Lake City, Utah. Every day, hundreds of people would meet for morning plenary talks and afternoon workshops on everything from algorithmic efficiency to the geometrical, algebraic, and fundamental aspects of numerical analysis. Cucker says:

> *Normally, a conference lasts three or four days. A month-long conference is enormous, really unheard of. It was almost like camping because we were all staying, dining, and organizing activities together. You could rent out these little cottages, so one night, we'd have dinner at the cottage of one person, and the next night, another person would host.*[11]

When the event came to a close, there was something in the air—a shared feeling that it had been an enormous success. Over lunch, a small group of six people, including Steve, Shub, Cucker, Arieh Iserles, Jim Renegar, and Narendra Karmarkar, agreed to establish the FoCM (Foundations of Computational Mathematics) society and continue organizing conferences every three years in different cities.[12]

Since then, FoCM has become an active international nonprofit organization with a cycle of conferences, the leading journal in the field, and the Smale Prize, which recognizes significant achievements that deepen connections between mathematics and computation.

It was also at this conference that Smale shared some positive news with Cucker, Blum, and Shub. He planned to move to Hong Kong to work at a new university: CityU. In his new role, Steve would have money to support their research, and he planned to invite all three collaborators to Hong Kong, where they could continue working on the book.

B. Working together at CityU

After his first semester in Hong Kong, Smale brought over Blum, Shub, and Cucker in February 1996. While working on the book, Shub would visit for short but frequent periods, Blum stayed for two years, and Cucker would remain the longest of all.

Lenore made the move to Hong Kong with another intellectual asset in tow. Her husband, Manuel Blum, was an eminent computer scientist who had just received the Turing Award, often regarded as the "Nobel Prize of Computing," in 1995 for his work on the foundations of computational complexity theory and its applications to cryptography and program checking. As one can imagine, CityU did not hesitate to offer Manuel a position as a Visiting Professor of Computer Science, and the situation worked out well for the couple.

During their time in the city, the Blums lived in a spacious three-bedroom apartment on the sixth floor of CityU's faculty housing, where they had a bird's-eye view of life in Hong Kong. In the mornings, Lenore recalls watching groups practice *tai chi* in a small park nearby while savoring views of CityU's leafy campus and the rolling mountains beyond. When planes descended into the old Kai Tak airport, they were so close Lenore could wave to passengers from her balcony. After the Blums left Hong Kong for positions at Carnegie Mellon, Steve and Clara moved into their spacious apartment and stayed for the next few years.[13]

Meanwhile, Cucker became a highly successful career professor at CityU. After moving to the Asian hub with his wife Antonia and their two children for a temporary position, he worked at the university for the next 27 years. As his career unfolded, he became a prolific author, chair professor, and the mathematics department head.

Cucker and his family enjoyed living in Hong Kong, where nature intermingled with everyday life, even in the middle of the city. "When I lived in Mid-Levels, the jungle began 100 meters [328 feet] from my apartment," says Cucker, referring to the affluent residential district on the northern slope of Hong Kong Island. "I used to take a walk every evening after dinner and would see wild boars. That side of Hong Kong was incredible."

He also appreciated the public services, such as healthcare and education, and relatively low crime and poverty rates compared with other major cities. "All in all, I had the impression that I had reached the land of milk and honey," he says.[14]

From 1996 to 1998, Smale, Blum, Shub, and Cucker worked on *Complexity and Real Computation.* Nearly 30 years younger than Smale, Cucker, who was nearing 40 then, was the kid in the group and eagerly soaked up the experience. "What I learned from working with Steve is that he is extremely patient," says Cucker. "He could have worked with many others, but he didn't. He is very patient and very generous."

While the mathematicians usually wrote alone, each working on their assigned chapters, they often met to work through technical details on a whiteboard or blackboard. Steve had idiosyncratic systems to keep track of all the moving parts. For example, in Steve's office, Blum recalls seeing legal pads full of handwritten notes and reams of paper organized with black binder clips.[15]

The authors did not know it then, but *Complexity and Real Computation* would become canonical. "It was a new subject, so it took a while for people to grasp what it was about," says Smale. "But it did well."[16]

As of 2024, the book is Steve's third-most cited work, with more than 2100 citations, according to Google Scholar.[17] Often used to teach graduate-level courses, *Complexity and Real Computation* has enabled greater collaboration between mathematicians and computer scientists. The cross-pollination has benefitted developing fields, though it is hard to draw a direct line to modern technology. "Eventually, almost all mathematics finds its way into applications, but the path from the pure mathematical result to an application is generally twisted," says Cucker. "Our book is no exception."

He suggests assessing the book's long-term impact in another way. At the FoCM conferences, some workshops disappeared due to a lack of interest, but a workshop on real number complexity has been held at every conference since 1995. "This measures how influential it has been," he adds.

While working together in Hong Kong, the mathematicians witnessed the university's early development. CityU was established as the City Polytechnic of Hong Kong in 1984 and achieved full university status in 1994. The campus had only just started opening in phases in 1990. It was early days—a time of rapid growth and heavy investment.

Many academic units, including mathematics, experienced swift expansion. The small but growing department had roughly 15–20 regular staff members, including tenure or tenure track, and three temporary staff members, such as research assistant professors. "There was a sense of excitement we all shared being involved in something new. For those of us working in the math department, there was more companionship," says Cucker of the collaborative environment.[18]

Smale raised CityU's profile considerably, attracting outstanding Ph.D. candidates and academics to the department. Blum recalls:

> *Roderick Wong* [then the head of the mathematics department] *really wanted to build it up as a top math department, so he went out of his way to hire great people. He hired Steve as an example to bring world-class people to CityU ... The students from mainland China were really spectacular. I think that also influenced the department. I don't know whether that had happened before, but I know students came down to work with Steve particularly.*[19]

As the CityU mathematics department evolved, Smale and Cucker continued working in real computation and complexity theory, with "Complexity estimates depending on condition and round-off error."[20] Published in the *Journal of the ACM* in 1999, the paper provides a theoretical foundation for solving systems of polynomial equations while accounting for the loss of precision produced during computation.

Around the same time, Smale penned what would become the fourth most-cited paper of his career: "Mathematical problems for the next century," which was published in *The Mathematical Intelligencer* in 1998.[21] He wrote the paper in response to the late Vladimir Igorevich Arnold, a renowned mathematician known for the Kolmogorov–Arnold–Moser theorem, who, like Smale, had traversed various areas, including dynamical systems theory, algebra, topology, geometry, differential equations, and many intersections therein.

Arnold was Vice President of the International Mathematical Union from 1995 to 1998. Inspired by "Mathematical problems"—a set of 23 unsolved problems posed by the famous German mathematician David Hilbert at the turn of the 20th century—Arnold wrote to several mathematicians, including Steve, suggesting they propose a list of great problems for the 21st century. In the introduction of his resulting paper, Steve wrote:

> *I have listed 18 problems, chosen with these criteria:*
>
> 1. *Simple statement. Also preferably mathematically precise.*
> 2. *Personal acquaintance with the problem. I have not found it easy.*
> 3. *A belief that the question, its solution, partial results, or even attempts at its solution are likely*

> *to have great importance for mathematics and its development in the next century.*

Smale summarized what he believed to be the three "greatest open problems of mathematics": the Riemann Hypothesis, the Poincaré Conjecture, and "Does P = NP?" which he describes as "a gift to mathematics from computer science." Smale's list stimulated steady work and new ideas as mathematicians chipped away at the problems.

As of 2024, 10 of Smale's problems remain unsolved, while five have been partially resolved and three fully resolved. In 2011, Cucker contributed to problem No. 17, "Solving polynomial equations in polynomial time in the average case." Working with Swiss mathematician and theoretical computer scientist Peter Bürgisser,[22] Cucker expanded on a partial solution provided by Carlos Beltrán and Luis Miguel Pardo. A few years later, Pierre Lairez, a young French algebraist, finished the problem.

C. The foundations of learning theory

After *Complexity and Real Computation*, Smale began shifting gears toward learning theory, an algorithmic discipline that enables computers to automatically perform tasks, learn from data, and make predictions or decisions. Many practical, real-world applications were already emerging around the world. For example, in 1997, the United States Postal Service launched a pioneering handwritten address interpretation system to streamline mail sorting. By 1999, the automatic technology had a success rate of 30–40 percent.[23] Today, the machines can read about 99 percent of all hand-addressed letters, according to the USPS.[24]

Fascinated by machine learning, Smale followed his curiosity. He observed that the discrete branch of learning theory was already quite developed and rigorous. However, the continuous side still seemed relatively experimental.[25]

"At the end of the 1990s, Smale visited my house in France for the first time," says Cucker, who renovated a winery in southern France to use as a vacation home. "We walked in the vineyards and talked about mathematics. It was always like that with Steve. He mentioned that people were starting to talk about learning theory, and he saw an opportunity."

Smale felt that the papers written on learning theory had intuition but lacked sound mathematical explanations. He invited Cucker to work on a paper that would lay the foundations. They started reading the latest literature on learning theory to get up-to-speed on the recent developments. In the meantime, the American Mathematical Society and the Hong Kong Mathematical Society held a conference in Hong Kong, where Smale and Cucker organized a workshop on the mathematics of learning theory.

"It was very beneficial for us because we got to meet some of the people at the forefront of learning theory and came to understand what they were doing," recalls Cucker. "After the conference, we had a very clear picture of what was happening in the field."

Smale and Cucker published their first paper related to learning theory, "On the mathematical foundations of learning," in the *Bulletin of the American Mathematical Society* in 2002. Like the BSS paper, it was a theory-building paper designed to reach a broad audience so others could develop and prove their results.

It was a huge success, becoming a highly cited work, meaning it performed in the top 1 percent concerning citations received compared to other papers published in that field in the same year. As of 2024, the publication has garnered more than 2000 citations, according to Google Scholar, and ranks in the top 10 percent of all research outputs ever measured by Altmetric, a data science company that tracks published research.[26]

Their work attracted another mathematician who would become one of Smale's long-term collaborators: Ding-Xuan Zhou. Then a young research assistant professor at CityU, Zhou attended a seminar on learning theory that Smale had given in the fall of 1999.

"It was the first time I had heard of mathematicians working in learning theory," says Zhou.[27] During the seminar, Zhou asked a question[*] that led Smale to seek him out a few months later. When Smale knocked on his office door in the spring of 2000, Zhou could hardly believe one of the greatest mathematicians of the era was standing before him.

Smale hoped to discuss learning for regression, posing another question on approximation by kernel convolutions to Zhou. In machine learning, kernel functions enable algorithms to operate in a higher-dimensional space without computing the data coordinates in that space. The young mathematician thought about it, then visited Smale's office a few days later to explain his answer. That exchange, Zhou says, would lead to a

* Zhou's question: "I asked him why he used covering numbers in the space of continuous functions instead of those in the L_2 space, which is used to measure the learning performance for regression."

A view of Tai Long Wan Bay in Sai Kung East Country Park. (Courtesy of Bosco Yip via Getty Images)

series of papers on learning theory with Smale and contribute to his fruitful career as an expert in deep learning, neural networks, approximation theory, and learning theory.

Zhou became a Chair Professor, the Head of the Department of Mathematics, the Founding Associate Dean of the School of Data Science, and the Director of the Liu Bie Ju Centre for Mathematical Sciences at CityU. In 2022, he left to join the University of Sydney as a Professor and the Head of the School of Mathematics and Statistics. He was rated as a highly cited researcher by Thomson Reuters in 2014 and 2015, as well as Clarivate Analytics in 2016 and 2017. This recognition was largely influenced by his seven joint papers with Smale on the mathematical foundations of machine learning.

D. Hiking, beaches, and minerals

During his downtime, Smale took long walks with Clara and explored the city's natural side. At the turn of the century, Hong Kong had, and still has, a reputation for rapid development. In 2000, infrastructure projects boomed, towers grew ever skyward, and the population ballooned to 6.7 million.[28]

Even as it expanded, the city safeguarded substantial green space. About 40 percent of the territory remains dedicated to country parks, where scenic mountains, beaches, farms, and islands provide a relaxing respite from the downtown density.[29] Within 20 minutes from his CityU apartment in Kowloon Tong, Smale could be immersed in nature. He spent many weekends tracing the coastline, lumbering up mountains, or clambering over volcanic rock formations, sometimes sharing the trail with wild buffalo, snakes, and boars.

Referring to Ham Tin, a remote crescent beach hugged by mountains in the northeastern New Territories, Smale says:

> *I liked to hike for some hours to this great beach—the most beautiful beach in Hong Kong. I would go there with my friends for a day or half a day and swim.*

To reach the coastal retreat, Smale usually followed the 6.6-mile-long Tai Long Wan Hiking Trail through Sai Kung East Country Park. He often started the hike by the Pak Tam Au Campsite on the park's northern end, about two hours by public transit northeast of Kowloon Tong. From there, Smale took the trail east to Tai Long Village, then curved south along Tai Long Wan Bay.[30]

When he reached Ham Tin, Smale enjoyed stopping for a swim, lunch, and a cold beer or two. Refreshed, Smale would start again, heading further south to another scenic stretch of sand called Sai Wan, then hike up into the mountains to reach the Sai Wan Pavilion, where a public bus offered rides back to Sai Kung town several times a day.

Smale also took advantage of another geographical benefit: proximity to the mainland's mineral world. He seized the opportunity to build his collection, flying to Changsha, in Hunan Province, to meet a mineral dealer named Zheng Jian-Rong, his wife Danna, and their daughter Cicy.[31]

After that first meeting, Zheng became Smale's primary source of minerals. Clara and Steve regularly took the hour-long train ride across the border to Shenzhen, a developing city north of Kowloon, to meet Zheng and other trusted dealers, such as He Xin-Jian. The Smales would usually stay at the Shangri-La Shenzhen in the Luohu District, examining specimens in their hotel room.[32]

As China's first special economic zone (SEZ), Shenzhen was on the front line of former Chairman of the Chinese Communist Party Deng Xiaoping's open-door policy, which encouraged economic growth via foreign direct investment and international trade in selected regions. Deng set up four initial SEZs—Shenzhen, Zhuhai, Shantou, and Xiamen—and all have developed rapidly over the past four decades, but none so fast as Shenzhen.

Today, Shenzhen is a powerhouse economic city and a hub of culture, design, and technology that is home to more than 15 million people—more than twice the population of Hong Kong. However, at the turn of the 21st century, the city was still considered a relatively dangerous backwater compared to its cosmopolitan neighbor to the south. Back then, Shenzhen was experiencing a severe crime wave. It was common to read of bag-snatching motorcyclists and worse: explosions, kidnappings, homicides, and sexual assaults. In Bao'an, the city's industrial district, crime rates jumped 70 percent from 2000 to 2004, according to an article in the *South China Morning Post,*[33] Hong Kong's primary English-language newspaper.

Smale had heard the cautionary tales, some of which hit close to home. His good friend and CityU colleague, William Shiyuan Wang, was forcibly detained in Shenzhen. A pioneer in language emergence and evolution, Wang was appointed Chair Professor of Language Engineering at CityU in 1995. He is now a Professor Emeritus of Linguistics at the University of California at Berkeley, where he first met Smale, and a Chair Professor of Language and Cognitive Sciences at the Hong Kong Polytechnic University.

In the 1990s, Wang often visited Shenzhen Book City, a massive bookstore containing millions of publications, to shop for mainland books and magazines. On one trip, the linguist was at the train station in Shenzhen when a group of men who claimed to be law officers took him into custody. Roughly a dozen men kept Wang in a nearby apartment for several days while they withdrew cash from his Hong Kong bank account and used his credit card. After the bank withdrawals stopped working, the gang disappeared, and Wang was free to return home.[34] While Wang's harrowing experience hardly deterred Steve, he exercised caution when visiting Shenzhen. He usually avoided traveling alone since he had to take thousands of dollars in cash to pay for minerals.

"He was careful because he knew Shenzhen could be dangerous," says Zhou. "He usually went there with Clara, and they would not stay there longer than needed."

During a solo trip in 2004, Zhou says Smale fell prey to pickpocketing, his wallet deftly plucked from his back pocket while he waited in a hotel lobby. Thankfully, Steve was unharmed, and his passport, kept in a side pocket, was overlooked.[35] To thwart any other attempts, Smale kept one hand—half-in, half-out—of his side pocket while heading home to Hong Kong, where he contacted Zhou to recount the ordeal and request a favor.[36] Zhou says:

> *When he returned to CityU, he asked to borrow some money—not much, maybe 1000 Hong Kong dollars* [$127]*—and a suitcase. I can't remember*

> *why he needed the luggage, but he likely bought some minerals and needed a suitcase to transport them back to Berkeley. I was happy to help. Then he pulled out a half-page notebook paper and wrote me* [an IOU]. *I said it was unnecessary, but he insisted.*

Zhou remembers Steve repaying him when they met at a conference in Barcelona in June 2004. The misadventure, which would have rattled many travelers, did not scare Steve away from the city. After all, he had a high tolerance for risk and had been in far more dangerous situations while acquiring minerals in remote areas of Pakistan and Brazil. During his time in Hong Kong, he regularly visited Shenzhen, collecting hundreds of specimens that elevated his collection. These included an unusually transparent scheelite from Sichuan, China, reminiscent of imperial topaz, a blood-red realgar with calcite from Hunan, and a green pyromorphite from Guangxi.

"It was [not long after] China opened, so it was a good chance to get minerals there. I bought beautiful quartzes and azurites ... many different things. I probably have the best Chinese collection in the world," Steve says. "I was in a very good position because Chinese minerals were just coming out, and I had a monopoly on them. At my house [in Berkeley], I have a whole wall filled with Chinese minerals."[37]

E. Tiger Leaping Gorge

Smale's appetite for adventure did not stop in Shenzhen. Seizing the opportunity to explore remote parts of China, he traveled to Tiger Leaping Gorge, or Hutiao Xia in Chinese, in Yunnan Province in 2000. According to local legend, the gorge earned its name from a tiger that leaped across the gorge at its narrowest point to escape from a hunter.

Cradled between the majestic Haba Snow and Jade Dragon Snow mountains, Tiger Leaping Gorge is considered one of the most picturesque places in China, owing to its alpine scenery and dramatic ravine. At its maximum depth, the chasm plunges over two miles. The size is difficult to comprehend until you are walking on the edge. To offer a sense of scale, it is more than twice as deep as the Grand Canyon and deeper than Mount Fuji is tall.

Steve organized the trek with several family members—Clara, Nat, Hermine, Dylan (Nat and Hermine's son), Celine (Nat and Hermine's daughter), and Laura—and a friend from work, Shuk Han Cheng.*

Before the trip, Nat and his immediate family spent a few weeks with Steve and Clara in Hong Kong. Laura joined them a day before their departure to the mainland, barely making the trip due to a missed flight connection. "I got there the morning they were taking off," Laura recalls. "My dad came to the airport to meet me and help me navigate Hong Kong's transportation maze."[38]

From Hong Kong, they flew to Kunming, the largest city and capital of Yunnan, where they spent the night and explored the city. The following day, they flew to Dali, an ancient metropolis on the banks of Erhai Lake. During

* As of 2024, Shuk Han Cheng is the Chair Professor in the Department of Materials Science and Engineering, a Chair Professor of Molecular Medicine in the Department of Biomedical Sciences, and the Associate Dean (Learning and Teaching) at the School of Veterinary Medicine, City University of Hong Kong.

their short stay, they toured the city's fabled Old Town and admired fine handicrafts made by the Bai people, an ethnic minority group that lives mainly around Dali.

Next, they drove to Lijiang, a historic city in northwestern Yunnan encircled by mountains. The city's Old Town was listed as a UNESCO World Heritage Site in 1997, just a few years before their visit. "It was just gorgeous. I mean, you look up, and there's the foothills of the Himalayas," says Laura.

On a day trip, they ventured north of Lijiang to Jade Dragon Snow Mountain. There, they took a cable car over highland grassland and forests, jagged peaks, and gleaming glaciers until they reached an observation point at nearly 15,000 feet. It was a breathtaking view, with spectacular peaks and valleys unfurling below.

Laura remembers Celine attracting attention for her blonde hair and blue eyes. "About six or seven Chinese people trying to touch her hair—it was such a novelty," recalls Laura to illustrate the area's remote nature. "It was different from anywhere I'd ever been."

When it came time to hike Tiger Leaping Gorge, Nat and Clara stayed back. Although a highly accomplished hiker and climber, Nat was suffering from a gastrointestinal illness and could not participate. Meanwhile, Clara preferred not to join longer hikes, at least not knowingly.

Along the main route, known as the High Trail Hike, trekkers follow a well-marked path that stretches a few feet wide at the narrowest, through pastoral scenery and leafy forests. One of the most challenging parts of the hike is known as the 28 Bends, where a series of switchbacks climb upward toward a verdant plateau.

"I remember hiking along this huge canyon—the deepest you could imagine," Steve recalls. "The walls go up a mile or more, and it's dangerous. There are paths, but you could fall off and tumble 10,000 feet."[39]

The Jinsha River swirled far below as they traced the canyon's unnerving ridge. At times, the path wove through bamboo groves and passed waterfalls cascading down the canyon. Occasionally, local farmers ambled by with goats or donkeys, though the Smales encountered very few, if any, other hikers.

Early in the hike, the group walked together. As they ventured further out, they diverged into smaller cliques, with Smale and Cheng hiking together for some time. At some point, the pair diverged, too, and Smale recalls walking alone for the last stretch. "I got to this tiny town at the end of the hike, and I was very happy to see Han there. The others came separately."[40]

While hiking alongside Hermine, Dylan, and Celine, Laura recalls staying overnight in a small hut-like guesthouse beside a farmhouse. During their stay, Laura fondly remembers three young Chinese children, roughly 5–10 years old, climbing up the hill after spending the week at school.

"It must have been a Friday. They climbed up and down the gorge alone, full of pep and looking very happy," she says. "After about 30 minutes of playing and talking to their parents, they pulled out their schoolbooks, sat down, and just started doing their homework on their own."[41]

Nat, who was on the mend, and Clara reconnected with the group in Zhongdian, a remote alpine city near Tibet that had just opened to visitors a few years prior in 1994. Today,

Clara and Steve after a graduation event at CityU.

it is known as Shangri-La (a Tibetan word meaning "a land of sacredness and peace") thanks to a 2001 tourism marketing campaign that renamed the destination after a fictional land of eternal peace and tranquility described by British author James Hilton in his 1933 novel *Lost Horizon*.[42]

At roughly 10,800 feet, Shangri-La is surrounded by mountains and rustic villages where travelers can experience Tibetan culture. While staying there, they visited the Songzanlin Lamasery (also called the Little Potala Palace)—the largest Tibetan Buddhist temple complex in Yunnan—ate yak cheese, and took a short hike nearby before driving northwest over a dramatic pass to Deqen (or Deqin), meaning "auspicious place" in Tibetan, a valley town in the heart of the Hengduan Mountains.

Though it looks close on a map, Nat recalls the long, winding drive taking about six to eight hours. "We got out of the car several times to look at the scenery. We had spectacular views of ice-covered peaks and deep, deep valleys. These valleys were extremely deep—maybe 8000 feet or deeper."[43]

F. Smale's final months in Hong Kong

When the journey ended, the Smales returned to Hong Kong, where Laura stayed with Steve and Clara for another week. The city's density, fast-paced way of life, and world-class dining made an impression on her.

It helped that her father knew where to find all of the best restaurants. Steve, who had a reputation for being a gourmet, sought two things when eating out: quality and simplicity. One of his favorite places to dine was the waterfront promenade in Sai Kung town, also known as Seafood Street. Along this lively corridor, fishmongers sell freshly caught fish, shrimp, cuttlefish, crabs, lobster, clams, abalone, and other delicacies. Diners can purchase their preferred catch directly from the fishermen and then pay for one of the waterfront eateries to prepare it. Alternatively, they can select their dish from one of the restaurants' large glass fish tanks and then enjoy it grilled, braised, steamed, or sauteed with salt and pepper.

Throughout his first appointment in Hong Kong, Smale returned to Sai Kung again and again with friends and visitors. All the while, he continued to keep a high profile, regularly attending banquets with influential government officials and speaking at international conferences. "When I first arrived in Hong Kong, my societal position was much higher than in Berkeley, and it only got stronger as time went on," he says.[44]

When his contract was coming to an end, Smale hoped to extend his appointment on similar terms. However, the university president at the time no longer wished to employ faculty older than 70. The university offered him a less desirable contract to encourage Smale to step down.

However, many people in the mathematics department wanted him to stay, Smale recalls. They even created a new job for him, albeit less prestigious, and bypassed the administration to fund his salary.[45]

"I don't remember Steve complaining, but he must have been unhappy with the move. This was not something he intended to do. It was decided by the university's president and was not restricted to Steve but to all staff above 70," says Cucker, adding that at least two other high-profile professors left simultaneously. "They were probably the three oldest people on the academic staff. It was imposed on him."[46]

Ultimately, he decided not to accept the lesser role at CityU. Instead, Steve took a high-paying position at the soon-to-open Toyota Technological Institute at Chicago (TTIC). Founded by the Toyota Technological Institute in Japan as an independent institution, TTIC started operations in September 2003, and Smale was one of the first people to be hired.

Any international move presents considerable logistical challenges. It was even more laborious for Steve because his extensive mineral collection had significantly expanded since his arrival. All told, Smale needed nearly 40 pieces of luggage to ship his mineral collection and had to rent a vehicle the size of a school bus to transport them to the airport.[47]

"The bus was not completely full, but almost," says Zhou, who accompanied Steve and Clara to the airport in 2001 to see them off.

Zhou remembers thinking Smale seemed disappointed about leaving Hong Kong. When asked about his feelings that day, Steve says he was probably influenced by Clara, who did not wish to relocate. "She surely wasn't so happy about moving to Chicago," says Smale. "And I wasn't that much happier either about the lifestyle. But that's not what mattered to me."[48]

Rather, he was attracted by the attractive salary and low teaching load—prerequisites for Steve at this point in his career. Moreover, he would be well-positioned at the intersection of mathematics and computer science to pursue his current research interests with his hallmark vigor and focus. And so, to Chicago, he went.

New Challenges in Chicago

14

A. On the cutting edge at TTIC

For the next six years, Steve split his time between his home in Berkeley and TTIC on the University of Chicago campus in Hyde Park, an architecturally and culturally rich neighborhood near Lake Michigan on the city's South Side.

With support from an approximately $255 million endowment from the Toyota Motor Corporation, the graduate school and research institute was ahead of its time in computer science theory, especially in areas like artificial intelligence and machine learning.[1] While technologies like computer vision, machine pattern recognition, and natural language processing are commonplace now, they were far from mainstream at the turn of the century.

Today, TTIC continues to provide exciting research opportunities in various areas within computer science, such as algorithms and complexity, computer vision and computational photography, robotics, computational biology, and speech and language technologies.

Smale joined TTIC in 2002 before it officially opened for operations, along with Chief Academic Officer and Professor David McAllester.[2] A respected computer scientist, McAllester's research areas cover machine learning theory, the theory of programming languages, automated reasoning, computational linguistics, and AI planning. Since the institute did not yet have a dedicated space for faculty, Smale worked in an office in Eckhart Hall, home to the Department of Mathematics at the University of Chicago, until TTIC moved into the grand Press Building.

TTIC was an unconventional appointment for Smale, reflecting his desire to work on cutting-edge problems at the intersection of mathematics and machine learning. While he was a Fields Medalist and a recipient of the National Medal of Science—a big name who likely raised the new institute's profile—some felt that Smale was out of place in the computer science atmosphere.

Steve recalls:

> *Some big bosses wanted me to have this job. But locally, those in computer science, including the director* [David McAllester], *weren't so crazy about me because I threatened the computer science atmosphere. He was supporting computer science. I was a mathematician. It wasn't a bad relationship, but he wasn't supportive … The reason was that I represented maybe a little different trend for TTIC, which had been purely computer science. Everything about it was the traditions of computer science. All these disciplines have their own narrow way of looking at things, and I intruded on that.*[3]

Their working relationship was professional and cordial, but McAllester would not have hired Steve if it had been up to him. "It is not a matter of turf. It is a matter of relevance to AI applications ... I wanted a computer science focus," says McAllester. "To the extent that we did theory, it was in theoretical computer science (algorithms and computational complexity). That was not really Steve's area. Although his work on kernels was related, his particular work was out of the AI mainstream."[4]

Smale had landed a high-paying role at TTIC without teaching duties but lacked the power to hire people and invite visitors without McAllester's approval. "I wouldn't say it didn't bother me. But I felt the difference,"[5] says Steve. That said, he worked with excellent people in the mathematics and computer science departments at the University of Chicago, so by Steve's calculation, it all balanced out.

While adjusting to new dynamics at work, Smale was also experiencing a degree of stress in his personal life. "That was one of the worst times in our marriage," he says.[6] "Clara wasn't so happy in Chicago."

In Hong Kong, where they lived in university housing, Clara interacted with other faculty members and their families throughout the day. It was a friendly environment where people greeted each other in the elevator, planned dinners, and joined clubs and other hobby groups. Many spouses also bonded over a shared experience of living in a foreign place far from home. By contrast, in Chicago, they lived on the 33rd floor of an apartment tower unaffiliated with the institute. It was harder to connect with others in the isolated environment, and the dramatically colder climate did not help.

Steve did not thrive socially in Chicago, either, but he had more opportunities to meet people at the office, which was about a 20-minute walk from the apartment. While working on campus, he developed close friendships with Shmuel Weinberger and Partha Niyogi, in particular. Smale recalls:

> *Shmuel had a reputation of being a whiz kid, a hot shot. He was a very sharp guy who knew a lot of math. He would come by my office because our offices were very close, and we often talked. He was a good friend.*[7]

Weinberger remembers those days very fondly.

> *One day, Steve knocked on my door and began asking me questions that, to him, were about machine learning and, to me, were about controlled topology, a theoretical branch of topology that I had been working on for the previous couple of decades. It was like Moses showing up to the synagogue and asking the local rabbi for help. It seemed unbelievable to me, so of course, I paid attention.*[8]

Niyogi, originally from India, was also one of Steve's best friends in Chicago. They worked on a few papers together and enjoyed dinner with their wives on many occasions. "We talked about everything, math especially, but other things, too," says Smale.[9] Tragically, Niyogi died of brain cancer in October 2010 at the age of 43. Specializing in machine cognition and learning, Niyogi had a brilliant mind and contributed fundamental work in many emerging fields.

During his first year at TTIC, Smale had many fascinating conversations with Niyogi about language, which helped him advance an earlier idea that he had started exploring with Cucker and Zhou at CityU.[10] Zhou recalls:

> *Steve was interested in the work of Partha Niyogi and Bill Wang on language and grammar evolution. Once, when we visited Tommy* [Tomaso] *Poggio at MIT* [the Massachusetts Institute of Technology], *I remember them talking about Chomsky's work on universal grammar (Chomsky likely joined a dinner during our visit) and the belief that everyone will speak the same language in the end.*[11]

Zhou is referring to renowned linguist Noam Chomsky's "universal grammar" hypothesis, which proposed that humans are innately capable of learning language and, as diverse as they may seem, languages share fundamental grammatical similarities.[12] While many competing theories exist on the origin of languages, Smale hoped to better understand their development using mathematical tools.

The resulting paper, "Modelling language evolution" by Cucker, Zhou, and Smale, explains how communication evolves through linguistic interactions with agents of society—meaning families, schools, peer groups, or social institutions. Central to Cucker, Zhou, and Smale's model was an algorithm for computing how languages change through interaction. Under certain conditions, they proved that multiple languages would converge to create new common languages over time. Cucker likens it to creole languages, which arose during the 1500s when European languages, such as English and French, merged with African languages on slave plantations to create new tongues. Their model also explained the mathematical formation of vowel systems in different languages.

"It was a model that essentially mixed two aspects: evolution following a fixed pattern of random interactions plus learning theory," says Cucker. "The pattern of random interactions explains how we interact with other people—different people have different degrees of influence on you. Based on those influences, there is a learning mechanism that makes you produce a new language."[13]

Looking beyond language evolution, Smale also suspected their findings could be applied more broadly to other systems, including economics, though they left the question open for others to pursue.[14]

Smale, Zhou, and Cucker's work on the mathematics of language development provided a glimpse into Steve's strategy when delving into less familiar areas. "Some people approach a new field like a student, not me. Oh, no, I wouldn't have patience," said Smale in a 2004 interview at the Institute of Mathematics, Academia Sinica in Taipei. "I just start looking at advanced research literature on the subject. It takes a while; I learn only by successful approximation—trying to do things almost immediately and usually failing but learning in parallel."[15]

Smale's top-down process enabled him to work across many fields of science where mathematics could be applied in new ways. It also helped him to focus on the big picture rather than get caught up in the details. Cucker says:

> *My memory of working with Steve is always the same. Steve has these wonderful ideas and shares them with me. Steve gets a general idea, which should lead to certain results. The idea is good, and indeed, it leads to the result he wants. But if you inspect the fine details with a magnifying glass, something may be wrong.*

Those small details can always be amended. More importantly, he's globally correct.[16]

B. Hiking Death Hollow

For recreation, Smale continued to enjoy surprisingly rigorous treks with friends and family. In June 2002, when he was 71, Steve and eight family members took on Death Hollow, a strenuous canyon trek in Utah's Grand Staircase-Escalante National Park. Given the demanding nature of the hike, Clara did not join.

As usual, Steve welcomed the challenge and looked forward to pushing himself on the five-day trek. He loaded his backpack with essentials, including a sleeping bag and food, and set off with the group. They followed a route Nat had done many times before. Soon into the trek, Steve struggled with the weight of his backpack, so Hermine and Nat each took about 10 pounds of gear to lighten his load. "After that, he did better. It is a hard hike, and he did well given his age," says Nat.[17]

They started at the Upper Death Hollow Trailhead, at around 9000 feet, on Hell's Backbone Road, a rugged, winding gravel road hugged by Boulder Mountain to the north and canyons to the south. The group descended through ponderosa pine forests and scrubby cactus brush to enter Death Hollow Canyon. There, they plodded along a dry streambed that evolved into pockets of muddy water, then a rushing tributary of the Escalante River.

Heading east, they navigated the natural obstacle course, dipping in and out of water at varying depths. Owing to millions of years of erosion, the canyon showcases spectacular slickrock scenery and rock formations. On either side of the narrow canyon, the vertical sandstone walls shoot upwards 800 feet, sometimes leaving just a sliver of blue sky overhead.

"It was pretty exotic," recalls Steve. "Sometimes you walked in the canyons, and they went so high up. Other times, we were forced to be in the water, which could be deep."[18]

They covered 11 miles on the first day alone, stopping just once for fresh water from the only natural spring along that leg of the route.[19] On the second day, they trekked through narrow stretches, navigating obstacles like drop-offs and pinch points before reaching an area of perennially flowing water and lush vegetation. Another long stretch took them through deep pools and waterways on the third day. By the end, they had traversed 35–40 miles with temperatures spiking around 90 degrees. "It was hotter than usual. But there's shade from thc canyon, and you're in the water for some time, so it wasn't that big a deal," says Nat.[20]

While traversing Death Hollow, hikers often remark on the canyon's colorful horizontal strata, comparing the scenery to a living painting that changes with the light. Morning blues turn the ribbons deep clay and salmon while the warm sunsets cast a magenta glow across the sandstone walls.

The canyon's beauty belies its perils. On the fourth day of the hike, Steve was wading through the tributary at the tail end of the group when the riverbed dropped from under his feet. Suddenly, his head was underwater. As water rushed around him, his backpack pushed him deeper. Luckily, Celine and Dylan raced to help. "My granddaughter Celine grabbed me and pulled my head above water," recalls Steve. "I don't know if I would have managed to get out."[21]

Steve's grandchildren unclipped his pack, which had been attached in such a way that it was floating to the top and pushing his body down. This distributed the weight better and helped him move through the water more easily.

After the scare, Steve took a few moments to catch his breath. Unharmed, he continued trekking through rushing water, climbing between large rocks, and floating across calmer pools. That was not the only misadventure to befall him. Before the hike ended, Steve developed a toe infection and could barely walk. But he finished. "He's a pretty tough cookie," says Dylan.[22]

C. Sailing the British Virgin Islands

The following year, Nat, Hermine, Dylan, and Celine flew to Chicago to spend a few days with Steve and Clara before departing for the British Virgin Islands (BVI) on a scuba diving trip. About a week later, Steve, Clara, Laura, and Kay met Nat and his family in Tortola, BVI's main island, where they chartered a boat and sailed around the archipelago.

The journey would prove tame compared to Steve's epic expedition from San Francisco to French Polynesia. But it was memorable in a different way. He was surrounded by his children and grandchildren, enjoying the beaches, snorkeling, and island culture.[23]

Steve chartered a Beneteau 50, a monohull sloop with two sails. Known for its reliability and performance, this 50-foot-long sailboat had a comfortable interior with four double-occupancy staterooms and a wood-clad galley. "They're set up

Steve and Clara with their family on the sailboat.

for cruising and spending the night in different places," says Nat. "It's a big boat, so anchorage can be tough."[24]

Steve took his role as captain seriously. He manned the helm and primarily relied on Nat to handle the sails. Though Kay was the most experienced sailor, having sailed across the Pacific, Steve largely excluded her on the trip, recalls Laura. However, Kay would be increasingly relied upon in later years.[25]

While island hopping in the British Virgin Islands, most sails took three hours or less. On a more extended, seven-hour journey to Anegada, a coral island known for its white sand beaches, Dylan recalls being surrounded by water as far as the eye could see. "That was the first time I was in a body of water and didn't see land," he says. "It was kind of spooky."[26]

Another day, they dropped an anchor by Virgin Gorda to see spectacular boulder formations, known as The Baths, and go hiking, dine, and snorkel. Snorkeling in the archipelago was highly rewarding, thanks to the islands' flourishing marine world full of parrotfish, yellow tails, clownfish, butterflyfish, and coral of every hue. Some coves even had nurse sharks, rays, and sea turtles gliding gently through the water.

"Every place we anchored was different from the previous one. In some, we'd swim and snorkel through almost like a maze of coral," recalls Laura. "In others, I remember snorkeling in more wide open spaces."[27]

At one point, they spotted Necker Island, Sir Richard Branson's private island. In another instance, close to Virgin Gorda, they snorkeled among dense clouds of tiny fish that would surround them. "There were probably hundreds of thousands of these fish on this coral reef," says Dylan. "And we saw seagulls picking off the fish cliffside. That was one of the most memorable moments."[28]

D. An era of big data

Between hiking and sailing, Smale continued laying the groundwork in developing fields like machine learning and data analysis. Among his pursuits, Smale was working on a paper with Zhou on learning theory and non-uniform sampling theory. Steve's idea for the project came a year earlier, in 2001, following a CityU conference that Zhou organized on computational harmonic analysis.

"After the conference, a few experts on sampling theory, such as [American mathematician] Akram Aldroubi from

Vanderbilt, talked about their work. Akram sent his *SIAM Review* paper to Steve," recalls Zhou.[29]

Steve shared the paper with Zhou and suggested they investigate the connection between learning theory and non-uniform sampling theory. "The essential difference is the sampling points in learning theory are random, while those in their study are deterministic," explains Zhou.

After visiting Berkeley in 2002 to discuss the project, Zhou and Smale collaborated on the paper by email and fax, since Smale preferred to print out documents and mark them up with handwritten comments, which he called "scribbles." The resulting paper, "Shannon sampling and function reconstruction from point values,"[30] was published in 2004.

They expanded on their study with "Shannon sampling II: Connections to learning theory." The sequel improved their error analysis and showed how their approach can be applied to learning theory.[31] Both papers influenced areas of study such as sampling theory, wavelets, signal processing, and approximation theory.[32]

Zhou recalls visiting Steve at TTIC later that year, where they pursued what would become one of their most influential works, "Learning theory estimates via integral operators and their approximations." Published in 2007, the foundational paper introduced the methodology of integral operators for studying learning theory problems.

"When Professor Smale proposed an approach to using integral operators for understanding learning theory, I did not think it was a seriously novel idea," says Zhou. "But later, it became obvious that the work became increasingly important."

At first, the reception was mixed. Some thought Smale and Zhou's results were too good to be true. "Some people even argued with me. But later on, they admitted that it was true. Many follow-up papers use this methodology for various purposes," says Zhou, adding that it is one of his most-cited papers.

Like many times before, Steve found himself in a familiar place—at the forefront of a nascent area adjacent to mathematics. Having observed the evolution of machine learning over the past few decades, from the Turing Test for determining AI in 1950 to master chess-playing computers by the late 1990s, Steve saw the technology's potential.

Working with Dr. Tomaso Poggio, the Eugene McDermott Professor in the Department of Brain and Cognitive Sciences at MIT who pioneered computational neuroscience, Smale explored the mathematics of pattern recognition. In their paper, "The mathematics of learning: Dealing with data," published in 2003, the co-authors wrote:

> *We believe that a set of techniques based on a new area of science and engineering becoming known as "supervised learning" will become a key technology to extract information from the ocean of bits around us and to make sense of it.*[33]

Supervised learning enables systems to train from a set of examples and data inputs. For example, a bank system could learn to process loan applications based on data and approve "good" ones. In another scenario, a car manufacturer could develop a system to detect people crossing the street by training using thousands of images.

To that end, they provided a foundation for learning theory and a simple regularization algorithm that is particularly effective when working with binary classifications (such as visual object recognition) and regression applications (such as the use of algorithms for predicting the temperature tomorrow morning based on the current measured weather data).

"I look at these things not just in terms of mathematical challenges—far from it. I want to see the development of important sciences of our era," explained Steve in a 2011 YouTube interview with MathemediaFilms. "Because I am a mathematician, I like to see how mathematics can be used to help develop these big questions."[34]

E. A breakthrough in flocking

While Smale leaned into the mathematics of data, an out-of-left-field idea fell in his lap. Cucker, Zhou, and Smale's work on the evolution of languages had reached the desk of Ali Jadbabaie, an Iranian-American systems scientist and decision theorist at MIT.

Having read the paper with interest, Jadbabaie reached out to Smale, suggesting that the model's matrix should change with time. To explain his point, he attached his own paper on stable flocking of mobile agents, which aimed to shed light on how a group of autonomous moving creatures, such as schools of fish or crowds of people, manage to form synchronized clusters moving at the same speed in the same direction. In his study, Jadbabaie provided a condition for flocking (synchronizing) involving the infinite sequence of positions and velocities of the birds.

The concept of flocking captured Smale's curiosity. It did not take long for him to formulate another compelling question: How does flocking emerge in the first place? "Imagine birds are drinking water, then they hear a noise. Suddenly, they start flying in all directions. That's time zero," explains Cucker. "Steve wanted to provide a condition at time zero that would guarantee eventual flocking."[35]

Motivated by this question, Cucker and Smale penned "Emergent behavior in flocks." Published in May 2007, the study offered a model, which would become known as the "Cucker–Smale flocking model," for interactions between birds that would lead to coordinated movement.[36] It could be a literal flock of birds, of course. But as usual, Smale also envisioned wider relevance—the emergence of a price system, common languages, or distributed computing, where reaching a consensus is essential.

"Emergent behavior in flocks" was a runaway success. As the fifth-most cited work of Smale's career, it stimulated a stream of new research that extended their results in many different directions. As of July 2024, there are more than 250 papers crediting the Cucker–Smale flocking model in their titles, including one by another Fields Medalist. "This is by far the most influential paper I ever wrote, and it was a complete outsider," remarks Cucker.[37]

Given the widespread interest, the duo quickly offered a follow-up, "On the mathematics of emergence," including several more examples. For instance, Smale and Cucker apply their flocking model to pairs of birds that do not communicate, a set of three birds and a leader, and emergence with discrete time, among other variations.

As was his way, Smale focused on flocking intensely while working on the subject, then moved on relatively quickly. "I tend to get obsessed with things. I can put in a lot of my energy for a while—and not just mathematics. I take advantage of different sources of information and practice," he shared in the 2011 MathemediaFilms interview.[38]

After he accomplished what he set out to do with flocking, Smale dropped the topic. "It was typical of him," says Cucker. "He explored the subject, realized something could be done there, wrote a paper, and then moved on to the next thing."

F. The Smale Collection, Wolf Prize, and Costa Rica

With the addition of specimens from China and India, Steve and Clara had built one of the world's finest international mineral collections. In 2006, the museum-worthy collection comprised roughly 2000 pieces ranging from two to seven inches in diameter. For the past 30 years, Steve had dreamed about publishing a coffee table book showcasing his stunning acquisitions. While splitting his time between Chicago and Berkeley, he finally tackled the project.

Working with editors Gloria Staebler and Günther Neumeier, who co-founded Lithographie publishing house in Germany, Steve's book *The Smale Collection: Beauty in Natural Crystals* was published in 2006. Primarily an artistic endeavor, the book contains mineral photography he had taken using a large format (8x10 inch) camera in his home studio years earlier, alongside beautiful shots by accomplished photographer Jeff Scovil.

Staebler and Neumeier hoped to highlight the development of the collection by presenting nearly 200 of the couple's most prized specimens in chronological order based on the date of acquisition, accompanied by captions written by Steve. The resulting book traces the collection's evolution and provides a map of the Smales' life together as they traveled the world for math-related jobs, conferences, projects, and other activities.

In the introduction, Steve details his approach to collecting. He attaches great importance to a few fundamental components—form, truth, and economy—and beauty above all else.

> *For me, beauty is not just a criterion but is the fundamental quality of a great mineral specimen. Beauty is more than skin deep, more than is observed when a specimen is displayed on a shelf or in a photograph—even when seen by a sophisticated collector. To me, beauty is also truth.*[39]

While Steve and Clara's collection showcases minerals of all colors and shapes, he says an ideal classic form exists: a single isolated crystal perched on a base of smaller crystals of the same or a different species. One can see variations of this classic shape throughout the collection. However, Smale also appreciates a departure when it results in a unique character. "At times, two different species become such an integral part of the specimen that it is not easy to distinguish a primary mineral; this type of specimen can be the ultimate in beauty."

Steve and Scovil bring the minerals to life through their fine art photography. Inspired by classical still-life paintings, they employ simple backdrops—often monochrome or complementary hues—and careful use of shadow to emphasize each crystal's beauty.

After the book was published, Steve received more good news. In January 2007, he was awarded the Wolf Prize in Mathematics. It is common for the Wolf Foundation to recognize two winners in mathematics each year, and Smale received the award alongside co-winner Harry Furstenberg of the Hebrew University of Jerusalem, with whom he split the $100,000 prize. The recognition honored Steve's "groundbreaking contributions that have played a fundamental role in shaping differential topology, dynamical systems, mathematical economics, and other subjects in mathematics," reads the prize citation.[40]

Since 1978, the Wolf Foundation has awarded the prestigious Wolf Prize to outstanding scientists and artists in six fields: Medicine, Agriculture, Mathematics, Chemistry, Physics, and the Arts, with winners selected by an international jury committee composed of renowned professionals in these fields. "I was happy to receive prizes, but they were not a big deal for me. I don't focus on awards too much," says Smale.[41]

Before attending the prize ceremony, which took place at Israel's Knesset (the country's unicameral parliament) in Jerusalem in May 2007, Steve helped organize another family trip, this time to Costa Rica. In December 2007, Steve and 13 others—a mix of family and friends—rented two large houses in the Gandoca-Manzanillo Wildlife Refuge, a less-explored national park along the country's southeastern coastline with diverse scenery, from golden beaches to swamps, lagoons, and rainforest.

Set between the coast and the jungle, the houses took some effort to reach. They drove along a rugged road in poor condition, then hopped out with their luggage and crossed a small river measuring about 20 feet across. Most days, the water was about ankle-deep. But it grew to chest-deep at one point due to a long rainstorm that poured five to ten inches.

"I remember crossing the river being tough for Clara. She could swim and everything, but the river was moving fast. Clara was nervous about that. The rest of us were more confident, and we managed," recalls Steve.

They walked about a quarter mile from the river to reach their lodgings, just 100 yards from the ocean. Peeking out from under the jungle canopy, the wooden houses were simple and rustic. One had an expansive outdoor deck, measuring about 500 square feet, where they spent much time. "It was covered and protected from the rain, so we would hang out on this deck often," says Nat.[42]

At night, Laura recalls hearing the booming, whooping barks of howler monkeys in the trees. "They're really loud! I remember the howling being unbelievably loud at night, around dawn or dusk," she says.[43]

The surrounding forests were also home to sloths, snakes, lizards, and colorful tropical birds like parrots and toucans. While they did not do as much trekking on this trip, the Smales spent their days in nature, exploring nearby beaches and snorkeling off the coast. The refuge includes a protected reef, home to brain coral and sea fans, as well as sea turtles and manatees.

One day, they experienced the thrill of ziplining through the jungle. "The first line went across a fairly steep, deep canyon, which was unnerving. I also remember my dad wanted to be the first to go," says Laura.

They took a tiny rocky road to reach the zipline course. Along the way, Steve and Nat were deep in conversation, working on a math problem together. "My dad had this idea. He wasn't sure where it would lead, but he told me about it, and we just got more involved in talking about it throughout the trip," says Nat.[44]

During that trip, Laura learned more about her father and his approach to math than at any other time in her life. He would come up with many ideas and run them by Nat, one after the next. "My dad would get excited, and my brother would sit there and think, then say, 'Nope, that doesn't work,' for one reason or another," she says.[45] "That was happening everywhere. No matter where we were, my dad would be thinking. He comes up with a gazillion creative ideas, most of which don't work. But some work really well."

G. Analyzing high-dimensional data

In 2007, Smale returned to CityU in Hong Kong for six months. While there, he attended the Liu Bie Ju Centre for Mathematical Sciences (LBJ)'s 10th-anniversary celebration in December and lectured on "Geometry on Probability Spaces."

His talk was based on a paper he had recently published with Zhou. The two mathematicians developed a theorem using partial differential equations with the motivation to understand natural phenomena in areas like vision, learning theory, and quantum field theory.[46]

"The paper introduced the problem and invited more people to think about it," says Zhou. "But the problem is too difficult even now. Although diffusion models are widely used in machine learning and large language models today, they are much less understood in theory."[47]

Back in Chicago, Smale continued collaborating with Niyogi and Weinberger. One of their more influential papers was published in *Discrete & Computational Geometry* in 2008. In "Finding the homology of submanifolds with high confidence from random samples," the authors provided algorithms to address common pain points in data analysis in high-dimensional spaces.[48] As Weinberger explains:

> *It was a proof of concept. There's a feeling that if high-dimensional data is unstructured, its analysis is essentially hopeless. As a result, people spent a lot of time trying to exploit "geometric structures" that could explain various data sets. We showed that, if indeed some geometry explained it, at least some aspects of that geometry could be inferred from samples taken from it.*[49]

It is like looking at a pointillist painting, he continues. The image comprises tiny dots, but you can still recognize the picture. The algorithms they presented encouraged the geometrically minded community and ultimately contributed to a standard toolkit for analyzing data in areas like graphics and pattern recognition.

Their probabilistic point of view has since become standard. "We adapted Valiant's* idea of 'probably

* Harvard computer scientist and computational theorist Leslie Valiant, who received the Turing Award in 2010 for his many significant contributions to theoretical computer science.

approximately correct' to our setting," says Weinberger. "At the time, I don't think there were other theorems like this in the geometry literature—there already were some very fine results on perfect and abundant data, but we allowed for imperfections."

While researching high-dimensional data in Chicago, Steve worked with Zhou on learning theory. They organized a week-long workshop, "Learning Theory and Approximation," with Kurt Jetter, a mathematics professor at the University of Hohenheim, from June 29 to July 5, 2008, at the Mathematisches Forschungsinstitut Oberwolfach (MFO), Germany. After a successful inaugural event, they would later be invited to co-organize similar conferences at the MFO in 2012 and 2016.

Upon returning from Germany, Smale resumed his collaborations with Niyogi and Weinberger, and the trio expanded on their findings on high-dimensional data with another influential paper. In June 2009, they submitted "A topological view of unsupervised learning from noisy data" to the editors of the *SIAM Journal on Computing*, which was published in 2011.[50] Though it was less cited than their initial paper on the topic, this one was more important because it enabled the data set to be noisy, meaning it could contain superfluous data. For example, gibberish input from a speech recognition program or meaningless data caused by programming errors. Weinberger explains:

> *It's like trying to see the road with blurry vision. We identified conditions under which it was possible to infer the geometry despite mirages and large errors in the data. It was obvious that this needed to be*

addressed. Some of the things we tried to do there, such as deal with noise without modeling the noise, are the right questions—and [those questions] *are not dealt with enough in later work.*[51]

Combined with machine learning, their work in geometric measurements has been indirectly influential in fields like medicine, material science, and social sciences. Remembering those days, Weinberger says working with Smale was at once rewarding and infuriating.

We think differently. Steve thinks very fundamentally about the key problems that must be confronted. I tend to rush toward solving a problem. If any depth develops in my way of thinking, it's usually the result of looking at a large set of questions so that the important ideas are so inevitable that they can't be missed, but with Steve, he seems to be there in his imagination immediately.

Smale approaches problems with a clear vision of the scientific questions and the confidence to stick to his vision—qualities Weinberger admires. But that determination also made him a bit of a stickler. Workarounds or partial answers never cut it. Even if Weinberger's solution had the same implications, Smale would reject it, saying something like, "That's not exactly the question I asked."

"He wouldn't like solutions that were technically correct but didn't penetrate to the core of the issue," says Weinberger. "He had a sense of why he wanted to know *exactly* what he asked, and those reasons governed whether he'd accept a solution."

As Smale broke fresh ground with Weinberger in Chicago, recently appointed administrators at CityU in Hong Kong were preparing to present a lucrative new offer to the famous mathematician with hopes of winning him back.

Back to Hong Kong

15

A. Smale returns to CityU

While Smale flexed his intellectual muscles in Chicago, changes were afoot on the CityU campus. Way Kuo, a pioneering engineer known for his work in systems reliability, became the university's fourth president in 2008. To elevate CityU's international reputation, Kuo knew it was essential to hire world-class professors—and Smale was at the top of his wish list.

Kuo enlisted Roderick Wong, concurrently the Vice President of Research & Technology and the Dean of Graduate Studies, to spearhead the recruitment process. Since Zhou was then the Head of the Department of Mathematics and still collaborating with Steve, Wong asked him to test the waters.

"In the fall of 2008, I called Prof. Smale to share the salary, teaching load, and terms. While we were on the phone, he calculated the salary [from Hong Kong dollars] into US dollars," recalls Zhou. "He said he would give it serious consideration."[1]

It was an attractive offer: HK$2.5 million ($320,000) a year for five years and no classroom teaching duties. In addition, Smale would resume his title as a distinguished university professor, a status reserved for a select few. "Steve's situation was really an exception," says Cucker. "Over the years, there have only been a couple of others with that title."[2]

As a distinguished university professor, he would have more prestige and freedom. For example, he would not have to report to anyone except the president. Moreover, he would have substantial funding to pursue projects and leeway to invite visiting professors.

"He's a big shot. You won't be able to recruit him unless you put in a lot of effort," says Kuo. "Smale can go to many big universities, you know? MIT, Princeton ... But if he returned to CityU, he could have a bigger impact on the academic community and propel the university's growth."[3]

Of course, moving around the world again was a big decision. Steve said he would discuss the proposal with Clara and weigh the pros and cons. Overall, he felt positive about the opportunity. Based on Smale's recollection, he accepted a contract with a $500,000 annual salary, rcquiring at lcast six months in Hong Kong per year and no teaching duties "Way Kuo made it much more attractive for me to return to CityU," says Smale. "He was very supportive."[4]

Steve and Clara returned to Hong Kong in August 2009. Just as he had seen them off in 2001, Zhou was there to pick them up again. "I remember him being excited because he was tired of the working arrangement at TTIC," says Zhou. "But to my surprise, Clara did not seem as happy. My impression was that she moved to Hong Kong simply because Steve wanted to go back."[5]

Laura also recalls her mother feeling less than enthusiastic about a second stint in Hong Kong. Nearing 80, Clara worried she was too old to relate to other faculty spouses and make new friends. "That second period wasn't as good for my mom," says Laura. "She played badminton a lot the first time and got to know all these other people. But when they returned, she didn't meet as many people and was less engaged."[6]

While Clara took longer to settle in, Steve adapted quickly to life at CityU. Walking across campus, he was easily recognizable from a distance with his thick white hair and de facto uniform: an open-collared shirt, pleated trousers, and supportive walking shoes. He did not put much thought into his wardrobe and preferred to keep things simple.

"By that time, I dressed informally. I hardly ever wore a tie or a suit," Smale says. "I usually wore something casual or a dress shirt without a tie. I think of it as more of a sporty shirt. When my shoes got worn out, I just bought new shoes—casual shoes, comfortable shoes."[7]

Kuo, who remained the president of CityU until retiring in 2023, recalls running into Smale on campus regularly. Occasionally, they would stop and enjoy lunch together. "He was an exceptionally energetic person. He exercised a lot. After hiking, he would come back to the office to work," recalls Kuo, adding that Steve had a natural affability, a trait not always seen in the upper ranks of academia. "He's not only a good mathematician, but he's also a good human. He always kept his door open for people to visit him. He also believes in academic freedom and supported me."

Kuo admires Smale's ability to change his professional interests every few years. "That's something that I always encourage youngsters to do," says Kuo. "Our lifespan is limited, but it's a long-term learning process. It's very encouraging."

Like his first appointment at CityU, Smale's presence attracted high-caliber talent to the university. "His impact on the mathematical community in Hong Kong has been tremendous. He has certainly left a fingerprint on the history of the City University of Hong Kong," says Kuo.[8] "Today, CityU has several excellent academic departments—and Smale influenced that."

A small but mighty roster of top-notch faculty has buoyed CityU's reputation over the past 15 years. "In 2008, the university was not as noticeable as it is today," says Kuo. "If you look at all kinds of rankings [today], we're doing well in almost every ranking."

For example, in 2008, CityU ranked No. 147 on the QS World University Rankings of top global universities. From 2020 to 2022, it hovered around No. 50 on the report[9] and was named "Most International University in the World" in the Times Higher Education World University Rankings 2024.[10]

Several other accolades underscore the university's growth and standing. In 2022, Stanford University included more than 180 CityU researchers—almost a quarter of the total faculty—on its list of the world's top 2 percent of most-cited scientists.[11] That same year, 29 CityU scholars were commended on Clarivate's annual Highly Cited Researchers list.[12] CityU ranked among the highest in Asia in both reports by faculty size.

"Smale joining us enabled us to attract a lot of good faculty members to CityU," says Kuo. "We have done that continuously. But certainly, Smale was an extremely successful case for us."

Respect, support, and warmth from the top brass made Steve feel valued in Hong Kong. He also had several fascinating mathematics problems to work on, including one at the intersection of neuroscience and mathematics. While at TTIC, Smale had been collaborating with Poggio to create a mathematical model of the visual cortex.

With financial support from CityU, Smale could continue working on pattern recognition with Poggio alongside Andrea Caponnetto from CityU, Lorenzo Rosasco, and Jake Bouvrie from MIT. The five researchers proposed a model of the neural response (how the brain receives and processes visual stimuli) using powerful machine learning techniques, such as learning algorithms and kernel methods.

"What we were doing is trying to, you know, recognize faces, and recognize cats and dogs. And what we did was to use the system in the brain—[these patches] in the brain—and they would fit together in different layers," explained Smale in a 2017 interview with *Plus* mathematics magazine.[13] "That gave us some kind of mathematical foundations for vision."

The resulting paper, "Mathematics of the neural response," published in 2009, provided a model that could test how computed results compare with how humans view images.[14] Setting the stage for further developments, their theoretical foundation could help explain how some artificial vision systems—such as advanced visual object recognition and rapid categorization of complex images—work and how accurate they are.

"The different levels of the brain that correspond to people seeing something, it goes through different levels, and it is not too much different from artificial vision," explained Smale in a 2021 interview with the American Mathematical Society (AMS) conducted by Indika Rajapakse, who would later become a close collaborator.[15]

Smale's study of vision exemplifies how he deployed mathematical tools to help make sense of natural phenomena. In a 2007 interview with the AMS, he said:

> *I consider* [mathematics] *as just one important thing to study. I see myself broader, as a scientist, even a little bit of an artist. So mathematics is not the sole motivating thing in my life—far from it. But I do see beauty in mathematics because of its elegance and its ability to idealize the things you see in everyday life. Understanding the things around you has been the motivating factor for me for the past 40 years.*[16]

B. Nat in Hong Kong, diving in the Philippines

In August 2009, Nat moved to Hong Kong for 10 months to work on a project with his father. Since their earlier discussions on the bumpy roads of Costa Rica, they had been speaking on the phone several days a week to volley ideas and hash out concepts. "I was overdue for a sabbatical [from The University of Utah's Department of Mathematics]," recalls Nat. "I had this good project to work on with my dad, so it made sense to go to Hong Kong."[17]

Steve and Nat were working on a problem related to Hodge theory, a powerful tool that connects algebra and geometry. Named after British geometrist W. V. D. Hodge, who introduced the concept in the 1930s, Hodge theory opened up many possibilities in mathematical research, especially related to the study of algebraic cycles. But newer applications of mathematics—such as statistics, data analysis, vision, and pattern recognition—called for a similar theory in a broader setting.

The two mathematicians hoped to develop a discrete version of Hodge theory to study metric spaces with a probability measure. "[Hodge] is a theory involving manifolds, a generalization of smooth surfaces in any dimension," explains Nat. "We were trying to develop an analogous theory for metric spaces, which are much more general objects—it could be a set of points or jagged edges glued together in weird ways."

For their first paper, "Hodge theory on metric spaces," which was published in February 2012, Nat and Steve worked with co-authors Laurent Bartholdi and Thomas Shick of the University of Göttingen in Germany to extend the classical theory into the more abstract setting of metric spaces.[18] Soon after, Nat and Steve extended the theory further with "Abstract and classical Hodge-de Rham theory," which was actually published a month earlier than their first paper. By taking these classical theories in new directions, Steve hoped to shed more light on the mathematical foundations of vision and pattern recognition.

It was Nat's first time collaborating professionally with his father. Though they had different styles of thinking and working, the two mathematicians worked well together.

Every day, we would talk about our ideas. He came up with a lot of ideas—a lot of them were wrong. Sometimes, they were obviously wrong. Sometimes they weren't obvious, but there was no reason to think they would be right. He's more of an optimist. Sometimes, though, he was right. I usually work through the details of figuring a lot of it out, I would say. He came up with some of the more important ideas.[19]

While in Hong Kong, Nat and Hermine lived in the Jockey Club Humanity Hall, an apartment building on campus. Nat could walk to work in a few minutes and appreciated the easy access to public transit. He recalls:

I spent most of the time working and talking with my dad. But my wife and I also did a lot of day trips and hiking trips because you could access just about anywhere in Hong Kong within an hour or an hour and a half. We took ferries to some islands and did a fair amount of climbing. I was serious about bouldering, so it worked out well.

They also took advantage of Asia's excellent diving with a trip to the Philippines. In December 2009, Steve, Clara, Nat, Hermine, Celine, Laura, and Kay made their way to Sangat, a remote island known for diving in Palawan, southwest of Manila. Not more than three square miles, the tiny island's jungled interior offers bird watching, rock climbing, hot springs, and some trekking.

"We stayed at a diving resort, which you could only access by boat," says Nat. "There were a few trails, but there's a lot of

Celine and Clara in the Philippines.

limestone rock and a fair amount of vegetation, so you can't go very far."

When it came to diving, the family wanted to explore the area's collection of shipwrecks, most of which are considered historically significant Japanese warships and supply vessels sunk in 1944 during the World War II battle of Coron Bay.

Before exploring the rich underwater world, Nat and Celine learned to dive using enriched air, known as nitrox. The term refers to a gas mix containing more oxygen and less nitrogen than normal air (78 percent nitrogen and 21 percent oxygen). When using nitrox, divers can enjoy longer submersions and shorter surface intervals because their bodies absorb less nitrogen (without adequate off-gassing, nitrogen accumulation can lead to decompression sickness, or "the bends").

During the trip, Nat, Celine, and Kay used enriched air to dive into four or five shipwrecks. "On one dive, we swam through a wreck, about 300 or 500 feet from one end of the boat to the other," recalls Nat. "We were swimming along this corridor, which was impressive."

Though out of practice, Smale, then 79, joined two or three dives during the trip, including a wreck that was a little more open.

"I remember diving into a sunken ship probably 60–80 feet deep, going from one end to the other and out. I had never done that before," says Steve. "Maybe I saw some sunken ships in the Caribbean. But in Sangat, it was more eventful because you could go into the ship and scuba along the inner side."[20]

C. Taking mathematics in new directions

In 2011, Steve began working with Berkeley Professor Moe Hirsch and Boston University Professor Robert L. Devaney—two of his former Ph.D. students who had become prominent mathematicians in their own right—on the third edition of their classic textbook, *Differential Equations, Dynamical Systems, and an Introduction to Chaos.*[21] Originally written by by Hirsch and Smale in the 1970s as *Differential Equations, Dynamical Systems, and Linear Algebra (Pure and Applied Mathematics)*, the textbook had become a foundational textbook for graduate courses—not to mention Smale's most-cited publication.

"Back then, many students did not have a strong exposure to linear algebra—a field of mathematics that was essential to solve the simplest differential equations, namely, linear systems of ordinary differential equations," says Devaney. "So Steve and Moe needed to include hefty sections in the book dealing with the basics of linear algebra."[22]

In the decades following, much had changed in terms of technology and dynamical systems. By the turn of the century, the book needed significant adjustments, and Hirsch and Smale entrusted Devaney to update the second edition from 2002 to 2004. "Of course, they both reviewed all that I wrote, approved all of the changes, and made some modifications," says Devaney.

Since most graduate students had a better idea of linear algebra by the early 2000s, he cut most of the basic information on this topic. In its place, he incorporated more contemporary research and applications related to discrete dynamical systems and chaotic behavior, which had greatly influenced math and science over the past few decades.

In the early 1960s, Edward Lorenz, a meteorology professor at MIT, discovered what became known as chaos theory. Lorenz attempted to predict future weather patterns using mathematical models. Instead, he realized that predictions varied dramatically depending on the accuracy of the starting condition. Just an extra decimal point could cause wildly varying results. His observations ultimately gave rise to the "butterfly effect," a thought experiment detailed in his 1972 paper, "Predictability: Does the flap of a butterfly's wings in Brazil set off a tornado in Texas?" Aside from weather patterns, other real-life examples of chaos theory include climate, heartbeat irregularities, fluid flow, and chemical reactions.[23]

The concept of chaos emerged in mathematics in the 1980s and became a major focus of Devaney's research. At the forefront of the field, he wrote two successful textbooks on chaotic dynamical systems and has since become a leading expert in the chaotic dynamics of complex functions. Devaney, who retired from Boston University in 2018, says:

> *Earlier on, it was hard to incorporate contemporary ideas and research topics into the mathematical curriculum. But now we can show our students many accessible topics in the field of chaotic dynamics using just the quadratic function, which really seems to impact their interest in pursuing mathematics.*[24]

He also included one of Steve's greatest discoveries: the "Smale horseshoe." "This was one of the first examples of a chaotic dynamical system that was completely understandable," adds Devaney. "Of course, back then, 'chaos' was not a subject of widespread interest, but the horseshoe map certainly began to evoke interest in this topic."

When it came time to work on a third edition, Hirsch and Smale once again agreed for Devaney to make the changes with their input and approval. Devaney expanded the book's theoretical approach to dynamical systems and chaos, weaving in more exercises and examples. The book, which provides a rigorous yet accessible introduction to differential equations, remains a standard text in undergraduate and graduate classrooms.

"I am pleased that the book is doing well. I think the combination of basic and important topics dealing with differential equations, together with the inclusion of many modern topics in this field, has a lot to do with this," says Devaney. "It is not so often that textbooks accessible to undergraduate students in mathematics can incorporate such modern topics into the curriculum."

While Devaney updated the textbook, Smale dove deeper into the mathematics of biology. Back at TTIC, Smale had attended several talks by computer scientists working in DNA. He realized he could transition from visual pattern recognition to genetics and immunology using a similar toolkit.

In a 2017 interview with *Plus* magazine Smale said:

> *At some point, it seemed to me that* [genetics] *was a lot easier than what I was doing because vision uses two-dimensional bunches of pixels, versus a one-dimension string ... So I figured what I was doing with pattern recognition, I could do much more with DNA.*[25]

He shared his ideas with a few biologists, who responded with encouragement. Smale continued:

> *When I got* [to Hong Kong], *I got a huge amount of support.* [CityU] *gave me everything. That's why I moved. In particular, I could put together a* [working] *group, and we learned biology that uses pattern recognition techniques to understand strings, so we were doing immunology and learning theory, machine learning. And that's the way it got started into biology.*

With support from CityU, Smale's group explored what he considered the century's most fascinating area: the intersection of intelligence, data, and biology. Observing the increasingly crucial role of data processing in emerging technologies, Smale first focused on computational biology and high-dimensional data in bioinformatics (the application of computational and analytical tools to capture and interpret biological data).[26]

Over time, his work evolved beyond algorithms toward more substantial questions about biological structure. In essence, Smale hoped to harness mathematics to construct the laws of biology. He was particularly interested in understanding the structure of genes and proteins, which he believed could have implications for diseases linked to gene mutations, such as cancer.

Back then, the mathematics of biology was a totally undeveloped area. "For one thing, biologists know statistics, computer science, and some math, but they don't know differential equations. Differential equations are the things that knit physics together with math, and I am trying to do that in biology," he said in a 2017 interview with the Heidelberg Laureate Forum. "So it's not a traditional field at all."[27]

Composed of Ph.D. students, postdocs, and assistant professors, Smale's working group met at least once weekly. Smale also invited visiting professors from all over the world, including collaborators like Poggio, Shub, and Rosasco, among many others. The working group became a magnet for bright minds in computer science, mathematics, statistics, and numerical analysis.

Many of the regular attendees became leaders in varied but interrelated fields. For example, Shuaicheng Li became a professor in CityU's Departments of Computer Science and Biomedical Engineering, studying bioinformatics, machine learning, and algorithms. Meanwhile, Massimo Andreatta went on to work at the Ludwig Institute for Cancer Research in Switzerland, applying innovative methods, like machine learning, to prevent and control cancer.

Yuting Wei, an assistant professor of Statistics and Data Science at Wharton, has been developing theoretical and algorithmic tools for learning from high-dimensional and structured data. On the commercial side, Quanwu Xiao is the Senior Director and Chairman of the Big Data Committee at Trip.com Group, where he applies his expertise in data analytics to real-world travel solutions like personalized flight recommendations and predictive modeling for airfare pricing and call center staffing.

D. A birthday conference in France

For Smale's 80th birthday in 2010, Cucker organized and hosted a math conference at his vacation home in southern France. Though happy to do it, Cucker says the plan took him by surprise during a lunch in 2007 with Wong and Jean-Pierre Bourguignon, then the Director of the Institut des Hautes Études Scientifiques.[28]

The conversation went something like this:

> ***Wong:*** *Oh, you know, Felipe has this large house in the south of France, and I always hoped we could organize a conference there.*

> ***Bourguignon:*** *Oh, that's a great idea. You know, in 2010, Steve will turn 80, and then we could organize a conference honoring him.*
>
> ***Wong:*** *We could have it at Felipe's house! You can handle it, right, Felipe?*

By the end of the lunch, the two bigwigs had organized the conference at Cucker's house and committed to contributing 5,000 euros (worth roughly $7,410 today) each to cover the expenses.

"I didn't say anything, but I thought, 'I only have six bedrooms there! What sort of conference do you want to organize?'" recalls Cucker, laughing. "I was thinking, 'What do I do?'"

Cucker did not waste any time, quickly assembling a team to help bring the conference to life. He turned to long-time friend and Argentine mathematician Teresa Krick, a professor at the University of Buenos Aires, to help on the organizational side. Meanwhile, Cucker's friend Estela Ponisio, a professional chef, generously offered to handle the meals.

His wife, Antonia, works in import-export logistics, so she sourced and shipped all the furniture and tableware they would need for the occasion. She bought cutlery, plates, furniture, a sofa, and even a blackboard on wheels. Completing the team was Cucker's long-term domestic worker, Floresca Nitafan, who joined him in France to assist with cleaning and cooking.

Cucker, Krick, Ponisio, and Nitafan arrived the Friday before the conference. On Saturday morning, they stocked up on essential supplies, cooking ingredients, and wine. "We bought three or four bottles of different red wines," recalls Cucker. "We had a tasting and decided unanimously that one was the best. The next morning, we returned to the supermarket and bought out the shop's entire stock of that wine."

When guests arrived, they were enthralled by the picturesque rural setting. Located in Les Tuileries d'Affiac, a remote hamlet about an hour from Narbonne, the house is surrounded by vineyards and mountains. Coincidentally, another mathematician, Max Dickmann, also co-owned a home in the same hamlet, so half of the party stayed with him while the rest slept at Cucker's.

Roughly 20–30 people attended the conference, doubling the hamlet's population for the week. Among them were Bourguignon, Fields Medalist Jean-Christophe Yoccoz, Lenore and Manuel Blum, Tomaso Poggio, Arieh Iserles, Michael Shub, James Renegar, and Nat Smale.[29]

Though the atmosphere felt casual, the conference had a relatively organized structure. Estela prepared a buffet breakfast in the courtyard at around 9 or 10 a.m. daily, followed by math talks until roughly 12:30 p.m. The presentations occurred in a large, open living room where Cucker set up the blackboard. After lunch, everyone reconvened for another few hours of talks, then came a happy hour of canapes and wine around 5 before dinner around 7:30 p.m. Cucker recalls:

> *The first day, Monday, went as planned. I realized we drank a lot, but that was fine. On the second day, the talks ended at 5 p.m. once again. At 5:15, I heard the sound of corks popping out of wine bottles. It turns out that someone in the audience noticed where we hid the wine. I tend to think of us mathematicians as a bunch*

> *of very moderate people. But no, no—there was no more wine after just two days!*[30]

For the remainder of the conference, the mathematicians drank wine that Cucker had made himself. Originally a 19th-century winery, his vacation home features winemaking facilities and two massive 20,000-liter wine barrels, one with a door that could be used as a makeshift bedroom in a pinch.

"Felipe renovated his place very nicely, and the food was delicious. It was like you see in the movies when French families get together with a table outside and bring out all this wonderful food," recalls Lenore Blum. "It was like a family party that had math talks."[31]

But there was a shadow hanging over the event. Steve was worried about Clara's health. "I had known Clara for 20 years by that point. And this was the first time I noticed something was wrong," says Cucker. "It was sad because, on the one hand, it was a celebration for Steve. On the other hand, he was apprehensive about Clara. You could see it. You could feel it."[32]

Nat also expressed concern about his mother. A day before departing for the birthday conference, he recalls a conversation with her that set off a red flag.

> *We stopped by my parents' place to pick up a few things before leaving for France, where we'd see them the next day. I said goodbye, and my mom replied: "Oh, are we going to see you again?" That shocked me because she knew we were all going to France.*

After the birthday conference at Cucker's, Steve and Clara spent some time driving around France. During their travels,

Steve grew increasingly concerned. Clara got lost, struggled to walk upstairs, appeared less engaged, and made decisions that seemed out of character. "It was in France that I knew something was wrong," he says.[33]

E. Losing Clara

When the Smales returned to Hong Kong, Clara's mental and physical deterioration accelerated. "I think it'd been happening slowly for at least a year before this, but it suddenly sped up," says Nat.

Just a month before the birthday conference in France, she had taken the MTR to Immigration Tower in Wan Chai, a district on Hong Kong Island, to renew their visas. "It's a complicated route," explains Nat. "She had to change MTR lines twice, walk 20 minutes, go up 20-plus floors to find the right office. She did all that by herself in June."

After Steve and Clara returned from France in July, Clara was hospitalized. "The immediate sickness was a leg infection, but the examinations revealed some hippocampus atrophy, serious body edema, hypothyroid, and especially an acceleration of protein malabsorption," Steve later wrote on a memorial website for Clara. "When Clara and I were married in 1955, she told me that her doctors had diagnosed the protein problem and were unable to treat it in spite of the danger of the condition."[34]

Steve believes the protein malabsorption problem was dormant for years, then, at the age of 80, suddenly overcame her. Mentally, she was getting worse and worse, too.[35] By September, she could not remember where they lived.

As Clara grew sicker, Steve started calling Nat and Laura more often—something he had not done before. "He would call me almost every day when my mom was sick," says Laura. "That suggested to me that he had a certain anxiety about what was happening and did not quite know what to do. He may also have been lonely since my mom was no longer able to be a companion."[36]

Worried about her parents, Laura flew to Hong Kong toward the end of August. During the visit, Clara could still walk and talk but struggled. For example, she had trouble getting in and out of the bathtub and often seemed confused. Steve took many precautions with Clara, but sometimes, his independent streak got the better of him. Laura remembers:

> *Once, we were walking back to their apartment, and my mom was not in a good state. She was losing her way, and I didn't know the city well. She ended up tripping and falling in the apartment's parking garage. He was so far ahead he didn't notice. He may have understood I was there and wouldn't leave my mom. It bothered me, but I don't know how he would have been if I hadn't been there.*[37]

Steve admits he can be impatient and prefers to walk at his own pace. He likes to be the first to arrive, the first on the plane, the first to finish a hike. He cannot pinpoint why; perhaps it is his competitive nature or singular focus, but it was not the first time he was admonished for forging ahead.

"My mother visited me in Hong Kong [in 1996] when she was 90. We were going upstairs to a restaurant, and I got ahead of her," Steve recalls. "She was walking slowly and got really upset with me because I didn't help her get up the stairs. She was probably right."[38]

Old habits aside, Steve did his best to navigate the challenging and painful situation. As his office was just a few minutes away, he could return home several times daily to check on Clara and spend time with her. "He was able to provide a lot of comfort," recalls Laura.

Soon after Laura left, Clara was hospitalized for the fourth and final time. It was hard on everyone. Clara tried to pull out the IV lines in her arm and fell out of the hospital bed trying to escape, recalls Steve. The hospital staff had to restrain her.[39] He consulted Clara's sisters (she had five sisters and two brothers) and decided to take her home, where she would be more comfortable.

"She hated the hospital," says Steve. "The doctors said we had to bring her back. But she, her sisters, and I, we all said no. Her sister Norma, who had a similar experience with her husband, said what I had done, taking her out of the hospital and bringing her home, was the right thing."[40]

Nat, who had left Hong Kong in June 2010, flew back to see his mom from December 2010 to early January 2011. On the memorial website, he wrote:

> *After I came back to the US last summer, her memory got rapidly worse, and she developed a number of physical problems. I, of course, heard regularly from my dad about her condition throughout the fall, but when I visited Hong Kong last Christmas, it was heartbreaking to see how much memory she had lost. However, she still had enjoyable moments, especially being with my dad, who could still make her laugh.*[41]

A few weeks later, in February, Laura visited her parents. By that point, Clara was bedridden and needed help dressing, eating, and bathing. Steve could not take care of Clara alone, so he hired Melissa Abon, a domestic worker from the Philippines. Melissa was the daughter of Floresca Nitafan, who worked with the Cucker family.

"That was one of the few occasions in which Steve asked me for help," recalls Cucker. "He described Clara's situation to me, and he was quite desperate. He really was lost. He realized he could not do the tasks that Clara's condition required and needed help."[42]

Cucker suggested he hire a domestic worker to help care for Clara, but Steve did not know where to start. While speaking to Floresca about the situation, Cucker learned that Melissa's employers had just left Hong Kong, so she was looking for a new position. "That's how the daughter of Floresca ended up working for Steve, and it turned out to be a very good recommendation," adds Cucker. "Clara was very happy with Melissa, and Steve was, too."

Well-educated with a degree in computer science, Melissa was gentle, caring, and calm. She was exactly what the Smale family needed during this challenging time. Laura says:

> *Melissa was amazing. She was just wonderful. She took great care of my mom. When I was on that trip, my mother needed to be fed by a spoon, and Melissa really wanted to do it. She was so much better than me, and seemed to be very, very caring as she did things like that.*[43]

Clara and Steve enjoying time together outdoors.

That visit was very hard for Laura. Her mother could not speak much then, and Laura did not think Clara knew her anymore. "There'd be about five minutes a day where I could get her to smile. But my dad was the one who she clearly wanted to have around," says Laura.

At that time, Clara still seemed to recognize Steve. "When he entered the room, she was clearly relieved and happy to see him. He was the one she turned to when she was especially anxious," adds Laura. "He was central to her."

Two months after Laura's visit, Clara Davis Smale died on April 28, 2011. She was 81 years old.

For Steve, her passing came as a relief. What happened to Clara was gradual, and she did not have a real life by the end. He says:

> *She had no awareness—nothing. She deteriorated. Her whole body, including her brain. So, in the last months of her life, she couldn't recognize anybody but me. And with me, it wasn't clear ... When I came into the room, she would follow me with her eyes. So it was a slight recognition, but she couldn't talk.*[44]

Steve and Clara were married for 56 years. Their relationship was complicated at times, but they had a strong bond and supported each other. "I think people always thought of her as the nicer of the two of us," says Steve.[45]

When asked about their relationship, family members describe it like opposites attracting. Clara was more cautious, while Steve took risks and thrived on adventure. Although she had lived all over the world, Clara felt anxious about traveling. Meanwhile, Steve could not get enough of it. Steve was highly ambitious, while Clara took less initiative. He could be very social, outgoing, and sometimes egotistical. Clara was more gentle, soft-spoken, and very humble.

"She was more timid, in general. Once, I remember when they visited Kay and me in Kenya, I took her to the river where we bathed, and a small cobra was moving across the path," Laura says. "When she saw that, she said she didn't really need a bath. After that, she'd get a bucket of water and clean herself that way."[46]

Clara was more adventurous when she was younger, as many people are. She worked on the *Michigan Daily* paper in college, actively participated in politics, especially the Vietnam anti-war movement, and moved around the world with Steve with two young children in tow. Not long ago, Laura recalls seeing a photograph that showed another side of her mother. "She was an actress in a play in Brazil in 1959 or so," says Laura, who was eight months old when her family moved to Rio de Janeiro. "I don't remember the play's name, but I was so surprised. That's the kind of thing that would have made her nervous later in life."

On reflection, Clara and Steve had much in common, too. Most importantly, their children. "When we were little kids, I don't think my dad was that interested in us," says Laura. "But as we got older, they both wanted to know how and what we were doing. Nat and I were something that brought them together."

They both valued independence and, as a result, had similar approaches to parenting, giving Nat and Laura wide latitude from a young age. "I always appreciated the freedom they gave us, which was something almost none of our friends had," says Laura, adding that she was allowed to take a European train trip with a friend at 13. "My mom was the youngest of eight kids, so her parents were probably very relaxed when they had her. She probably grew up with a lot of freedom, too."

Clara and Steve were also good listeners, patient, intelligent, and kind. "My mom was the kind of person who liked to listen to other people rather than talk," says Laura. "She was warm and wonderful but didn't spoil us. She set limits in a gentle way."

Nat shares a similar sentiment, writing on the memorial website that he missed his mother's kind, comforting presence the most. "When I was young and had a bad day at school, it was a relief to come home to her," wrote Nat. "Even when I was grown up and no longer living at home, it was always a great pleasure to visit and just be around the house with her."[47]

Their home was a warm one, full of life and music. Steve and Clara had a penchant for jazz and opera music, which they regularly played at home. The couple shared a sense of humor and laughed together often.[48]

"My parents had a common interest in politics and the news and *The New York Times*, and they spent a lot of time talking about what was happening in the world," says Laura. "With the minerals, my mother started just being too nervous about spending so much money—sort of gambling on them—but that changed over the years, and she became more involved and appreciated them more."

Long before Clara fell ill, Laura remembers her parents taking long walks around the Berkeley Marina, where a 4.8-mile loop traces the waterfront. They would start together, then Steve would move ahead faster, and Clara would continue at her slower pace. They would meet again at the end of the loop and head home together. "That's how they took walks in the Berkeley area," says Laura. "I remember that being part of their routine."[49]

Every once in a while, Steve and Clara would hold hands and talk while they took walks. For Laura, one moment stands out in her mind. In the summer of 1979, her parents visited her in the high Sierras, where she worked as Kay's research assistant, to meet Kay. "They were walking hand in hand on some trails through the wildflowers, and I remember Kay saying that it seemed very pleasant and romantic," says Laura.[50]

Most of all, Laura remembers her mother's laugh. When living in Berkeley, Clara would read *The New York Times* and *The New Yorker*, starting with the comics. She says:

> *I remember hearing her just laughing and laughing by herself from the big living room* [at their home in Berkeley] *as she read the comics. If she really liked one, my mom would bring it over and show me. She had a good sense of humor. It wasn't that she would make other people laugh, but she laughed so easily, genuinely and wholeheartedly, at so many things.*

Biology, Berkeley, and a Lasting Legacy

16

A. Clara's memorial

Watching Clara deteriorate physically and mentally over the past seven months caused Steve deep sorrow. She had been by his side for over half a century, and now, he would have to learn to live without her. Encouraged by friends and family to look after himself, Steve continued to lead a full life after her death.

Melissa continued working for Steve, taking care of the shopping, cleaning, and cooking, and providing welcome company and support. About a month after Clara passed away, Nat and Laura visited Steve in Hong Kong to spend time with their father and, in Nat's case, attend the Fourth International Conference on Computational Harmonic Analysis, organized by Steve and Zhou from May 23 to 27, 2011. During their trip, Laura recalls going out to dim sum together and visiting a beach out in the countryside, while Nat did some climbing.[1]

That summer, Steve accepted several opportunities for work and travel. In July, he attended the 2011 FoCM conference in Budapest, where he gave a plenary talk on the mathematical foundations of data analysis. It was a notable conference, celebrating the society's 15th anniversary and the inaugural Smale Prize. Named in Steve's honor, the award recognizes young mathematicians whose work furthers the integration of mathematics and computation.

Professor Snorre Christiansen of the University of Oslo, Norway, was the first to receive the Smale Prize for his pioneering work in modern scientific computing. The award summary commends Christiansen's use of Calderón's formulas to construct preconditioners for the electric field equations and contributions to the development of finite element exterior calculus, among other notable achievements.[2]

After the conference, Steve flew home to California to host a memorial he had organized for Clara, whose cremated ashes had been buried in Hong Kong. Laura and Kay volunteered to take care of the flower arrangements, and Kay also created a photo slideshow of Clara that played throughout the event. "People appreciated the photos. It grounded and centered things on my mom," adds Laura.[3]

On August 1, the day of the memorial, around 60 to 80 people[4] gathered at the DoubleTree by Hilton Hotel Berkeley Marina. Among them was Zhou from CityU, as well as many other friends from the mathematics and mineral realms. Roughly a dozen people shared heartfelt reflections about Clara, including one of her closest friends, Jean Thomas, who was married to the late Berkeley mathematician Emery Thomas.

Visibly emotional, Steve stood up and recounted a few memories about his late wife. "My dad said some very nice things and wept. He had a hard time speaking," says Laura. "That's the main thing I remember about him from the event. But overall, it was a cheerful scene full of friends we hadn't seen in a long time."[5]

B. New results in the mathematics of biology

Following the memorial, Smale immersed himself in the mathematics of biology. He returned to Hong Kong for the fall semester and embarked on a three-year project entitled "New Directions in Hodge Theory," funded by the Research Grants Council of Hong Kong. Collaborating with several members of his working group, Smale developed a kernel-based algorithm to compute the similarities of amino acid sequences and predict their binding affinity.

In "Introduction to the peptide binding problem of computational immunology: New results," published in the October 2014 issue of *Foundations of Computational Mathematics*, Smale and his co-authors—including CityU computer science professor Hau-San Wong and computer science Ph.D. student Wenjun Shen—presented a cutting-edge algorithm called KernelRLSpan, which offered a simple and promising methodology for studying immunology and amino acid sequences using string kernels. According to the paper's introduction, the authors hoped their results would enhance scientists' understanding of biological and immune processes:

> *Large scientific and industrial enterprises are engaged in efforts to produce new vaccines from synthetic peptides. The study of peptide binding to appropriate MHC II molecules is a major part of this effort. Our goal here is to support the use of a certain "string kernel" for peptide binding prediction as well as for the classification of supertypes of the major histocompatibility complex (MHC) alleles. We hope that this will contribute to the prediction of vaccines and the understanding*

> *of organ transplants. This work actually could be applied to any other receptor-binding system where large binding sets are available.*[6]

Teaming up again with Shen, Wong, and a few other researchers, Smale expanded on the initial findings with a follow-up, "MHC binding prediction with KernelRLSpan and its variations," published in the *Journal of Immunological Methods* in April 2014. The authors adjusted their algorithm to predict the binding affinity of a broader class of molecules called MHC I.[7]

C. Sailing around St. Martin

In the summer of 2012, Steve embarked on another sailing trip alongside Nat, Hermine, Laura, Kay, Cucker, Michael (Kay's ex-husband), and Sarah (Michael's wife). They met in St. Martin, a Caribbean island governed by two nations: the northern half of the island (Saint-Martin) is part of France, while the southern half (Sint Maarten) is an autonomous region of the Netherlands. The island has had an open border for centuries, and both sides offer excellent cuisine, multicultural heritage, pristine beaches, lagoons, and other natural treasures.

After provisioning the boat, another Beneteau 50, the group set off on a 10-day voyage, beginning with a short sail north to the neighboring island of Anguilla. It was a smooth passage, taking only a few hours in calm, sunny weather. While exploring the laidback British Overseas Territory and its outlying islands, Steve recalls dining at local restaurants and meeting other sailors.

On the next leg, a longer sail southeast to St. Barts, Steve says conditions worsened as they passed through an unprotected stretch of water. It was not raining, but strong 25–30 knot easterly trade winds kicked up 10-foot waves. "The sailboat heeled [tilted] a lot," says Smale. "It heeled so far, you could have slipped into the sea. We wouldn't have been able to recover someone in those conditions. It is scary when I look back on it, but somehow, I wasn't too scared at the time."[8]

Upon arrival in St. Barts, they explored the small volcanic island, mostly on foot. Steve invited more people to help sail, including Kay and Michael, than he did on the British Virgin Islands trip. Due to the ship's large dimensions, they had trouble anchoring, which caused stress and frustration, but overall, Nat says they worked pretty well together.

Cucker enjoyed the scenery and company but admits he was not an asset on deck. "Even though I knew nothing about sailing, Steve was still happy to have me there," says Cucker. "That points to his generous nature because, honestly, I was dead weight."[9]

As they cruised around the islands, Cucker recalls spending long, leisurely hours sitting in the cockpit chatting with Steve and the other sailors, with occasional interruptions when it came time to trim the sails or prepare to jibe. When they were anchored or the conditions were calm, Cucker had time to speak with Steve about mathematics.

"He has an unorthodox way of working," says Cucker. "Most mathematicians like to work at a large institute with huge blackboards and a big library. Steve would also consider that an excellent place to work. But why not also do mathematics while sailing around the Caribbean?"

As he had on the beaches of Rio, along the hiking trails of Hong Kong, and in the vineyards of France, Steve successfully merged travel, family, friends, and math. "My work is global. It does not depend on my location or job," says Smale. "I can do it anywhere. It's just more pleasant to do mathematics on the beach."[10]

D. Exploring the human genome

Back on dry land, Steve's work on computational immunology and amino acid chains sparked some interest in the scientific community. His alma mater, the University of Michigan,* invited Smale to speak on the subject in 2013. During that four-day trip to Ann Arbor, an enthusiastic visitor knocked on his door.

"When I heard Dr. Smale was going to be here [the University of Michigan], I told the organizers I wanted to spend at least a half day with him," says Indika Rajapakse. "When I walked into the room, he must have wondered, 'Who is this guy?'"[11]

With a big smile and even bigger ideas, Rajapakse could not wait to meet Smale. At the time, Rajapakse held dual appointments at the University of Michigan, Ann Arbor, as an Assistant Professor of Computational Medicine and Bioinformatics in the Medical School and an Assistant

* Steve earned his bachelor's degree in mathematics in 1952, master's in 1953, Ph.D. in 1957, and an honorary degree in 1996 from the University of Michigan.

Professor of Mathematics in the College of Literature, Science, and the Arts. As of 2023, he became a full professor and remains one of the few biologists at the university with an engineering and mathematical background, enabling him to study the human genome from an interdisciplinary perspective.

"Dr. Smale was trying to understand the dynamics," says Rajapakse, who was born in Sri Lanka. "I was, and still am, interested in reprogramming cells. You can reprogram biological systems in biology, but you need to understand the dynamics, so I was eager to talk to him."

In his work, Rajapakse focuses on the human genome—the complete set of DNA instructions for our cells, which enables our bodies to develop and function. In humans, the genome comprises 23 pairs of chromosomes crammed into the cell's nucleus and a smaller set of mitochondrial DNA. Increasingly, biologists believe that changes in nuclei organization can lead to uncontrolled cellular division and, in turn, mutations that can cause diseases like cancer. By better understanding the 4-dimensional nucleome (the 3-dimensional nucleus over the fourth dimension, time), Rajapakse hopes to identify aberrations and reprogram cellular systems to resume healthy functionality.

Rajapakse was familiar with Steve's vast cannon of work and hoped to collaborate. By the end of that first meeting, Steve was open to working with Rajapakse, too, as he felt that ordinary differential equations through dynamical systems could do great things for science.

At that time, the Defense Advanced Research Projects Agency (DARPA) funded Rajapakse's research. Dedicated to discovering breakthrough technologies, DARPA has supported several innovations, including developing spacecraft, orbital satellites, stealth military technology, ARPANET (considered the precursor to the internet), automated voice recognition, and GPS receivers small enough to fit into personal devices.

When Rajapakse told his program manager about the prospect of working with Smale, DARPA invited them for a meeting. "About three months later, Dr. Smale and I went to DARPA together and got funded for our first paper," recalls Rajapakse. "We wanted to apply mathematics to answer biological questions and find solutions in human medicine."

Soon after, they began work on their first paper together, "Functional organization of the human 4D nucleome," with seven other co-authors, including Lindsey Muir, a Research Assistant Professor of Computational Medicine and Bioinformatics at the University of Michigan, who is married to Rajapakse. In the paper, published in June 2015, the researchers provided a mathematical framework for studying the 4-dimensional nucleome,[12] where linear DNA compresses and folds over time to regulate gene expression and organ development.

"We used data and ideas from dynamical systems, which Dr. Smale studied a long time ago, to get a hint about how the genome folds," says Rajapakse. "Every second in your body, millions of cells die, and new cells replace them. Behind that process is the genome, so we wanted to know: How is the genome organized? What is its topology? How do genes express functions?"[13]

Suppose they could decode how cellular frameworks restructure over time. In that case, Rajapakse continues, scientists could get closer to understanding how and why various diseases and biological processes develop, as well as how to target and treat them.

Their initial work stimulated a string of projects to establish the mathematical underpinnings of gene function. Their article "Mathematics of the genome," published in 2017 in *Foundations of Computational Mathematics*, provides a mathematical foundation for bifurcation from a stable equilibrium in the genome. In biological systems, a bifurcation event occurs when minor changes to inputs at equilibrium eventually impact the system's observable behavior.

Consider the flow of water. As the stream rate increases, the water transitions from a stable flow to an unstable one, ultimately forming a whirlpool. The point at which the system changes from stable to unstable, impacting the system's behavior, is the bifurcation point. In the context of cellular behavior, Smale and Rajapakse investigated bifurcation points in biological circuits that affect gene expression.[14] In the paper's introduction, they write:

> *Development of a single cell into a multicellular organism involves a remarkable integration of gene expression, molecular signaling, and environmental cues. Emergence of specific cell types depends on selective use of genes, a phenomenon known as gene regulation, at critical times during development.*
>
> *Our goal is to build a mathematical foundation for gene regulation and its dynamics. We believe this could lead to new insights in biology.*

As Smale investigated the mathematics of the genome, CityU invited the famous mathematician to become a senior fellow of the university's new Institute for Advanced Study (IAS) when it was established in November 2015. He accepted the prestigious appointment alongside several other high-profile scientists, including Nobel Laureates Jean-Marie Lehn and Serge Haroche, and fellow Fields Medalist Pierre-Louis Lions.

The IAS takes inspiration from the Princeton Institute for Advanced Study, where Smale was a member from 1958–1959 and again in the fall of 1966. With a vision for open intellectual exchange and specialized research that addresses critical global challenges, the IAS at CityU invites internationally recognized fellows in areas like mathematics, physics, chemistry, materials science, and biology to investigate new research areas and present their results during courses, workshops, lectures, and seminars.

Kuo says:

> *We could recruit many world-class mathematicians* [to IAS] *because of Smale. We have the best. We are a heavy-duty institute with several Nobel Prize winners and many mathematicians from France, Canada, the US, the UK, and Germany. Many influential people are associated with the institute, and Smale was one of the first. We're so honored to have him.*[15]

As part of his appointment, Steve had more funding to pursue research initiatives and host workshops on the mathematics of biology. For example, he and Rajapakse organized a five-day gene dynamics and chromosomes

Steve with a group of friends hiking in Hong Kong.

workshop in August 2016. They invited leading scientists to IAS, where they explored the interaction between genes and chromosome conformations in the nucleus.

During the visit, Smale showed Rajapakse around Hong Kong. "He is such a generous person. Dr. Smale took me out to lunch and dinner at Sen-ryo, this great sushi restaurant in Festival Walk," recalls Rajapakse. "He is an amazing host. He always wants everything to be simple but of the highest quality."[16]

Over the weekend, Steve took his guest hiking around Tai Long Wan in Sai Kung. They planned to hike to Ham Tin and catch a boat back to town. But when they arrived, there was no ferry in sight. Rajapakse recalls:

> *I was shocked. I knew I could walk back, but Dr. Smale was in his mid-80s. We had to walk back a few*

> *miles to another beach to find a boat. I was worried something was going to happen to him. But he's so determined and cocky. He said, "I can do it"—and he did.*

E. "Retiring" again and sailing to Grenada

While studying the human genome with Rajapakse, Smale moved home to Berkeley in 2015. Though he would remain an honorary professor at CityU indefinitely, Smale, then 84, was ready to be closer to his family.

"Living in Hong Kong was a big part of my life. I was happy there. I had good friends," he says. "I still have friends in Hong Kong, but returning to Berkeley was good because of my family."[17]

He also looked forward to spending more time in the house he and Clara had bought decades earlier. "It's a good house. Not flamboyant, not fantastic," he says. "The main good characteristic is the backyard. We have an extensive backyard. A beautiful backyard."[18]

Growing up, Nat and Laura spent their free time climbing on boulders or zip-lining across the backyard under the California sunshine. "When the kids were young, we installed a zip line, about 50 feet long, from one tree to another," says Smale.

When Smale returned to Berkeley, his grandson, Dylan, moved in with him. "Dylan indicated that he wanted to move here and help me out with cooking and doing errands," says Steve. "He does a lot for me. He cooks, shops, and goes to the grocery store. He's saved me some trouble."

At first, the arrangement was temporary because Steve returned to CityU for six months in 2016. When he retired from the university later that year, Smale was paid $1 million in a lump sum pension. Of that sum, he gifted $500,000 to each of his children to help them buy houses in the expensive Berkeley area.

In January 2021, after Nat retired as the Associate Department Chair of Mathematics at the University of Utah, he and his family moved into a house less than a mile from Steve. In 2022, Laura, a Professor Emeritus of Behavioral Neuroscience at Michigan State University, and Kay spent a year in Berkeley while on sabbatical. They moved to Point Richmond, about a 20-minute drive northwest along the San Francisco Bay, in 2023.

"I always wanted to move home to Berkeley, and my dad made it possible for Nat and I to move back to where we grew up," says Laura. "That's partly because he wanted to see us more. We were really happy to move back—it made things better for us and him."[19]

In the summer of 2016, when Steve was 86, the Smale family and Cucker took another sailing trip. This time, they explored the West Indies, starting in Grenada, home to the pastel-painted capital of St. George's and numerous spice plantations, before heading north to St. Vincent and the Grenadines. They chartered another Beneteau 50, cruising from island to island, stopping to enjoy the snorkeling, scenery, and local cuisine. Of the three Caribbean sails, it was the most pedestrian in some ways but more challenging in others.

"Coming and going from Grenada, you sail along the west side of the island, which is protected from the trade winds, so it's very slow going," says Nat. Later on, he says, there was more tension, and Steve made more mistakes than on previous sails. "Nothing disastrous happened. It all worked out. But my wife and I decided we wouldn't go on another sailing trip with Steve after that. He eventually decided that was fine."[20]

Steve, who has not been sailing since Grenada, glows with pride as he shares a letter from The Moorings, the world's largest yacht chartering company, which he received in January 2012, before their trip to St. Martin, and has kept all this time. It reads:

> *Congratulations, we have received and approved your request to skipper a Moorings yacht on a future sailing vacation. We are pleased to inform you that you're qualified to sail any Moorings bareboat yacht, monohull, or catamaran at any of Moorings' 42 bases worldwide, etc.*

F. Genome architecture and reprogramming

Back in Berkeley, Steve settled in at home and focused on the mathematics of biology. Meanwhile, Dylan moved in permanently. Aside from vacations and hiking trips, they had not spent much extended time together before since Dylan primarily lived in Utah and France, while Steve moved between Berkeley, Chicago, and Hong Kong.

While getting to know Dylan better, Smale worked with Rajapakse on "Emergence of function from coordinated cells in a tissue," published in January 2017. Building on the theoretical scaffolding they had erected in earlier studies, the pair aimed to unpack the mathematics of tissue dynamics.

In their paper, they presented an idealized mathematical foundation for understanding dynamical forces—both within the genome and between cells—that lead to the emergence of cellular function. In other words, they explored how information from a single genome brings about the cell's identity and role, shedding more light on normal and dysfunctional tissue development. Rajapakse says:

> *People are still building on the ideas Dr. Smale and I introduced in that paper. Dr. Smale has been studying this for a long time, inspired by Alan Turing, who wrote about the chemical basis of morphogenesis* [the biological process that causes a cell, tissue, or organism to develop its structure and function] *in 1952 and defined a model about how the cells interact. Our work is like an extension of Turing.*[21]

The developments also link to Smale's earlier studies with Cucker on synchronization and flocking, Rajapakse continues. "There are more universal ways to think about flocking—it happens in your body, too," he says. "We are increasingly realizing that your genome, inside the nucleus, has organization principles and experiences a form of synchronization. As you age, you lose those robust organization principles inside the nucleus and are more likely to develop disease."

At the same time, Smale and Rajapakse continued to explore pitchfork bifurcations, which they had investigated earlier in "Mathematics of the genome." In their follow-up study, "The pitchfork bifurcation," published in 2017, the pair presented the development of a new theory. While previous literature on pitchfork bifurcation required symmetry and a third derivative, Smale and Rajapakse tried to build a mathematical model of cell differentiation with neither. They succeeded, presenting a novel bifurcation theory that provides a better understanding of genome dynamics and cell division.[22]

At the 2017 FoCM conference, which took place in July in Barcelona, Smale and Rajapakse collaborated with Mike Shub to organize a workshop, entitled "Computational Mathematical Biology with Emphasis on the Genome." In addition to hosting the seminar, Smale also presented their work on the mathematics of cell division.

The duo then dove into another related topic, "Genome architecture mediates transcriptional control of human myogenic reprogramming," with a team of 10 co-authors, including Muir. Published in *iScience* in August 2018, the paper examined the interaction between genome architecture and cellular transcription. A critical biological process, transcription defines cell identity during development, enables cells to respond to signals, and coordinates cell activity.

After reprogramming fibroblast cells into muscle cells, they generated a rich data set that captured the dynamics. Upon analyzing the data, the researchers observed that organizational changes in genome architecture preceded the transition, and once the process started, some muscle cells synchronized with biological rhythms, shedding even more light on how our cells function and develop.[23]

"The paper provided a foundation for how genome architecture and cell reprogramming relate when a cell changes from one type to another. We observed that a bifurcation occurs from fibroblasts to skeletal muscle," says Rajapakse. "We also found some hints about entrainment—meaning, how the myogenic master regulators couple, or work together, with our biological rhythms."[24]

G. The beauty of synchronicity

Just as the emergence of flocking in birds had piqued his interest a decade earlier, biological synchronization would capture Smale's attention for the next few years. The study of synchronization dates back more than 350 years ago, when Dutch mathematician and physicist Christiaan Huygens studied two pendulum clocks ticking in harmony on a single wooden structure. While they started at separate paces, the pendulums eventually swayed in unison.[25] Huygens' discovery laid the groundwork for understanding synchronization in oscillating systems, including the human heart.

In the heart, billions of muscle cells beat in unison. When placed in a petri dish, muscle cells oscillate independently at first. Over time, they start syncing with neighboring cells. Fascinated by this phenomenon, Smale began working with Rajapakse and an old friend, Charles Pugh. A professor in the Berkeley mathematics department who researches geometry, topology, and dynamical systems, Pugh had joined Steve on the heroic voyage to the Marquesas on *Stardust* in the summer of 1987.

The three men worked on a mathematical model of the human heart using graph theory, geometric analysis, Morse theory, and the Kuramoto model. One of the most useful and flexible tools for studying the emergence of synchronization, the Kuramoto model is a dynamic system that works in phases to

Rajapakse and Steve at the IAS Extended Workshop in July 2018.

Steve at IAS in July 2019.

model the behavior of coupled oscillators. At first, each oscillator has an intrinsic natural frequency. But if an attraction is strong enough, the system achieves synchronicity.

Smale spoke on the topic during the 2019 Heidelberg Laureate Forum lecture:

> *My thoughts have gone back to this subject for a few decades ... Turing gave some examples of morphogenesis, which are just the other direction from what I am looking at. He saw morphogenesis as a break in symmetry, and I want to see how symmetry is created.*[26]

Using computer science and dynamical systems tools, the researchers measured the chemical and electrical reactions within and between cells. At time zero, the oscillation will be different, explains Smale. "But as time gets bigger and bigger, because of the diffusion term, one could expect these to all coalesce at about the same point, and that's synchrony."

Keen to share their findings and discuss the possibilities with other mathematicians and scientists, Smale and Rajapakse returned to CityU, where they held another workshop in July 2018. The second iteration, IAS Extended Workshop on Genomes, Cells, and Mathematics, lasted a month and attracted almost 400 participants across 10 workshop sessions. Smale delivered the opening lecture, "Beating in Unison," while Rajapakse gave a lecture titled "Algorithm for Cellular Reprogramming."

Throughout the event, they exchanged views on genome-wide sequencing studies with over a dozen other experts in mathematics, computer science, cellular biology, and human genetics. Smale returned the following July to give another talk on the same subject, "The Heart, Beating in Unison, and the Mathematics of Synchronization," at IAS.

"We gave Professor Smale the royal treatment whenever he came to IAS," says Kuo.[27] "I just got an email from a French scholar, Jean Salençon [a recognized leader in continuum mechanics], who was the French Academy of Sciences president and is returning [as a visiting fellow] at IAS, so we have quite a few good mathematicians because of Smale."

H. More biology and a second mineral book

Smale may have retired from CityU in 2016, but his ambition and creativity continued to work overtime. Toward the end of 2018, Smale embarked on a more in-depth investigation of Kuramoto oscillators with Rajapakse.

Within a few months, Rajapakse invited Cleve Moler—the renowned American mathematician, computer programmer, and co-founder of MathWorks—to contribute to the project. Moler created a Kuramoto program to simulate oscillations, enabling more experimentation on the numerical side, while Smale focused on the mathematical perspective.

They later welcomed Pugh to the team in 2019 to investigate a greatly simplified Kuramoto model that leads to synchronization. "We want to understand its global dynamics from a mathematical point of view," says Pugh. "Not just the fact that synchronization occurs for most initial conditions, but exactly what they are, and exactly what the other initial conditions lead to. This [is what] the modified Morse Theory is about."[28]

When the COVID-19 pandemic halted travel, Steve continued to work on the problem at home, meeting with Rajapakse and Pugh every week on Skype, with the addition of Gregorio Malajovich (one of Smale's former students) in 2022.

"We've been working on it for a few years now. The main thing about Dr. Smale is he gets obsessed with problems," says Rajapakse. "Once it's in his head, he focuses on it. He wants to establish a deeper, more global understanding of this notion of synchrony."[29]

Rajapakse believes Smale's interest in synchronization links back to a deep appreciation for simplicity. "Synchronization implies simplicity because it means everything is coordinated," says Rajapakse. "There could be 100 variables, but if they are all doing the same thing, they are reduced to one."[30]

Speaking with Smale daily, Rajapakse has gotten to know him well over the past decade. In addition to being collaborators, they have also become good friends. In July 2020, Rajapakse began organizing virtual birthday conferences for Smale, starting with his 90th. Dozens of friends, colleagues, collaborators, and former students attended the event, which was held online due to the pandemic. More than 15 people shared thoughtful remarks, traversing personal memories, humorous anecdotes, and tributes to Smale's accomplishments.

"Dr. Smale is different," says Rajapakse. "While other mathematicians focus on the answers, Smale cares more about the questions. He's guided by real questions that can build foundations to use the language of math to humanize science."

But meaningful questions should not be complicated, he continues. "Dr. Smale uses simple tools and asks good, simple questions. I think that's what we need to encourage when it comes to younger generations—don't obsess about the answers, try to formulate and simplify your questions."

Smale elaborated on his thought process during a joint interview with Dr. Leland H. Hartwell, who won the 2001 Nobel Prize in Physiology or Medicine for discovering key regulators of the cell cycle, conducted by Rajapakse. According to the resulting article, "Conversation with Dr. Steve Smale and Dr. Lee Hartwell," which appeared in the *Notices of the AMS* in 2021, the experts saw eye-to-eye on the matter:[31]

> ***Smale:*** *I think a lot of science has to do with formulating problems rather than answering them. Sometimes answering problems can be revolutionary, but it's more important to understand what the problems are in a subject. It's my feeling that formulation of problems is pretty key.*
>
> ***Hartwell:*** *I agree entirely with Dr. Smale, in fact, what I am trying to emphasize in science education is the questions rather than the answers. I think the formulation of the question really defines the problem, it puts boundaries on what we're trying to figure out.*

While applying dynamical systems to biological questions, Smale worked on another personal project close to his heart. In 2021, he published the second coffee table book about his mineral collection, *The Smale Collection II,* accompanied by striking photography by Jeff Scovil and brief captions about each mineral's type, size, and acquisition. Once again, Steve gladly worked with editor Gloria Staebler of Lithographie publishing house.

Whereas Steve's first book shared his and Clara's philosophy toward collecting minerals, the sequel spotlighted select specimens featured in *The Mineralogical Record* and the Virtual Museum of Chinese Minerals. Created and hosted with support from CityU, the latter showcases some of the finest crystals China has produced over the last decade.

In the book's introduction, Steve shares his thoughts on how his mineral collection mirrors his life and changes in the world.

> *The 1980s and 1990s saw the beginning of a political and cultural shift from West to East that continues to this day. This shift extended to science and mining and was thus relevant to me and my story, as they influenced where I lived, my jobs, and the character of my mineral collection.*[32]

I. Family, health, and a full social calendar

Just as his collection has evolved over the years, so has Steve. His sister Judy, who has known him longer than anyone, offered some observations. Earlier in life, she says, he was not always as considerate as he could have been and was domestically traditional. "Because he's so driven, so determined to accomplish his goals, sometimes that takes precedence over things and people he might have neglected in different ways," says Judy, who spent Christmas with her brother and the family in 2023.[33]

Though it was typical in his generation, Steve rarely cooked or cleaned, leaving those domains to Clara. "I never remember him sharing the preparation of a meal. He was always called for dinner," Judy says, adding that he could be cutting or dismissive at times. "If we were sitting and I was visiting Steve and Clara, and we were talking about something he wasn't interested in, he'd get up and leave and go to his study. After dinner, he worked until nine o'clock."

But any lingering sharpness has rounded with age. These days, Steve is more patient in conversation, capable in the kitchen (he usually makes his own breakfast and lunch, while Dylan handles dinner), and more involved with the family. "Steve enjoys meals with family and can be very generous," says Judy. "He used to take us out to dinner at nice restaurants in Berkeley, and he'd pay for it. He also pays for half my airfare when I visit."

While Steve is generally upbeat and mild-tempered, he can be tough, having learned to stand up for himself at an early age, says Judy. When they were in elementary school at the Horton School, a one-room country schoolhouse in a rural area south of Flint, Steve was bullied, especially by a child named Calvin.

> *He would beat him up every day ... My brother would go home to my father and say, "Well, Calvin beat me up, and that's why I've got this cut on my face. What can I do?" And my dad would always say, "Just walk away." And* [Steve] *would say, "Well, I tried to do that." I used to feel so sorry for him because he didn't know what to do.*

One day, Steve had enough. He hit Calvin in the face, hard enough to break his glasses. Calvin never bothered him after that. "Steve did what he had to do. He had to figure it out for himself," says Judy.

As he progressed through life, Steve continued to figure things out on his own. Instead of confronting childhood bullies, he boldly tackled new ideas and questions in his work. He was not one to shy away from a challenge—a trait that enabled him to become a passionate political activist, exceptional mathematician, top mineral collector, and avid traveler. Naturally self-directed and independent, he preferred to teach himself rather than learn from instructors. By the same token, he avoids doling out advice and influencing other people's choices. "He does things in his own time and of his choosing," says Judy. "That's his way."

At 94, he remains highly disciplined and health conscious. He wakes up most days around 6 a.m. and reads *The New York Times* in print, and eats a breakfast of cold muesli with milk and berries alongside freshly ground Illy coffee.

Lunch is the same every day: creamy French cheeses, Wasa crackers, mixed greens with olive oil, and a glass of red wine. He prefers wines from the Châteauneuf-du-Pape region of southeastern France because they are high quality but less expensive than Burgundy or Bordeaux, he says, adding: "I want to keep things simple, so I can host people and not think too much about it."

Dinner is usually straightforward, too. Dylan often prepares foods like black beans, one of Smale's favorites, lentils, or chickpea-crusted pizza. "I am on a self-imposed diet. It has gotten more intense as I have gotten older because I want to avoid disease," says Smale, adding that he takes roughly 30 supplements and vitamins each day. "I don't eat much red meat and avoid sugar; except I make one exception: Lindt 90% dark chocolate."

Even on special occasions, Smale does not budge. When colleagues at CityU presented him with a birthday cake during a visit in 2018, he refused to take a bite, recalls Rajapakse, who was there at the time.[34]

While he does not hike anymore, Smale stays active at home. He walks on the treadmill for 12 minutes almost daily, uphill at 15 degrees at 2.1 miles per hour. He is doing well overall when it comes to his health. Steve had a TAVR (transcatheter aortic valve replacement) operation a few years ago and a Mohs operation to remove squamous cell carcinoma, a type of skin cancer, from his lip in early 2024.

He continues to work on projects he enjoys, such as a third coffee table book about his mineral collection, which is scheduled to be published in late 2024. Between mathematics, minerals, reading, music, and exercise, Smale welcomes several weekly visitors, from former students to collaborators and friends. In addition, Laura says the family typically gets together once a week for lunch or dinner.

"Nat's wife, Hermine, jokes that he has a busier social calendar than the rest of us," says Laura. "We talk once or twice a week on the phone, and I rarely go a week without seeing him."[35]

Steve travels some, but mostly shorter trips within the U.S. In February, he went to the Tucson Gem and Mineral Show 2024 with Nat, where he purchased a brown tourmaline and a green fluorite.[36] "Both are about three inches long and are quite pretty," adds Nat. During the show, he spent time with four or five collectors, including Sandor Fuss and Wayne

Thompson. By the fourth day, he was ready to leave, but not before meeting with an old friend from the University of Michigan, Art Rose, who hitchhiked to San Francisco with Steve in the summer of 1952.

He also keeps in touch with close friends like Lenore and Manuel Blum, Mike Shub, Felipe Cucker, and Indika Rajapakse. "Lenore and Manuel are among my oldest friends—they are like family to me," says Smale. "It's a close relationship that's persisted for 40–50 years since Lenore was a teacher at Berkeley."

Lenore, Rajapakse, and Michael Xuan (one of Smale's Berkeley Ph.D. students and now the co-founder of big data company UniData Technology in Shanghai) are organizing "Smale 95," a conference in honor of Smale's 95th birthday in July 2025. Scheduled to take place at Berkeley's Simons Institute for the Theory of Computing, the two-day math conference will celebrate his legacy as one of the most versatile, bold, and unconventional mathematicians of his era.

While still in the proposal stage at the time of writing, Rajapakse also hopes to set up the Stephen Smale Institute at the University of Michigan. He aspires to build an interdisciplinary research institute that bridges applied mathematics, computational science, and bioinformatics. "In recent years, the integration of applied mathematics, computational science, and bioinformatics has emerged as a powerful approach to unraveling the complexities of biological systems," writes Rajapakse in his proposal. "The mission of the Smale Institute will be to provide the next generation of mathematicians with the resources and infrastructure needed to revolutionize these fields."

In terms of research, the institute would revolve around Smale's 18 problems, which he proposed at the turn of the 21st century, and the most challenging problems facing medicine and biomedical research. From network-based modeling of biological pathways to optimization algorithms for drug discovery and pharmacogenomics, the institute would be a world-class hub of interdisciplinary innovation with a long-term goal of addressing complex biological challenges and improving human health.

J. Stephen Smale's legacy

Most people in mathematics choose an area, like algebra or geometry, and spend their careers building a reputation as an expert in that field. But Steve does not think linearly—he thinks in multiple dimensions, and his career has unfolded with a series of unexpected moves. Perhaps foretold by his success as a skilled chess player in his youth, Smale so often had the anticipatory insight, imagination, and deftness required to dip in and out of emerging fields.

Of course, it was not always easy for Smale. After a lackluster early academic career, he finally found his footing in algebraic topology, thanks to an advanced graduate course taught by the late Raoul Bott, who was also Smale's doctoral thesis advisor at the University of Michigan. With his mathematics talent finally unleashed, a young Smale achieved a series of successes in differential topology and dynamical systems. Cucker says:

> *When Steve started working in dynamical systems, he changed the way people dealt with the field so dramatically that people say the most important person*

in dynamical systems in the first half of the 20th century was French mathematician Henri Poincaré, and in the second half of the 20th century, it is Steve, This gave him as much visibility as his work on topology but in a completely different community.[37]

While topology and dynamical systems continued to inform his later work, Smale did not pigeonhole himself. Instead, he moved on to mathematical economics, then computation, complexity, learning theory, probability, emergence, synchronicity, and computational biology. Cucker says:

While the problems would shift, some of the core tools he used to solve them stayed the same. Steve is essentially, at heart and, in mind, a geometer. He's extremely strong in geometry—he has this capacity to see things. When he attacks a problem in any area, he spends a lot of time trying to see how to translate this problem into geometry. And then, at some point, things click. Once he rephrases things in geometric terms, he gets amazing results.

Whether in topology or economics, learning theory or biology, Smale hoped to use mathematics to understand the world and explore broader phenomena impacting society. In a 2007 interview with the AMS, he said:

Mathematics is so effective because one can look for universal laws more easily with mathematics than without. It enables us to abstract the main ideas. With formalization and symbols one is able to see what is universal. The abstraction allows us to see universal ideas. I have been very inspired by Newton who could see a falling apple and the motion of planets and recognize them as part of the same phenomenon.[38]

When entering a new area, he had the confidence to pursue cutting-edge problems, the curiosity to learn, and the discipline to do the work. Few mathematicians have moved between fields with such agility. Some may see it as a curious choice. However, Steve wanted to keep moving and evolving and, perhaps most importantly, contribute to understanding the problems of our era.

"He built the foundations from topology and dynamical systems, then applied that understanding to different fields of economics, biology, and learning theory," says Rajapakse. "That's why Dr. Smale is Dr. Smale. That's why he can move into different fields and ask meaningful questions guided by a desire to unravel a deeper understanding of phenomena, not just to prove a theorem."[39]

Many brilliant mathematical minds have been impacted, directly or indirectly, by Smale as well. During his 30 years at the University of California at Berkeley, Smale supervised the dissertations of nearly 50 doctoral students, including Moe Hirsch, Mike Shub, James Renegar, Jacob Palis, Nancy Kopell, Robert L. Devaney, and many more. Beyond California, his legacy expands far and wide, from the halls of the University of Michigan to the Instituto de Matematica Pura e Aplicada in Rio de Janeiro, the Foundations of Computational Mathematics society, and of course, the City University of Hong Kong.

Even in his 90s, Smale still has a mind for math and the energy to experiment. Over the past seven decades, his work has advanced our understanding of the world and pioneered a different approach to mathematics—one that champions simplicity and flourishes at the intersection of disciplines. As one of the greatest mathematicians of the era, Smale's adventurous approach to his work and life will undoubtedly inspire for generations to come.

Mathematical Appendix

Mathematical Appendix A: Smale's Thesis

Smale's Thesis Problem: *Classify, up to regular homotopy, regular closed curves in an arbitrary manifold.*

To explain the problem we must give meaning to the terms: classify, regular homotopy, regular closed curve, and manifold. Often mathematics proceeds by starting with one result, and then generalizing it in some fashion to a deeper theorem that includes the original and more. Steve's work was preceded and inspired by a paper[1] of Hassler Whitney in 1937 addressing the following:

Whitney Problem: *Classify, up to regular homotopy, regular closed curves in the plane.*

Notice that the only difference is that the familiar term *the plane* replaces *an arbitrary manifold.* As we shall see, the plane is a manifold, and Smale's result included that of Whitney. We will begin by explaining the mathematical jargon in the Whitney problem.

To describe a *closed curve in the plane,* think of a standard unit circle and the two dimensional coordinate plane (denoted R^2). Consider a function whose domain is the circle and which assumes values in the plane. View the independent variable as a parameter t, determining a point on the circle given by its angle (in radians from 0 to 2π). For each such domain value the function gives a corresponding point, $f(t) = (x(t), y(t))$ in the plane. The graph of all these points $f(t)$ in R^2 will be a curve. The curve is closed in the sense that $f(0) = f(2\pi)$ are the same points, and thus the curve starts and ends at the same spot. Examples of closed curves and their graphs are shown in Figure A.1.

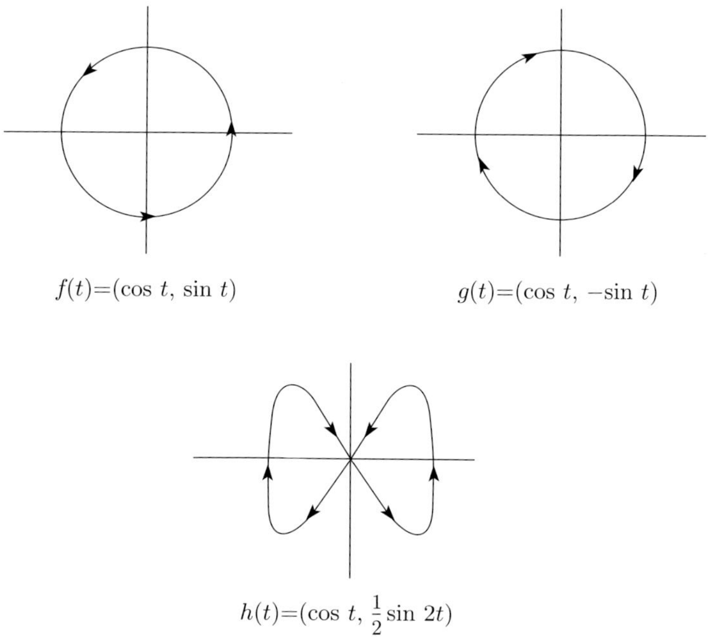

Figure A.1. Examples of closed curves in the plane.

The curve $f(t) = (\cos t, \sin t)$ is a circle because $x = \cos t$, $y = \sin t$, and $x^2 + y^2 = \cos^2 t + \sin^2 t = 1$. As t increases, points in the plane move counterclockwise along the circle beginning at (1, 0) when $t = 0$. In g the motion begins at the same point, but is clockwise. For the figure eight in h the motion proceeds through the upper right section and then past the origin to the lower left, continuing to upper left, and finishing in the lower right.

For a closed curve in the plane to be *regular*, there must be a tangent vector at each point on the curve that varies continuously as the point moves along the curve. In Figure A.2 these vectors have been drawn at selected points for f, g, and h. Note that the vectors point in the direction

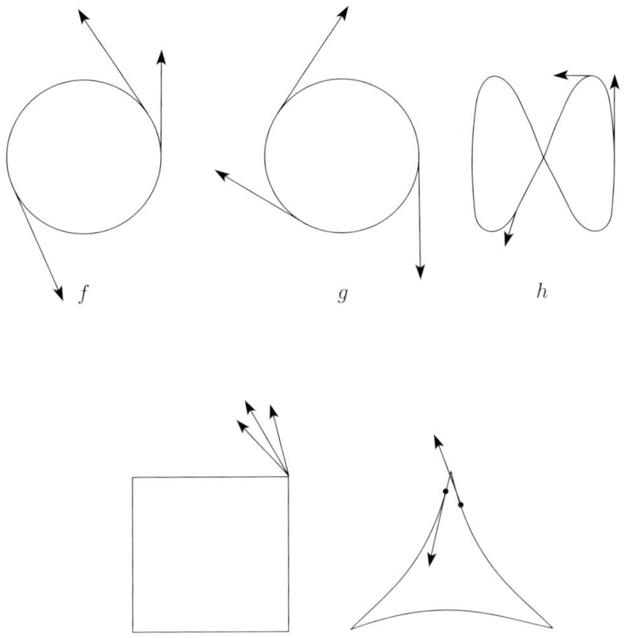

Figure A.2. *f*, *g*, and *h* are regular, but the square and cusp are not regular.

of motion. The lengths of the vectors depend on the function and must be positive.

The square curve is not regular because there is no well defined tangent at the corners. Similarly the *cusp* in the next curve precludes regularity because the tangents point in opposite directions on each side and thus do not vary continuously. To obtain a rigorous definition of regular curve requires calculus.

Definition. A closed curve in the plane f is regular provided its derivative $f'(t) = (x'(t), y'(t))$ exists, is continuous, and that the values of $x'(t)$ and $y'(t)$ are never simultaneously 0.

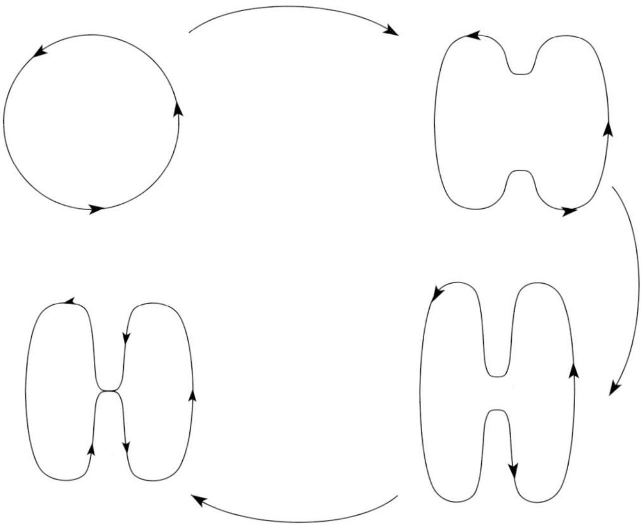

Figure A.3. A regular homotopy of curve *f*.

For two regular curves in the plane, *f* and *k*, to be *regularly homotopic* requires that they share the relationship that one image can be reshaped into the other in a "nice" manner, without breaking or kinking in the process. Roughly speaking, *f* and *k* are regularly homotopic provided *f* can be deformed to *k* through other regular closed curves so that the tangent vectors of the intermediate curves change continuously with the deformation.

For example begin with *f*, the counterclockwise rotation of the circle. If we carefully push down at the north pole and push up at the south pole, the resulting curve is also regular. The amount the poles are displaced toward the origin affects the curve, but not the regularity. As shown in Figure A.3, both poles can be pushed to the origin to obtain a regular curve *k*. For each s between 0 and 1, there is a regular curve g_s passing through the point $(0, s)$ (e.g., $f = g_1$ and $k = g_0$). The tangent vectors of the curve change continuously with both t and s. Thus *f* and *k* are regularly homotopic.

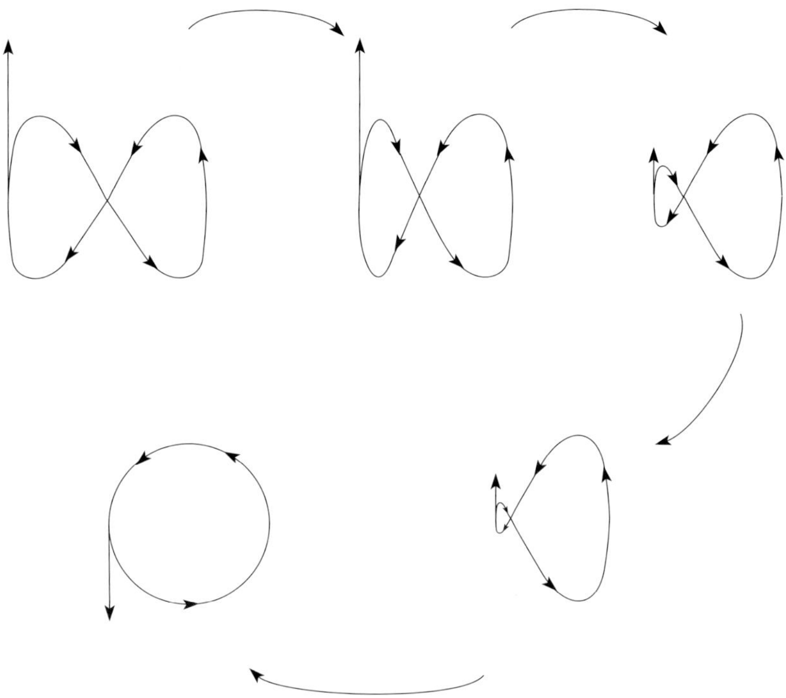

Figure A.4. A homotopy of the figure eight curve *h* to *f* that is not regular.

The figure eight curve, *h*, is homeomorphic to the image of *k*, however the direction of motion is different as the curves pass through the origin. Are *f* and *h* regularly homotopic? We can start with *h* and deform by regular curves to *f*, moving the intersection point to the left (see Figure A.4).

For a somewhat subtle reason this does not satisfy the requirements of a regular homotopy. Notice that the tangent vector at $f(\pi) = (-1, 0)$ points down. If we look at the tangent vectors of each of the preceding curves at this point, the direction is up. That means that there is a sudden change (discontinuity) in the direction as we get to *f*. This violates the definition. (Rigorous mathematical aside: The precise condition is that $\frac{d}{dt} f_s$ be continuous in *s* and *t*.)

The question remains as to whether *f* and *h* are regularly homotopic. Certainly our first attempt failed to establish anything about their relation. However, perhaps some more complicated deformation might work. This is the essence of the classification question. To classify, up to regular homotopy, we would like some conditions on the curves from which we could readily determine whether they are regularly homotopic, without actually constructing the homotopy.

In his solution, Whitney introduced the concept of the *winding number* of a curve in the plane. If we follow the angles of the tangent vectors, as we move from the initial to the terminal point, the winding number is the signed net number of revolutions made in the journey. In *f* the vector points up at the beginning (when $t = 0$) and rotates counterclockwise one time. We define counterclockwise as positive and thus the winding number is 1. In *g* the motion is clockwise and the winding number is -1.

The figure eight in *h* is more complicated. It begins pointing up, but never actually completes a revolution since the direction is reversed each time it passes through the origin. For this curve the winding number is 0. As a final example, verify that 1 is the winding number of *k*. It is not a coincidence that *f* and *k* are regularly homotopic and have the same winding number. Whitney proved the following result, acknowledging the contribution of Graustein.[2]

Whitney–Graustein theorem. *Two regular closed curves in the plane are regularly homotopic if and only if they have the same winding number.*

As an immediate consequence of this theorem, no two of the curves *f*, *g*, and *h* are regularly homotopic.

Returning to Smale's problem, it remains to explain the concept of a manifold. This is a generalization of Euclidean space. One dimensional Euclidean space is the real number line in which every point corresponds to a real number. Similarly Euclidean 2-space, R^2, is the plane where each point is determined by a pair of real numbers, (x, y). Next in Euclidean 3-space, R^3, points correspond to triples of numbers, (x, y, z). Notice that we may think of each Euclidean space geometrically, or simply work with the number representations. We can continue the latter analogy to Euclidean 4-space, R^4, in which points are quadruples of numbers, (x_1, x_2, x_3, x_4). Following in this manner, for any positive integer n, there is a Euclidean n-space, R^n, consisting of n-tuples of numbers.

In R^2 we are familiar with the unit circle, the locus of all (x, y) with $x^2 + y^2 = 1$. The (open) unit disk in R^2 consists of the points it encloses, $x^2 + y^2 < 1$. In R^3 there is a unit sphere, $x^2 + y^2 + z^2 = 1$, and a unit ball, $x^2 + y^2 + z^2 < 1$. Proceeding analogously to R^4, the unit sphere consists of the quadruples (x_1, x_2, x_3, x_4) with $x_1^2 + x_2^2 + x_3^2 + x_4^2 = 1$ and the unit ball is defined by $x_1^2 + x_2^2 + x_3^2 + x_4^2 < 1$. The definitions of spheres and balls extend to R^n.

Consider the unit sphere in R^3. We say that it is locally Euclidean in the same sense that, prior to Columbus, people thought the earth was flat. If you are at a point on the surface of a sphere and possess limited vision, you would think that you are on R^2. Thus the sphere is locally Euclidean and we denote it by S^2, with the 2 signifying the dimension of the corresponding Euclidean space. In R^n the unit sphere, S^{n-1}, is n – 1-dimensional.

The unit ball in R^3 is also locally Euclidean, but it is 3-dimensional and denoted D^3. We will use the notation D^n for the unit ball in R^n. An *n-manifold* is a structure M which is locally like D^n. More specifically, each point $x \in M$ has a neighborhood (set of nearby points) that is homeomorphic to D^n. Thus at each point on M, it looks like R^n, but globally it might be quite different.

Thus far our inventory of 2-manifolds include S^2, D^2, and R^2. Another example is given by a hollow doughnut, known to mathematicians as a torus, T^2. To a population living on a large doughnut, locally it is indistinguishable from a sphere or plane. For each positive integer n, there are n-manifolds S^n, D^n, R^n, and T^n. One motivation for the manifold concept arises from the notion that the universe is locally Euclidean, but globally unknown.

Another way to view a manifold is as a space patched together from Euclidean pieces. To actually do calculus on a manifold, we would require a technical assumption that the pieces are sewn together in a smooth manner. When referring to an arbitrary manifold we often omit the n-prefix. A closed curve on a manifold M is a function from the circle to M. Regularity requires the existence of tangent vectors varying continuously along the curve. The definition of a regular homotopy is analogous to the plane setting, except that Smale used the additional stipulation that the curves maintain the same initial point and that the corresponding tangent vector remain the same throughout the deformation.

Smale's problem was a generalization of Whitney's in the sense that the curves could live on an arbitrary n-manifold rather than the plane. On the other hand the plane setting was central to the concept of a winding number in the Whitney–Graustein solution. For example if the curve meandered in a 3-dimensional space, it is unclear how to count revolutions of the tangent vector. To distinguish the regular homotopy classes Smale employed a mathematical structure that is substantially more complicated than the winding number, but applicable to arbitrary n-manifolds. He showed that his problem could be translated to this other setting where more was known.

Describing the new structure requires two giant steps. First consider the space of all vectors that are tangent to the manifold and have length one. This *unit tangent bundle*, T, is itself a $2n - 1$-manifold. Next, on this space he applies an algebraic topology structure denoted by π_1 and called the fundamental group. Putting the steps together yields the fundamental group of the unit tangent bundle $\pi_1(T, x_0)$, where x_0 is a designated element of T. Smale proved that, in the manifold setting, this structure distinguishes the curves that are regularly homotopic, analogous to the winding number in the plane. Smale's theorem was published in the *Transactions of the American Mathematical Society* and is reproduced below, verbatim.[3]

Theorem A. *Let x_0 be a point of the unit tangent bundle T of a Riemannian manifold M. Then there is a* 1-1 *correspondence between the set π_0 of classes (under regular homotopy) of regular curves on M which start and end at the point and direction determined by x_0 and $\pi_1(T, x_0)$.*

While this reformulation of the problem might seem highly complex, $\pi_1(T, x_0)$ was already resolved for many manifolds. For example, with S^2

the set was known to consist of two elements. Combining this with Smale's theorem, it follows that there are exactly two regular homotopy classes of regular closed curves on S^2.

The proof of Theorem A required 13 pages of mathematical development, employing a variety of techniques from algebraic topology and analysis.[4] Steve skillfully utilized concepts from the paper of Serre that he had studied in Bott's course. Most of the work involved establishing a deep relationship between T and the set of regular (not necessarily closed) curves beginning at x_0. Bott had suggested pursuing this course, but envisaged quite different techniques than those employed by Smale. The following year he was amazed when Steve pushed the concept to an even more general setting.[5]

Mathematical Appendix B: Everting the Sphere

In 1957 Smale submitted a manuscript to the *Transactions of the American Mathematical Society*. The ten page paper, entitled "A Classification of Immersions of the Two-Sphere," appeared in the February 1959 issue. Among the results was the following theorem:

Theorem B. *Any two C^2 immersions of S^2 in E^3 are regularly homotopic.*

A consequence of Theorem B is that the sphere can be turned inside out. To understand why this is the case requires an examination of the mathematical terms. The symbol E^3 is just another notation for R^3, Euclidean 3-space. Different mathematical symbols or terminology sometimes arise for the same structure. There is no clearinghouse for mathematical jargon, and duplication may result from independent discovery or an effort at improvement. The symbol S^2 represents the surface of the unit sphere.

A C^2 immersion of S^2 in R^3 is a function with domain S^2 that assumes values in R^3 and satisfies some additional conditions. Prior to describing the extra constraints, consider some examples of functions from S^2 into R^3. Figure B.1 shows the resulting images in R^3. The functions may be described as follows:

a. Identity: each point of S^2 is mapped to itself.

b. Circumscribed box of side length 2: each point of S^2 is mapped out to the point of the box that lies on the same radial ray from the origin.

c. Teed up football with pointed ends: another radial projection.

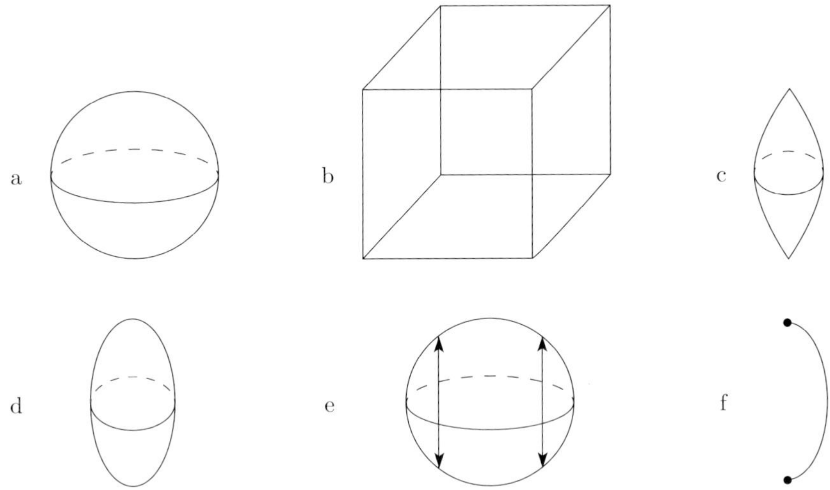

Figure B.1. Some images of functions from S^2 to R^3.

d. Teed up football with rounded (ellipsoidal) ends: yet another radial projection.

e. Reflection about equator: fixes equator and switches northern and southern hemispheres.

f. Semicircle: all points of a given latitude map to the point on the prime meridian at that latitude.

Notice that at each point on the graph of a, d, and e, there is a well defined tangent plane. However, at the corners of the box and at the pointed ends of c, there is no single tangent plane. In example f there is a tangent line to each point, but no tangent plane. To be a C^2 immersion of S^2 in R^3 requires that there exist a tangent plane at each point and more. Certainly b, c, and f fail. The adjective C^2 imposes a technical condition that nearby tangent planes relate to each other in a smooth manner. We stipulate that a, d, and e meet all the criteria of a C^2 immersion of S^2 in R^3.

The following approach generates a family of immersions. Beginning with the identity map, deform the image by symmetrically pushing down at the north pole and up at the south pole. With a bit of care, the resulting map is a C^2 immersion. Let g_s denote the result of pushing each pole by s units. Figures B.2a and B.2b show $g_{.2}$ and $g_{.5}$. With g_1, the north and south poles are mapped to the origin. Immersions permit the image to cross itself, provided all the other conditions are met. With $g_{1.2}$ and $g_{1.9}$ there is a circle of intersection, where points from the northern and southern hemisphere are mapped inside the equator. When $s = 2$, g_2 is the reflection from B.1e.

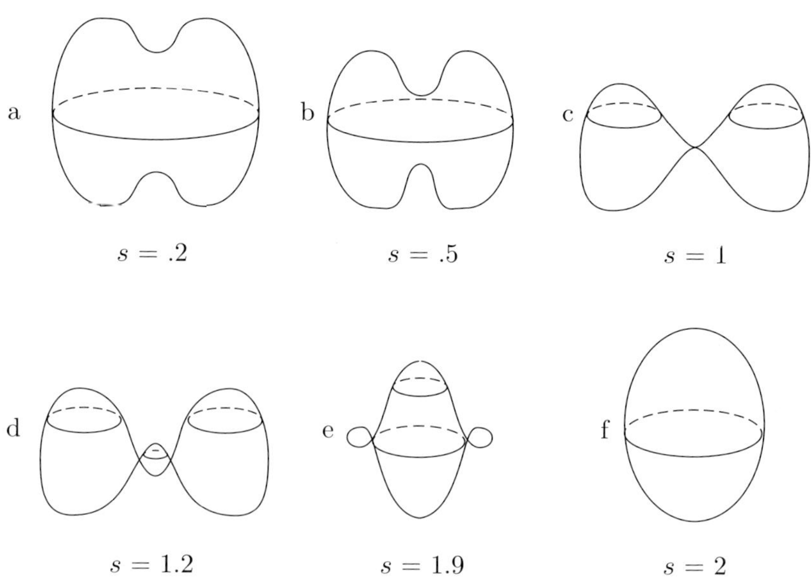

Figure B.2. A homotopy from the identity to reflection on the sphere that is not regular.

Now consider the family of maps g_s as s goes through the real numbers from 0 to 2. Each g_s is a C^2 immersion. For g_0, the identity from B.1a, think of the inside of the sphere having the color red and outside as blue. As s increases the sphere is deformed. When s passes 1 and the sphere goes through itself, portions of the red push outside. By the time $s = 2$, the outside is red and the inside is blue. In some sense this process, called a homotopy, turned the sphere inside out. However, it does not meet the mathematical standard for a regular homotopy. The difficulty lies with the crimping that occurs near the equator as s gets close to 2. To be a bit more precise, consider what happens to the tangent planes near the west pole. The tangent plane at the west pole is vertical for each g_s. Nearby, there are points (at the top of the crimp) where the tangent plane is horizontal. As s gets closer to 2, a point with a horizontal tangent plane approaches the west pole. When $s = 2$, the only horizontal tangent planes are at the north and south poles. It is the sudden evaporation, at $s = 2$, of the horizontal tangent planes that violates regularity of the homotopy. To be a regular homotopy each g_s must be an immersion, which they are, but additionally the tangent planes must change continuously with the deformation. We can not have horizontal tangent planes move into vertical ones.

Summarizing the previous paragraph, the family g_s is a homotopy of immersions from the identity to the reflection. It is not a regular homotopy. The mathematical criterion for turning the sphere inside out is that there exist a regular homotopy between the identity and the reflection. The exercise with g_s gives some of the flavor of mathematical research: a promising attempt that fails to evert the sphere. Next the mathematician might attempt to rehabilitate the homotopy in some manner so as to address the regularity problem. Remember, the failure of g_s to evert the sphere does not preclude the existence of a different homotopy that is regular. When the g_s example resisted remedy, a mathematician might seek a different, likely more complex, deformation. Another possibility to consider, prior to Smale's work, was that no eversion existed (i.e., the identity and reflection are not regularly homotopic). Some support for this point of view came from the analogous lower dimensional problem of regular curves in the plane (immersions of S^1 in R^2) where the Whitney–

Graustein theorem was applicable. There the identity and reflection have different winding numbers and are not regularly homotopic.

With no candidate for a regular homotopy, the conventional wisdom was that none existed. Most mathematicians would have pursued that course. Smale's Theorem B established that the identity and reflection, as well as any other C^2 immersion of S^2 in R^3, were regularly homotopic. However, the statement of the theorem provides absolutely no clue as to the nature of the deformation. Sometimes that type of information is present in the proof. To follow this thread we first note that Smale's Theorem B was essentially a corollary of his more complicated Theorem A.

Theorem A resolved the problem of regular homotopy for C^2 immersions of S^2 in R^n. The solution involved taking a certain set of $n \times 2$ matrices (rigorous mathematical aside: those with rank 2), known as Stiefel manifolds, $V_{n,2}$. Next he considered an algebraic topology structure on $V_{n,2}$, denoted $\pi_2(V_{n,2})$. Theorem A established a 1-1 correspondence between $\pi_2(V_{n,2})$ and the distinct regular homotopy classes of C^2 immersions of S^2 in R^n.

In the Theorem B situation, $n = 3$. From Theorem A there is a 1-1 correspondence between $\pi_2(V_{3,2})$ and the number of different regular homotopy classes of C^2 immersions of S^2 in R^3. Now for the interdisciplinary beauty of mathematics. The algebraic topology question as to the nature of $\pi_2(V_{3,2})$ was already answered. The set $\pi_2(V_{3,2})$ consisted of just one element. Thus there was only one regular homotopy class of C^2 immersions of S^2 in R^3, implying that the identity is regularly homotopic to the reflection.

The enterprise of obtaining an explicit eversion by backtracking through Smale's mathematics is already daunting. At the very least, it requires understanding some algebraic topology of Stiefel manifolds as well as Smale's proof of Theorem A. In any event, a careful examination of these topics[1] provides little, if any, insight into the nature of a comprehensible eversion. However, consider the impact of Theorem B on a mathematician seeking to construct an eversion. Prior to Smale's work, the existence of an eversion was a long shot that had not yet been ruled out. After Theorem B, there definitely was one out there, just waiting for discovery. Despite the drastic increase in the expected return of the problem, it was several years before Arnold Shapiro devised his complicated solution.

Most conceptualizations of the eversion involve the *antipodal map*, rather than reflection across the plane of the equator. The antipodal immersion is defined by mapping each point of the sphere to its diametrically opposite companion. In functional notation, antipodal $(x, y, z) = (-x, -y, -z)$ and reflection $(x, y, z) = (x, y, -z)$. Note that performing a sequence of three reflections in z, y, and x yields the antipodal map. By Smale's theorem each of these immersions is regularly homotopic to the others.

Understanding an actual eversion remains a challenge. A variety of visual aids* have been produced including pictures,[2] films,[3] and wire models.[4]

In mathematics the proof is the substance, but the theorem is the hook. Although Smale's proof of theorem A employed tools and followed a choreography similar to that of his thesis, the two works elicited entirely different reactions. This was because Steve's thesis appeared to be a mundane extension of Whitney–Graustein, but the existence of the eversion was an intriguing shock that provoked interest in exploring the methods. Soon Smale extended the work from the two sphere S^2 to higher dimensional spheres. In his article, "The classification of immersions of spheres in Euclidean spaces,"[5] he classified immersions of S^k into R^n, for $k < n$, in terms of the algebraic topology of the Stiefel manifold $V_{n,k}$.

Over the years a common misconception has developed among mathematicians that Smale turned the sphere inside out in his thesis. History is rarely a vital part of mathematical education and the folklore is replete with inaccuracies and incorrect attributions. This theme is explored further in Chapter 4.

* The image on the book cover (from 2000) is a frame from the video "Outside In" generated at the Geometry Center of the University of Minnesota and distributed by A K Peters, Ltd.

Mathematical Appendix C: Chaos and the Horseshoe

A more in depth analysis is available in Devaney's book, *An Introduction to Chaotic Dynamical Systems.*[1] To define a dynamical system requires a set and a function. As a warm-up for the horseshoe, we examine the asymptotic behavior of two simpler sets and functions. In both cases the set is the closed unit interval I consisting of all numbers between 0 and 1, inclusive. The first function is $f(x) = .5x(1 - x) = .5x - .5x^2$. This is a second degree polynomial whose graph is a parabola (see Figure C.1). Note that each point of I is mapped into I. We are interested in the asymptotic behavior that results from iterating an initial state by f. If the initial state is x_0, consider the orbit

$$x_0, f(x_0), f(f(x_0)), f(f(f(x_0))) = f^3(x_0), \ldots$$

What can be said about trends in this orbit sequence?

If $x_0 = 1$, then $f(1) = 0$, $f^2(1) = f(f(1)) = f(0) = 0$, and for all subsequent n, $f^n(1) = 0$. The initial state 1 reaches and remains at the target 0.

If $x_0 = \frac{1}{2}$, then $f(\frac{1}{2}) = \frac{1}{8}$ and $f^2(\frac{1}{2}) = \frac{7}{128}$. Continuing, the orbit appears to proceed toward 0. This is also the case for any x_0 with $0 < x_0 < 1$. To obtain a more rigorous argument, notice that $f(x) = .5x(1 - x) < .5x$ and $f^n(x) < \frac{x}{2^n}$. If $0 < x_0 < 1$, then $f^n(x_0) < \frac{x_0}{2^n} < \frac{1}{2^n} \rightarrow 0$. The dynamical behavior of f is simple: every orbit is asymptotic to 0.

The second example is the quadratic polynomial $g(x) = 4x(1 - x) = 4x - 4x^2$. The parabola maps I into I and its graph is shown in Figure C.2.

If $x_0 = 1$, then $g(1) = 0 = g^2(1) = g^3(1) = \ldots$.

If $x_0 = \frac{1}{2}$, then $g(\frac{1}{2}) = 1$, $g^2(\frac{1}{2}) = 0 = g^3(\frac{1}{2}) = \ldots$.

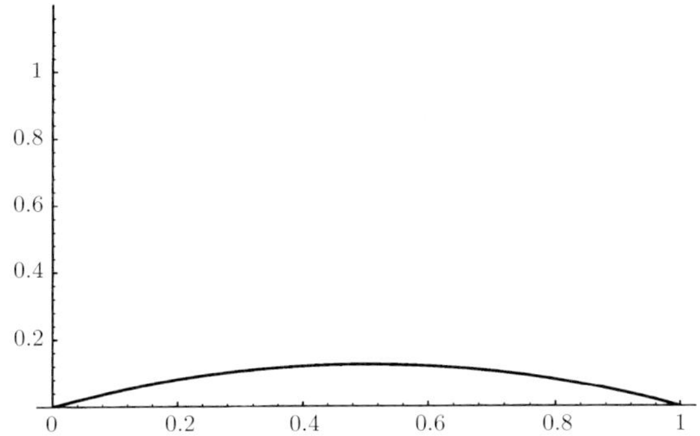

Figure C.1. Graph of $f(x) = .5x(1 - x)$

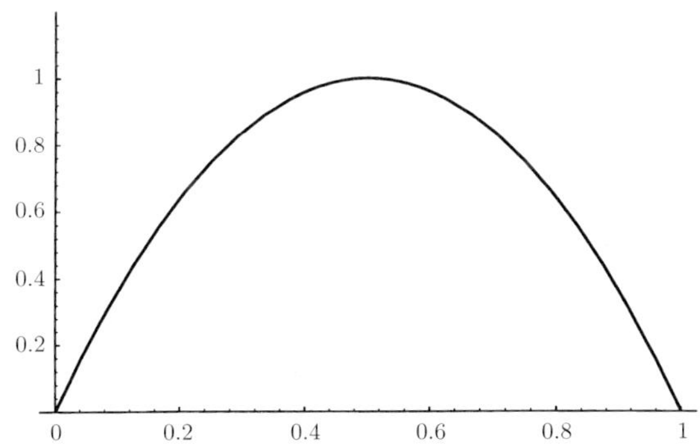

Figure C.2. Graph of $g(x) = 4x(1 - x)$

The initial states $\frac{1}{2}$ and 1 both reach and remain at 0. However, unlike f, 0 is not the only target. If $x_0 = \frac{3}{4}$, then $g(\frac{3}{4}) = \frac{3}{4} = g^2(\frac{3}{4}) = \ldots$. The initial state $\frac{3}{4}$ remains fixed at $\frac{3}{4}$ forever. Geometrically the existence of such a solution is apparent from superimposing the graphs of $g(x)$ with $y = x$ as in Figure C.3. The graphs intersect at 0 and another point. The actual value $\frac{3}{4}$ can be obtained by algebraically solving $x = 4x(1 - x)$.

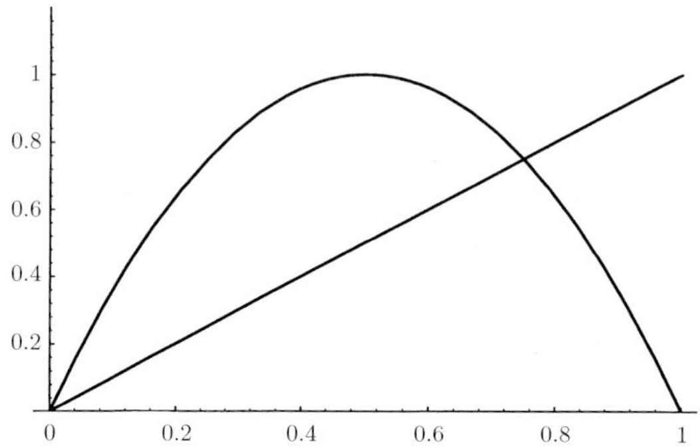

Figure C.3. Finding the fixed points of g.

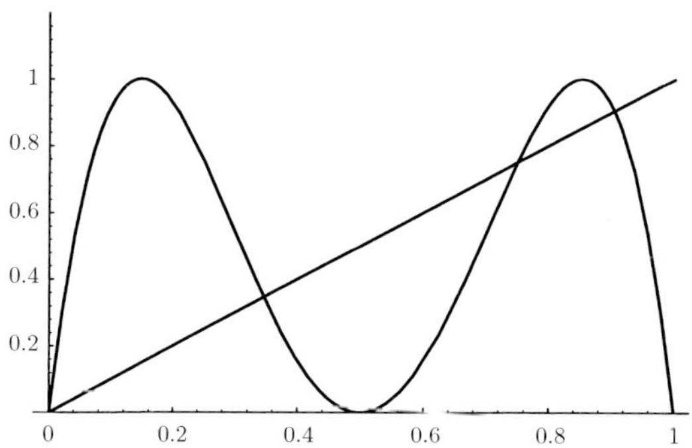

Figure C.4. Finding a period two orbit of g.

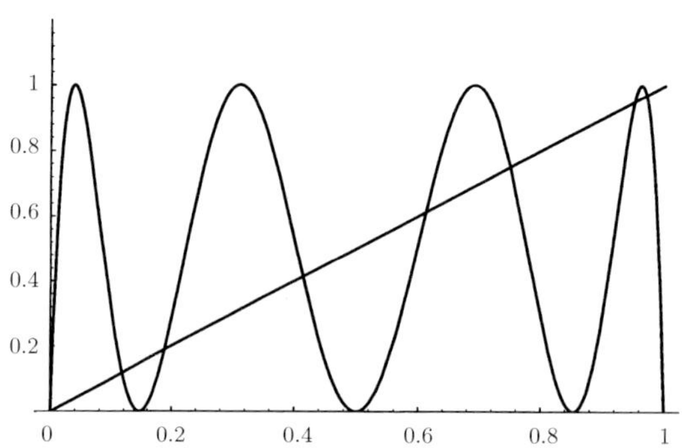

Figure C.5. Finding period three orbits of g.

Superimposing the graphs of g^2 with $y = x$ as in Figure C.4 reveals four intersections. Two of these are states 0 and $\frac{3}{4}$ which remain fixed through all iterations. The other two points must form an orbit of period 2 (they cannot be fixed). Similarly g^3 has eight points of intersection (see Figure C.5), the two fixed points and two orbits of period three. Continuing in this manner reveals that g has periodic points of every period. Consequently g has an infinite number of asymptotic targets.

To see the chaotic elements of g, fix a state x_0 and consider its future orbit and that of states close to x_0. Notice that the graph of g^n consists of 2^{n-1} humps from 0 to 1 and back down to 0. By choosing n sufficiently large, the width of a hump can be made arbitrarily narrow. For the states in this narrow interval, the values of the nth iterate are spread over the entire spectrum from 0 to 1. Since x_0 lies in one of these narrow intervals, the orbit values depend sensitively on the initial state.

Another feature of chaos is randomness. If state values greater than $\frac{1}{2}$ are considered *upper* (U), and state values less than $\frac{1}{2}$ are considered *lower* (L), we will see that any specified sequence of high and low values may be obtained. Depending upon whether $x_0 < \frac{1}{2}$ or $x_0 > \frac{1}{2}$, the orbit begins with an L or U. Now look at the graph of g in Figure C.2. There is a unique point c_1 between 0 and $\frac{1}{2}$ such that $g(c_1) = \frac{1}{2}$. Similarly there is a unique point c_2 between $\frac{1}{2}$ and 1 where $g(c_2) = \frac{1}{2}$. For an initial state between 0 and c_1, the orbit begins with an L and then produces another L, since $g(x_0) < \frac{1}{2}$. For initial states between c_1 and $\frac{1}{2}$ the orbit commences LU.

Initial State	Orbit
$0 < x_0 < c_1$	LL
$c_1 < x_0 < \frac{1}{2}$	LU
$\frac{1}{2} < x_0 < c_2$	UU
$c_2 < x_0 < 1$	UL

Initial State	Orbit
$0 < x_0 < c_3$	LLL
$c_3 < x_0 < c_1$	LLU
$c_1 < x_0 < c_4$	LUU
$c_4 < x_0 < \frac{1}{2}$	LUL
$\frac{1}{2} < x_0 < c_5$	UUL
$c_5 < x_0 < c_2$	UUU
$c_2 < x_0 < c_6$	ULU
$c_6 < x_0 < 1$	ULL

In Figure C.4 let c_3, c_4, c_5, and c_6 denote the four values that g^2 maps to $\frac{1}{2}$. These are spliced between 0, c_1, $\frac{1}{2}$, c_2, and 1. Above we list the first three itinerary values for initial states in these intervals. Each of the eight possible sequences of L and U is realized.

From Figure C.5 there are eight states which g^3 maps to $\frac{1}{2}$. Combining these with the previously identified states yields 16 subintervals and all the possible strings of 4 U's and L's. It can be proven that for any infinite sequence of U's and L's there is an initial state whose orbit values meet the specifications. This is quite a change from f where all orbits are asymptotic to 0. With g there are orbits of every period and every random design.

While the system g exhibits features that are currently associated with chaos, neither the set nor the function qualified for consideration in Smale's 1960 study. The mathematical defect with the set is the *boundary points* 0 and 1. The sets for Smale's systems were not permitted to possess such points. The circle and sphere are examples of sets without boundary points. The difficulty with the function g is that different points have the same image, such as 0 and 1 which are both mapped to 0. For Smale's systems there was a stipulation that the functions be homeomorphisms (plus a little more). With a homeomorphism, no two domain points can have the same image.

We reiterate the reasons for examining g in this text. The domain and function are simpler than those of the horseshoe, making the mathematics more accessible. Furthermore g shares several of the chaotic properties possessed by the horseshoe. However, g does not meet the technical requirements that Smale associated with dynamical systems and, with further argument, it is not structurally stable. We have used g as a transitional aid to understand the horseshoe, rather than as historically relevant to Smale's 1960 research.

In Figure 3 of Chapter 4 we began the construction of the horseshoe by defining a function h whose domain was the rectangle R. Further development is required to remedy two deficiencies. The edges of the domain rectangle consist of undesired boundary points. Moreover, some points, such as A, are mapped outside the domain and their orbits are undefined (what is $h^2(A)$?). To address the second issue, add semicircular caps to the left and right sides of R, and extend the definition of h so that each cap is mapped into the cap on the right, as in Figure C.6. Notice

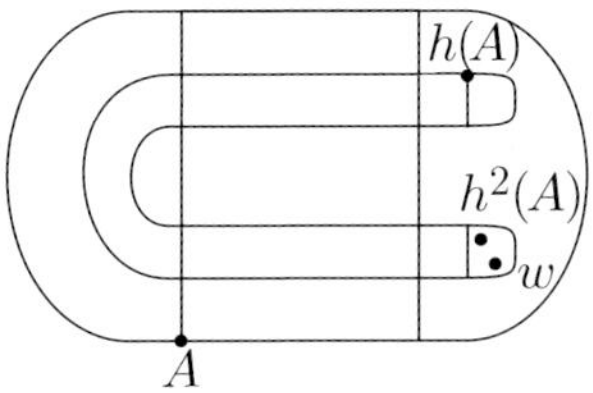

Figure C.6. Extending the horseshoe so that its image is inside the domain.

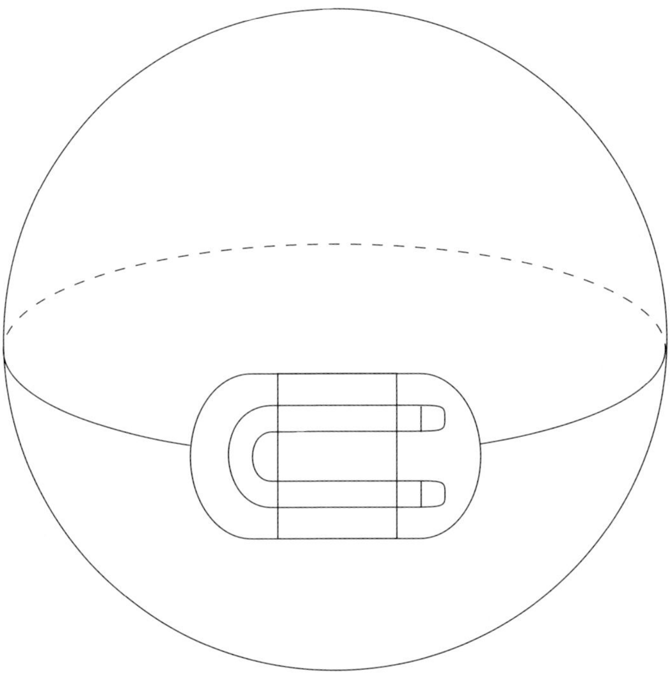

Figure C.7. Extending the horseshoe to the sphere.

that the right cap is contracted to a smaller cap, inside itself. With a bit of care, the extension map may be defined so that the right cap has a single attracting fixed point. This fixed point, denoted w, is the target for all orbits that enter the right cap. Now follow the orbit of A. Since $h(A)$ is in the right cap, the orbit remains trapped forever and is asymptotic to w.

To avoid boundary points in the domain, think of Figure C.6 embedded in the front of a sphere as in Figure C.7. Extending the map to a homeomorphism of the sphere requires defining h on a set which is the complement of R and the two caps. This set is homeomorphic to an open disk. The map can be constructed so that there is a single fixed point on the back of the sphere which repels its neighbors in a manner so that all orbits eventually reach R or a cap.

The map h is a homeomorphism that maps the sphere back onto itself. Actually it can be made to satisfy a higher standard of smoothness and is

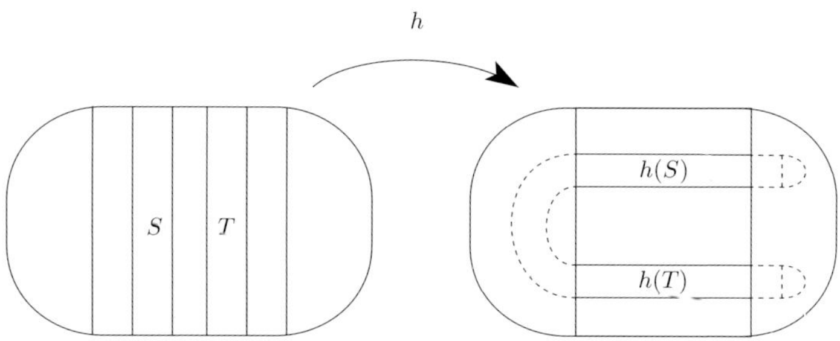

Figure C.8. Locating the portions of R that remain in R under the horseshoe function.

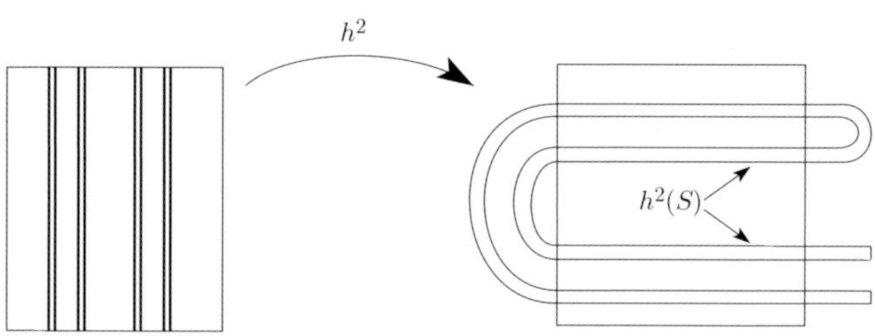

Figure C.9. Locating the portions of R that remain in R under two iterations of the horseshoe function.

called a diffeomorphism. The interesting dynamics occur inside R. For the points beginning in R, we are primarily concerned with identifying those whose orbits remain in R for all time. If an orbit ever leaves R, then it is trapped in a cap and is asymptotic to w.

Consider the $h(R)$ as shown in Figure C.6. Portions of this horseshoe image lie in the caps, but there are two rectangular pieces inside R. The next task is to determine which initial states map to the rectangular sections in R. Thinking back to the construction of $h(R)$, we see that the upper and lower rectangles are the images of vertical rectangles, S and T, located on the left and right sides, respectively, of R (see Figure C.8). The rectangles S and T contain all the initial states whose orbits remain in R, but some of the states in these rectangles move into a cap on the second or subsequent iterations. The intersection of $h^2(R)$ with R consists of four rectangles. Looking back again to the corresponding initial states produces four skinny vertical rectangles, two each in S and T (see Figure C.9). The third iteration yields eight vertical skinnier rectangles and so on. The set of initial states, for which the entire orbits remain in R, is an infinite union, Λ^+, of vertical lines.

Our study of the asymptotic behavior of the horseshoe map is now reduced to an analysis of the initial states in Λ^+. Does Λ^+ have fixed points? periodic points? points with random orbits? Identifying fixed and periodic points is more complicated for h than it was for g. Recall that with g, we superimposed the graph of $y = x$ with g. Intersections gave the fixed points. Points on the parabolic graph of g have coordinates $(x, g(x))$, including both the domain state x and image state $g(x)$. In Figure C.6, points in the domain and image of h have separate representations, each in a two dimensional plane. To combine these plots in the graph for h requires an impractical four dimensional picture. The search for fixed and periodic points of h, within R, necessitates a new tool.

To analyze the asymptotic behavior of the orbits that remain in R, Smale applied techniques from a mathematical subspecialty known as *symbolic dynamics*. Putting aside the horseshoe for a moment, consider the set Σ of all infinite sequences of Ss and Ts. Each integer position holds an S or a T. Thus there is an S or T in positions $\ldots -3, -2, -1, 0, 1, 2, 3, \ldots$. Formally we can think of these sequences as functions from the set of integers to the set consisting of the two elements S and T. Here are some examples specifying entries in positions –3 to 4:

$$\ldots SSSSSSSS \ldots$$

$$\ldots STSTSTST \ldots$$

$$\ldots STSSTSSS \ldots$$

Next, define a function σ on the set of sequences. The function σ takes a sequence and shifts each entry to the left by one position. Applying σ to each of the sequences above gives the following images in positions –3 to 4:

$$\ldots SSSSSSSS \ldots$$

$$\ldots TSTSTSTS \ldots$$

$$\ldots TSSTSSST \ldots$$

Notice that the first sequence is a fixed point and the second has period two. Viewing σ and Σ as a dynamical system, there are points of every period. The last example is not periodic and is not asymptotic to any other sequence. Exploring asymptotic behavior requires a notion of distance between sequences. We measure this distance by counting the positions from 0 in which the two sequences agree. If they match at all positions within some proximity of zero, then the sequences are defined to be close to each other. The sequence with all *S*s is close to the sequence with *S*s in positions –2000 to 2000 and *T*s elsewhere. Applying σ to each sequence produces images that are slightly further apart than the originals since they disagree in position 2000. After 2000 iterations, the orbits are distant. This is sensitive dependence on initial states, a feature of chaos. Since any sequence of *S*s and *T*s is present in Σ, the dynamical system can produce random behavior.

Smale showed that all the dynamical elements of σ are embedded in the horseshoe. Consider an initial horseshoe state z whose orbit remains in R. We will associate a sequence $K(z)$ to z. Since each point of the orbit of z is in R, its predecessor is in S or T. Thus each element of the orbit is either in S or T. The sequence elements of $K(z)$ in positions 0, 1, 2, … are determined by the locations of z, $h(z)$, $h^2(z)$, … in S or T. We would like for $K(z)$ to be a sequence in Σ. It remains to define the sequence at the negative integer positions. Since h is a homeomorphism of the sphere, it has a well defined inverse function h^{-1} which undoes the action of h (for h^{-1} to make sense it is crucial that no two domain points of h have the same image). The existence of h^{-1} permits construction of the backward orbit z, $h^{-1}(z)$, $h^{-1}(h^{-1}(z))$, … . Let Λ^- consist of the initial states whose backward orbit remains in R. With the additional stipulation that $z \in \Lambda^+ \cap \Lambda^-$, we can complete the sequence $K(z)$. The choice of S or T in the -1 position depends upon whether $h^{-1}(z)$ lies in S or T. Continuing with $h^{-1}(h^{-1}(z))$ for the -2 position and so on produces a function K mapping $\Lambda^+ \cap \Lambda^-$ to Σ. Smale proved that K is a homeomorphism that relates h to σ by the composition relation $K(h(x)) = \sigma(K(x))$. Mathematically this means that K translates the dynamics of $\Lambda^+ \cap \Lambda^-$ in the horseshoe to the sequence space. Among the shared properties are the infinite number of periodic points and the existence of initial states with essentially random orbits. Structural stability of the horseshoe is a separate argument which is beyond the scope of this appendix.

We close with some further historical perspective. Smale's early 1960 work on dynamics was motivated by the letter from Levinson. Steve realized that his own vision of dynamics was too simplistic. Out of his study of Levinson's paper on differential equations, Steve produced a diffeomorphism that was both structurally stable and possessed an infinite number of periodic points. The actual geometric picture in Smale's Rio de Janeiro construction was somewhat more complicated than the horseshoe of Figure 3. (See his paper, "On how I got started in dynamical systems."[2]) When Smale described the original picture in a talk during the summer of 1960 at Berkeley, Lee Neuwirth suggested the simplification to Figure 3 and the horseshoe was born. Smale's primary exposition of the work appeared several years later in the proceedings of a symposium honoring Marston Morse.[3] The results went well beyond the sphere, applying to all manifolds of dimension greater than one. First he showed that all such spaces admit diffeomorphisms containing the dynamics of σ and that this property persists under perturbation of the map. Next he analyzed the impact of a homoclinic point, a state with the same periodic target in both future and past time. Smale extended the analysis of Poincaré and Birkhoff, showing that the presence of a homoclinic point is associated with σ type dynamics.

Mathematical Appendix D: The Higher Dimensional Poincaré Conjecture

Smale's proof of the Higher Dimensional Poincaré Conjecture applies to manifolds of dimension ≥ 5. Higher dimension is a daunting concept. To provide some of the flavor of Smale's breakthrough, we will focus on a key aspect of his development, but examine the technique within the friendly confines of 2-manifolds. The essence of proving the Poincaré Conjecture is to show that certain manifolds are homeomorphic to the sphere. Consider the 2-manifold M shown in Figure D.1. We may think of this surface as situated in 3-space so that the bottom z-coordinate is 0, the highest z-coordinate is 2, and the *saddle points* have z-coordinate 1. To obtain a homeomorphism to a sphere, simply extend the middle hill upward, while pushing the saddles above the left and right peaks (Figure D.2). Some further rounding deformation yields a big sphere (Figure D.3) which is homeomorphic to S^2.

The previous approach was fine for producing a homeomorphism between a given 2-manifold and S^2, but the Higher Dimensional Poincaré Conjecture contains two additional challenges. Rather than working with a specific manifold, we merely know that it satisfies certain abstract conditions. The other problem is the difficulty of coping with manifolds of higher dimension.

We now return to M in Figure D.1 and devise a more general homeomorphism approach. Let the real valued function $f: M \to R$ give the height (or z-coordinate) of each point on M. Now think of the points of M as flowing along M under the influence of gravity, that is in the direction of steepest descent or greatest decrease in f. There are six *critical points* which remain stationary because there is no single path of steepest descent. These are the three maxima, two saddles, and the minimum. Consider the set

Figure D.1. A surface in 3-space.

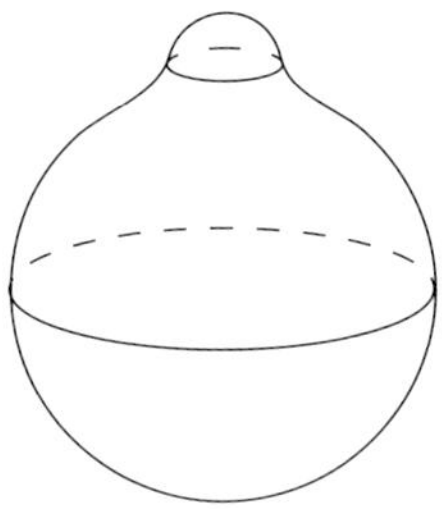

Figure D.2. Deforming the surface by pushing the middle hill upward.

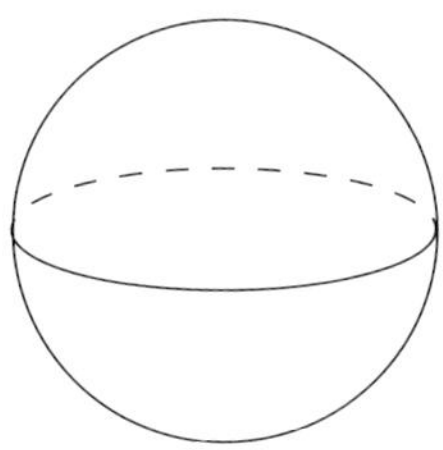

Figure D.3. Rounding the deformation into a big sphere.

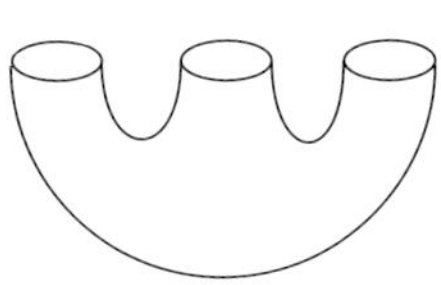

Figure D.4. X_s with $1 < s < 2$.

of points asymptotic to each critical point. For the minimum this set is 2-dimensional. The saddles each attract a one dimensional curve, and the maxima are isolated by themselves. Each critical point has an *index* which is defined to be the dimension of the set of points it *repels* (rather than attracts). The respective indices of maxima, saddles, and minima are 2, 1, and 0. Notice that M was contrived so that the function f returns the value of the index at each critical point. This is called *self-indexing*.

The critical points play an important role in identifying topological changes. Let X_s denote the set of points from M with f-values (or z-coordinate) $\leq s$. If $0 < s < 1$, then X_s is homeomorphic to a closed disk. For $1 < s < 2$, X_s is homeomorphic to the surface shown in Figure D.4. The critical value $s = 1$, distinguishes where X_s changes its topological nature. Similarly at $s = 0$ and $s = 2$, the homeomorphism class also changes.

Recall the deformation of M from Figure D.1 to Figure D.2. In effect the right maximum cancels with the right saddle and the left maximum cancels with the left saddle, leaving just the minimum and middle maximum. Let g be the height function for the resulting manifold in Figure D.2. Examine the flow for Figure D.2 that moves points in the direction of greatest decrease of g. There are just two critical points, one each of index 0 and index 2. The X_s analysis can be used to show that the manifold in Figure D.2 is homeomorphic to S^2. The reason is that X_s is a closed disk for each value of s less than the maximum of g. Adjoining the maximum essentially joins the boundary of the disk into a single point, making a sphere.

This suggests an alternative method of showing that a compact 2-manifold is homeomorphic to S^2. If flowing down yields just two critical points, a minimum and a maximum, then the 2-manifold is homeomorphic to S^2. Next we abstract the notion of the flow. If $h : M \to R$ is a smooth real valued function on M (not necessarily a height function), then calculus methods can produce a flow in the direction of greatest decrease in h, with critical points defined as the stationary points. In Figures D.1 and D.2 the flows were down in the direction of greatest decrease of f and g, respectively. Suppose that instead of deforming the manifold in D.1 to D.2 we could change the function f to a new function h on M so that the corresponding flow has just two critical points. It could then be argued that

M is homeomorphic to S^2. The central idea is to change the real valued function, rather than the manifold, so that the corresponding flow cancels pairs of critical points whose indices differ by 1.

For any compact n-manifold N, Smale proved[1] that there exists a real valued function k with certain technical properties, including self-indexing of the critical points. The values of k range from 0 to n with each critical point assuming an integer value. The algebraic topology of S^n and principles of Morse theory place restrictions on the number of critical points of each index. In his article "Generalized Poincaré's conjecture in dimensions greater than four,"[2] Smale proved that if $\dim(N) \geq 5$, then k could be modified so as to cancel pairs of critical points with indices that differed by 1. Combined with the Morse restrictions, it is possible to obtain a function for which the only critical points of the flow are a single minimum and a single maximum. Thus N is homeomorphic to S^n. Further mathematical details are available in Smale's papers and the Hirsch survey article in the book, *From topology to computation: Proceedings of the Smalefest.*[3]

Notes

Chapter 1. One Room Schoolhouse

1. James Gleick, *Genius*, Pantheon, New York, 1992.
2. Norman Macrae, *John von Neumann*, Pantheon, New York, 1992.
3. Steve Smale, "Some Autobiographical Notes", *From Topology to Computation: Proceedings of the Smalefest* (M. W. Hirsch et al., eds.) Springer-Verlag, New York (1993), p. 4.
4. Helen Smale interview, 5/23/93.
5. Helen Smale interview, 5/23/93.
6. Information on this period obtained from the 5/23/93 interview with Helen Smale; the 9/30/93 interview with Judy Smale; the 5/23/93 interview with Steve Smale; and from "Some Autobiographical Notes," by Steve Smale, in *From Topology to Computation: Proceedings of the Smalefest* (M. W. Hirsch et al., eds.) Springer-Verlag, New York (1993), pp. 3–21.
7. Bill Carpenter interview, 5/2/96.
8. At the time of the 7/15/93 interview, Norm Doorenbos was dean of the graduate school at Auburn University.
9. Norm Doorenbos interview, 7/15/93.
10. Steve Smale, "Some Autobiographical Notes," *From Topology to Computation: Proceedings of the Smalefest* (M. W. Hirsch et al., eds.), Springer-Verlag, New York, 1993, p. 5.
11. Joe Jewett interview, 10/2/93.
12. While all of the quotes are unattributed, this appears to come from *Fathers and Sons* by Turgenev.
13. Steve Smale interview, 5/23/93.

Chapter 2. Marxism and Mathematics at Ann Arbor

1. Jack Scruggs interview, 7/22/93.
2. Norman Doorenbos interview, 7/15/93.
3. Descriptions of the Thrall class are derived from an interview with Steve Smale on 5/23/93, an interview with Vincent Giuliano on 3/19/94, an interview with Art Rose on 10/3/93, and a telephone interview with Robert Thrall on 1/6/94.

4. Jack Scruggs interview, 7/22/93.
5. Steve Smale interview, 5/23/93.
6. David Caute, *The Great Fear*, Simon and Schuster, New York, 1978.
7. David Caute, *The Great Fear*, Simon and Schuster, New York, 1978.
8. Ellen Schrecker, *No Ivory Tower*, Oxford University Press, Oxford, 1986.
9. Ellen Schrecker, *No Ivory Tower*, Oxford University Press, Oxford, 1986, p. 96.
10. Ellen Schrecker, *No Ivory Tower*, Oxford University Press, Oxford, 1986.
11. Ellen Schrecker, *No Ivory Tower*, Oxford University Press, Oxford, 1986.
12. For the history of the HUAC, see *The Committee*, by Walter Goodman, Farrar, Strauss, and Giroux, New York, 1968.
13. Ellen Schrecker, *No Ivory Tower*, Oxford University Press, Oxford, 1986.
14. Levinson's testimony can be found in *Communist Methods of Infiltration* (*Education—Part* 4), Hearings before the Committee on Un-American Activities, Washington, D.C.: U.S. Government Printing Office, 1953. Differing reactions to his testimony are given in *A Beautiful Mind*, by Sylvia Nasar, Simon & Schuster, New York, 1998, pp. 153-154, and "The Purge" by Chandler Davis in *A Century of Mathematics in America, Part I* (Peter Duren, ed.), vol. 1, AMS, Providence, RI, 1988, p. 415.
15. Steve Smale, "Some Autobiographical Notes," *From Topology to Computation: Proceedings of the Smalefest* (M. W. Hirsch et al., eds.), Springer-Verlag, New York, 1993.
16. Steve Smale, "Some Autobiographical Notes," *From Topology to Computation: Proceedings of the Smalefest* (M. W. Hirsch et al., eds.), Springer-Verlag, New York, 1993, p. 7.
17. *New York Times*, 8/13/51, p. 1.
18. Vince Giuliano interview, 3/19/94.

19. This description of the controversy relies on material from the Smale archive, reports from the *Michigan Daily* in March, April, and May, 1952, and material on the McPhaul Case in Box 58 of the Ruthven Papers at the Bentley Historical Library.

20. The group had assisted the Communist leaders who had been convicted in 1949 of conspiracy charges. When four of these individuals jumped bail, the government sought their whereabouts from the trustees of the fund. In addition investigators attempted to identify the contributors of the bail money. Green's failure to respond before the grand jury led to a contempt citation (*New York Times*, 7/28/51 and 7/31/51).

21. Testimony of Stephen Smale, Investigation, March 18, 1952, in Box 58 of the Ruthven Papers at the Bentley Historical Library.

22. Joint Judiciary transcript of the Case of Stephen Smale, April 15, 1952, in Box 58 of the Ruthven Papers at the Bentley Historical Library.

23. Appeal hearing of Stephen Smale at the meeting of the Sub-Committee on Discipline, May 7, 1952, in Box 58 of the Ruthven Papers in the Bentley Historical Library.

24. *New York Times*, 11/15/60; and *The Committee* by Walter Goodman, Farrar, Strauss, and Giroux, New York, 1968.

25. Art Rose interview, 10/3/93. Rose is now a physician in the Detroit area and remains a devotee of chamber music.

26. Steve Smale, "Some Autobiographical Notes," *From Topology to Computation: Proceedings of the Smalefest* (M. W. Hirsch et al., eds.), Springer-Verlag, New York, 1993, p. 9.

27. Raoul Bott interview, 3/22/94.

28. Hans Samelson interview, 5/28/93.

29. J.-P. Serre, *Homologie singulière des espaces fibrés*, Annals of Mathematics, **54** (1951), 425-505.

30. *Nomination for Stephen Smale*, Notices Amer. Math. Soc., **38** (1991), 758-760; and Raoul Bott interview, 3/22/94.

31. James Munkres interview, 3/19/94.

32. Luncheon talk by Raoul Bott in *From Topology to Computation: Proceedings of the Smalefest* (M. W. Hirsch et al., eds.) Springer-Verlag, New York, 1993, pp 67-69.

33. "Steve Smale" in *More Mathematical People. Contemporary Conversations*. Albers et al., eds., Harcourt, Brace, Jovanovich, Boston, MA, 1990, p. 310.

34. *Investigation of Communist Activities in the State of Michigan—Part* 6, Hearing before the Committee on Un-American Activities, May 10, 1954, Washington, D.C.: U.S. Government Printing Office, 1954.

35. *Investigation of Communist Activities in the State of Michigan—Part* 6, Hearing before the Committee on Un-American Activities, May 10, 1954, Washington, D.C.: U.S. Government Printing Office, 1954, p. 5362.

36. Lionel Lewis, *Cold War on Campus*, Transaction Books, New Brunswick, NJ, 1988; and "The Purge" by Chandler Davis in *A Century of Mathematics in America, Part I* (Peter Duren, ed.), vol. 1, AMS, Providence, RI, 1988; and Ellen Schrecker, *No Ivory Tower*, Oxford University Press, Oxford, 1986.

37. *New York Times*, 8/27/54 and 12/8/59; and Ellen Schrecker, *No Ivory Tower*, Oxford University Press, Oxford, 1986.

38. Gardner Ackley to Ralph Sawyer, 12/13/54, Marvin Niehuss Papers, Box 4, Folder S, Michigan Historical Collections, Bentley Historical Library, University of Michigan, referenced in *No Ivory Tower* by Ellen Schrecker, Oxford University Press, Oxford, 1986.

39. M. E. Sharpe interview, 10/17/94.

40. Michigan State Police Report in Smale archive, 10/5/54.

41. Steve Smale, "Some Autobiographical Notes," *From Topology to Computation: Proceedings of the Smalefest* (M. W. Hirsch et al., eds.), Springer-Verlag, New York (1993), p. 9.

42. Frank Raymond conversation, 4/26/96.

43. Steve Smale interview, 5/23/93.

44. Clara Smale interview, 5/27/93.

45. *New York Times*, 6/24/93.

46. For a contemporary report see K. Rubin and A. Silverberg, *A report on Wiles' Cambridge lectures*, Bull. Amer. Math. Soc. (N.S.), **31** (1994), pp. 15-38.
47. A. Wiles, *Modular elliptic curves and Fermat's Last Theorem*, Annals of Mathematics, **141** (1995), 443-551; and R. L. Taylor and A. Wiles, *Ring theoretic properties of certain Hecke algebras*, Annals of Mathematics, **141** (1995), 553-572.
48. Raoul Bott interview, 3/22/94.
49. Saunders MacLane telephone interview, 6/30/94.
50. Wilder's 3/14/56 letter to A. W. Tucker at Princeton for Smale and his letter to Caltech are among the Raymond Louis Wilder Papers in the Archives of American Mathematics at the Center for American History at the University of Texas at Austin.

Chapter 3. Early Mathematical Audacity

1. Steve Smale, *The story of the Higher Dimensional Poincaré Conjecture*, The Mathematical Intelligencer, **12** (1990), pp. 44-51.
2. Raoul Bott interview, 3/22/94.
3. Clara Smale interview, 5/27/93.
4. William McNeill, *Hutchins' University*, The University of Chicago Press, 1991.
5. Marshall Stone, *Reminiscences of mathematics at Chicago*, The Mathematical Intelligencer, **11** (1989), pp. 20-25.
6. Felix Browder, *Stone age of mathematics at the midway*, The Mathematical Intelligencer, **11** (1989), pp. 22-25.
7. Edwin Spanier interview, 7/20/94.
8. Raoul Bott interview, 3/22/94.
9. Anthony Phillips, *Turning a sphere inside out*, Scientific American, **223**, May 1966, pp. 112-120.
10. Morris Hirsch, "Reminiscences of Chicago in the Fifties," *Paul Halmos. Celebrating 50 Years of Mathematics* (John H. Ewing and F. W. Gehring, eds.), Springer-Verlag, New York (1991), 109-118.

11. Morris Hirsch, "Reminiscences of Chicago in the Fifties," *Paul Halmos. Celebrating 50 Years of Mathematics* (John H. Ewing and F. W. Gehring, eds.), Springer-Verlag, New York (1991) p. 118.

12. Moe Hirsch interview, 7/18/94.

13. Clara Smale interview, 5/27/93.

14. Clara Smale interview, 5/27/93.

15. Courtesy of Edwin Spanier from his files.

16. See *Who Got Einstein's Office* by Ed Regis, Addison-Wesley, Reading, 1987; and "The School of Mathematics at the Institute for Advanced Study" by Armand Borel in *A Century of Mathematics in America. Part III* (Peter Doren, ed.), History of Mathematics, vol. 3, American Mathematical Society, Providence, RI, 1989, pp. 119-147.

17. "Oswald Veblen" by Deane Montgomery in *A Century of American Mathematics. Part I* (Peter Duren, ed.), History of Mathematics, vol. 1, American Mathematical Society, Providence, RI, 1988, pp. 119-129; and *John von Neumann* by Norman Macrae, Pantheon, New York, 1992.

18. John W. Dawson, Jr., *Logical Dilemmas*, A. K. Peters, Wellesley, MA, 1997.

19. Ed Regis, *Who Got Einstein's Office*, Addison-Wesley, Reading, 1987.

20. Philip Stern, *The Oppenheimer Case*, Harper and Row, New York, 1969.

21. Steve Smale, *Obituary of Marston Morse*, The Mathematical Intelligencer, **1** (1978/79), no. 1, pp. 33-34.

22. Steve Smale interview, 5/27/93.

23. Steve Smale, "On How I Got Started in Dynamical Systems," *The Mathematics of Time*, Springer-Verlag, New York, 1980, p. 148.

24. M. M. Peixoto, "Some Recollections of the Early Work of Steve Smale," *From Topology to Computation: Proceedings of the Smalefest* (M. W. Hirsch et al., eds.), Springer-Verlag, New York (1993), pp. 73-75.

25. Courtesy of Edwin Spanier from his files.

Chapter 4. On the Beaches of Rio

1. Steve Smale, "On How I Got Started in Dynamical Systems," *The Mathematics of Time*, Springer-Verlag, New York, 1980, pp. 147-151.
2. *New York Times*, 12/23/59.
3. *Chaos: Finding a Horseshoe on the Beaches of Rio*, by Steve Smale, The Mathematical Intelligencer, **20** (1998), pp. 39-44, and as a preprint, to appear in the Proceedings of the International Congress of Science and Technology—45 years of the National Research Council of Brazil.
4. See *Chaos* by James Gleick, Penguin, New York, 1987, p. 45; and "On the Contributions of Smale to Dynamical Systems" by Jacob Palis in *From Topology to Computation: Proceedings of the Smalefest* (M. W. Hirsch et al., eds.), Springer-Verlag, New York, 1993, p. 167.
5. *Communist Methods of Infiltration* (*Education—Part* 4), Hearings before the Committee on Un-American Activities, Washington, D.C.: U.S. Government Printing Office, 1953.
6. From *Chaos: Finding a Horseshoe on the Beaches of Rio*, by Steve Smale, The Mathematical Intelligencer, **20** (1998), pp. 39-44, and as a preprint, to appear in the Proceedings of the International Congress of Science and Technology—45 years of the National Research Council of Brazil.
7. T.-Y. Li and James A. Yorke, *Period three implies chaos,* American Mathematical Monthly, **82** (1975), pp. 985-992.
8. See *Chaos* by James Gleick, Penguin, New York, 1987.
9. *Chaos: Finding a Horseshoe on the Beaches of Rio*, by Steve Smale, The Mathematical Intelligencer, **20** (1998), pp. 39-44, and as a preprint, to appear in the Proceedings of the International Congress of Science and Technology—45 years of the National Research Council of Brazil.
10. Steve Smale, *The Story of the Higher Dimensional Poincaré Conjecture*, The Mathematical Intelligencer, **12** (1990), 44-51.
11. Steve Smale, *The Story of the Higher Dimensional Poincaré Conjecture*, The Mathematical Intelligencer, **12** (1990), 44-51.

12. John Stallings, "How Not to Prove the Poincaré Conjecture," *Topology Seminar Wisconsin*, 1965 (R. H. Bing and R. J. Bean, eds.), Princeton University Press, Princeton, NJ, 1966, pp. 83-88; and Gary Taubes, *What Happens When Hubris Meets Nemesis*, Discover, July 1967, pp. 66-77.
13. Steve Smale, *The Story of the Higher Dimensional Poincaré Conjecture*, The Mathematical Intelligencer, **12** (1990), 44-51.
14. Moe Hirsch interview, 7/18/94; and James Munkres interview, 3/19/94.
15. Steve Smale, *The Story of the Higher Dimensional Poincaré Conjecture*, The Mathematical Intelligencer, **12** (1990), 44-51.
16. John Stallings interview, 6/3/96.
17. John Stallings interview, 6/3/96.
18. Steve Smale, *The Story of the Higher Dimensional Poincaré Conjecture*, The Mathematical Intelligencer, **12** (1990), 44-51.
19. Steve Smale interview, 1/4/95.
20. John Stallings interview, 6/3/96.
21. Steve Smale, *The Story of the Higher Dimensional Poincaré Conjecture*, The Mathematical Intelligencer, **12** (1990), 44-51.
22. John Milnor, "The Work of M. H. Freedman," *Proceedings of the International Congress of Mathematicians* 1986 (Andrew M. Gleason, ed.) American Mathematical Society, Providence, RI, 1987, vol. 1, pp. 13-15.

Chapter 5. Berkeley to Columbia and Back to Berkeley

1. Ellen Schrecker, *No Ivory Tower*, Oxford University Press, Oxford, 1986.
2. *New York Times*, 11/20/60.
3. Steve Smale interview, 5/23/93.
4. J. Milnor, "Differentiable manifolds which are homotopy spheres," mimeographed notes, Princeton, 1958.

5. For a mathematical summary see M. W. Hirsch, "The Work of Stephen Smale in Differential Topology," *From Topology to Computation: Proceedings of the Smalefest* (M. W. Hirsch et al., eds.), Springer-Verlag, New York, 1993, pp. 83-106.

6. Steve Smale, *On the structure of manifolds*, American Journal of Mathematics, **84** (1962), pp. 387-399.

7. John Milnor, *Lectures on the h-cobordism theorem*, Princeton University Press, Princeton, NJ, 1965.

8. Steve Smale interviews, 5/23, 24, 25/93; and "Steve Smale" in *More Mathematical People. Contemporary Conversations* (Albers et al., eds.), Harcourt, Brace, Jovanovich, Boston, MA, 1990.

9. Clara Smale interview, 5/27/93.

10. Steve Smale interview, 5/23/93.

11. Steve Smale, "On How I Got Started in Dynamical Systems," *The Mathematics of Time*, Springer-Verlag, New York, 1980.

12. Jacob Palis, "On the Contribution of Smale to Dynamical Systems," *From Topology to Computation: Proceedings of the Smalefest* (M. W. Hirsch et al., eds.), Springer-Verlag, New York, 1993, pp. 165-178.

13. Mike Shub interview, 10/27/95.

14. Steve Smale, "Dynamical systems and the topological conjugacy problem for diffeomorphisms," *Proceedings of the International Congress of Mathematicians* 1962 (V. Stenstrom, ed.), Institut Mittag-Leffler, Djursholm, Sweden, 1963, pp. 490-496.

15. Jacob Palis, "On the Contribution of Smale to Dynamical Systems", *From Topology to Computation: Proceedings of the Smalefest* (M. W. Hirsch et al., eds.), Springer-Verlag, New York, 1993, p. 171.

16. Mike Shub, "Appendix: Personal Reminiscences", *From Topology to Computation: Proceedings of the Smalefest* (M. W. Hirsch et al., eds.), Springer-Verlag, New York, 1993, pp. 296-299.

17. Mike Shub interview, 10/27/95.

18. Mike Shub, "Appendix: Personal Reminiscences," *From Topology to Computation: Proceedings of the Smalefest* (M. W. Hirsch et al., eds.), Springer-Verlag, New York, 1993, pp. 296-299.

19. Mike Shub interview, 10/27/95.

20. Robert S. McNamara, *In Retrospect*, Times Books, New York, 1995.
21. Steve Smale, "Some Autobiographical Notes", *From Topology to Computation: Proceedings of the Smalefest* (M. W. Hirsch et al., eds.), Springer-Verlag, New York, 1993, p. 10.
22. Steve Smale interview, 1/4/95.
23. Anthony Tromba, "Smale and Nonlinear Analysis: A Personal Perspective," *From Topology to Computation: Proceedings of the Smalefest* (M. W. Hirsch et al., eds.), Springer-Verlag, New York, 1993, pp. 481-492.
24. Steve Smale, *Morse theory and a non-linear generalization of the Dirichlet problem*, Annals of Mathematics, **80** (1964), 382-396.
25. Steve Smale, *An infinite dimensional version of Sard's theorem*, American Journal of Mathematics, **87** (1965), 861-866.
26. Steve Smale interview, 1/4/95.
27. For a more complete history of the Free Speech Movement see the early account *Berkeley: The Student Revolt* by Hal Draper, Grove Press Inc., New York, 1965; the scholarly *Berkeley At War* by W. J. Rorabaugh, Oxford University Press, Oxford, 1989; and the retrospective *The Free Speech Movement* by David Goines, Ten Speed Press, Berkeley, 1993. See also Mark Kitchell's documentary film "Berkeley in the Sixties", and *Revolution at Berkeley* by Michael Miller and Susan Gilmore, The Dial Press, New York, 1965.
28. Jack Weinberg interview, 11/13/94. At the time of the interview Weinberg worked for Greenpeace.
29. W. J. Rorabaugh, *Berkeley At War*, Oxford University Press, Oxford, 1989, pp. 21-22.
30. This interview was done by Marston Schultz and Burton White, and was quoted in *The Free Speech Movement* by David Goines, Ten Speed Press Berkeley, 1993, pp. 93-99.
31. Clark Kerr, *The Uses of the University*, Harvard University Press, Cambridge, MA, 1960; and *The Frantic Race to Remain Contemporary*, Daedalus, Fall 1964.
32. See W. J. Rorabaugh, *Berkeley At War*, Oxford University Press, Oxford, 1989, for more on the Berkeley political dynamic.

33. David Goines, *The Free Speech Movement*, Ten Speed Press, Berkeley, 1993, pp. 103-111.
34. W. J. Rorabaugh, *Berkeley At War*, Oxford University Press, Oxford, 1989.
35. Hal Draper, *Berkeley: The Student Revolt*, Grove Press, Inc., New York, 1965, pp. 31-32.
36. From contemporary KPFA tape in *The Free Speech Movement* by David Goines, Ten Speed Press, Berkeley, CA, 1993, p. 152.
37. Jack Weinberg interview, 11/13/94.
38. For the text see *Berkeley: The Student Revolt* by Hal Draper, Grove Press, Inc., New York, 1965, pp. 56-57.
39. Jack Weinberg interview, 11/13/94.
40. Jack Weinberg interview, 11/13/94.
41. Steve Weissman interview, 1/26/95.
42. Steve Weissman interview, 1/26/95. At the time of the interview Weissman operated a small publishing house specializing in boating and fishing.
43. David Goines, *The Free Speech Movement*, Ten Speed Press, Berkeley, CA, 1993, pp. 303-326.
44. Hal Draper, *Berkeley: The Student Revolt*, Grove Press, Inc., New York, 1965, pp. 35-37; and David Goines, *The Free Speech Movement*, Ten Speed Press, Berkeley, CA, 1993, pp. 332-333.
45. David Goines, *The Free Speech Movement*, Ten Speed Press, Berkeley, CA, 1993, pp. 339-340, 348-349.
46. Draper and Rorabaugh argue that the action was directed by Kerr and the regents while Goines makes the dubious suggestion that it came from Sherriffs.
47. Steve Weissman interview, 1/26/95.
48. From contemporary KPFA tape in *The Free Speech Movement*, by David Goines, Ten Speed Press, Berkeley, CA, 1993, p. 361.
49. W. J. Rorabaugh, *Berkeley At War*, Oxford University Press, Oxford, 1989, p. 32.

50. W. J. Rorabaugh, *Berkeley At War*, Oxford University Press, Oxford, 1989, p. 33.
51. Steve Weissman interview, 1/26/95, in which Weissman stressed his dislike for the phrase "role model."
52. Steve Weissman interview, 1/26/95.
53. Steve Weissman interview, 1/26/95.

Chapter 6. The Lone Ranger of the Antiwar Movement

1. David Goines, *The Free Speech Movement*, Ten Speed Press, Berkeley, CA, 1993, pp. 484-489.
2. W. J. Rorabaugh, *Berkeley At War*, Oxford University Press, Oxford, 1989; and David Goines, *The Free Speech Movement*, Ten Speed Press, Berkeley, CA, 1993.
3. Robert McNamara, *In Retrospect*, Times Books, New York, 1995, p. 31.
4. Robert McNamara, *In Retrospect*, Times Books, New York, 1995.
5. Robert McNamara, *In Retrospect*, Times Books, New York, 1995.
6. Robert McNamara, *In Retrospect*, Times Books, New York, 1995.
7. Robert McNamara, *In Retrospect*, Times Books, New York, 1995, p. 138 (Dean Rusk statement in a Senate hearing); and *New York Times*, 8/8/84, with Fulbright's response to Gaylord Nelson's proposed amendment.
8. Tom Wells, *The War Within*, University of California Press, Berkeley, CA, 1994.
9. *New York Times*, 2/20/65.
10. Robert McNamara, *In Retrospect*, Times Books, New York, 1995, pp. 165, 176.
11. Anatol Rapaport, "Dialogue or Monologue" *Teach-Ins: U.S.A.* (Louis Menashe and Ronald Radosh, eds.), Frederick A. Praeger, New York, 1967, pp. 4-8.
12. *Daily Californian*, 3/24/65, 3/25/65.
13. Robert McNamara, *In Retrospect*, Times Books, New York, 1995, p. 179.

14. *New York Times*, 4/18/65.

15. *San Francisco Chronicle*, 4/18/65, reported 2500.

16. Steve Smale, "Some Autobiographical Notes," *From Topology to Computation: Proceedings of the Smalefest* (M. W. Hirsch et al., eds.) Springer-Verlag, New York, 1993, p. 13.

17. Telephone interview with Barbara Gullahorn Holecek, an independent filmmaker at the time, 6/11/95.

18. Barbara Gullahorn Holecek interview, 6/11/95.

19. Jerry Rubin, *Growing (up) at thirty seven*, M. Evans and Company, New York, 1976.

20. Barbara Gullahorn Holecek interview, 6/11/95.

21. Jerry Rubin, *Growing (up) at thirty seven*, M. Evans and Company, New York, 1976.

22. Barbara Gullahorn Holecek interview, 6/11/95.

23. Barbara Gullahorn Holecek interview, 6/11/95.

24. Steve Smale, "Some Autobiographical Notes," *From Topology to Computation: Proceedings of the Smalefest* (M. W. Hirsch et al., eds.), Springer-Verlag, New York, 1993.

25. Steve Smale, "Some Autobiographical Notes," *From Topology to Computation: Proceedings of the Smalefest* (M. W. Hirsch et al., eds.), Springer-Verlag, New York, 1993, p. 14.

26. Louis Menashe and Ronald Radosh (Editors), *Teach-Ins: U.S.A.*, Frederick A. Praeger, New York, 1967, p. 118.

27. *Time*, 5/14/65.

28. *San Francisco Chronicle*, 5/21/65.

29. Unpublished manuscript by Steve Smale.

30. CU 149, Box 65, Folder 133, University Archives, University of California, Berkeley.

31. Marilyn Milligan interview, 7/21/94.

32. Barbara Gullahorn Holecek interview, 6/11/95.

33. S. Smale, *Structurally stable systems are not dense*, American Journal of Mathematics, **88** (1966), 491-496.

34. Staughton Lynd, "Nonviolent Alternatives to American Violence," *Teach-Ins: U.S.A.* (Louis Menashe and Ronald Radosh, eds.), Frederick A. Praeger, New York, 1967, pp. 54-59.

35. Jerry Rubin interview, 7/22/94.

36. Steve Smale, "Some Autobiographical Notes," *From Topology to Computation: Proceedings of the Smalefest* (M. W. Hirsch et al., eds.), Springer-Verlag, New York, 1993, pp. 20-21.

37. Unpublished manuscript by Steve Smale.

38. Marilyn Milligan interview, 7/21/94.

39. Robert McNamara, *In Retrospect*, Times Books, New York, 1995 (Westmoreland memo 187-8, poll 190).

40. Robert McNamara, *In Retrospect*, Times Books, New York, 1995.

41. Tom Wells, *The War Within*, University of California Press, Berkeley, CA, 1994, p. 51.

42. *San Francisco Chronicle*, 8/6/65.

43. Unpublished manuscript by Steve Smale.

44. W. J. Rorabaugh, *Berkeley At War*, Oxford University Press, Oxford, 1989, p. 94.

45. Steve Weissman interview, 1/26/95.

46. Steve Weissman interview, 1/26/95; Jack Weinberg interview, 11/13/94.

47. Unpublished manuscript by Steve Smale.

48. Unpublished manuscript by Steve Smale.

49. Marilyn Milligan interview, 7/21/94.

50. Unpublished manuscript by Steve Smale.

51. Steve Smale interview, 7/20/94.

52. *San Francisco Chronicle*, 8/25/65.

53. Steve Weissman interview, 1/26/95.

54. Barbara Gullahorn Holecek interview, 6/11/95.

55. *San Francisco Chronicle*, 8/26/65.

56. Tom Wells, *The War Within*, University of California Press, Berkeley, CA, 1994.
57. Steve Smale interview, 1/4/65.
58. Marilyn Milligan interview, 7/21/94.
59. Frank Bardacke interview, 7/23/94.
60. *Life*, 12/10/65.
61. *Life*, 12/10/65.
62. *San Francisco Chronicle*, 9/27/65; and *Daily Californian*, 9/27/65.
63. Unpublished manuscript by Steve Smale and interview, 7/19/94.
64. CU 149, Box 65, Folder 134, University Archives, University of California, Berkeley.
65. Letter, 10/2/65, CU 149, Box 65, Folder 134, University Archives, University of California, Berkeley.
66. Kerr to Heyns, 10/3/65, CU 149, Box 65, Folder 134, University Archives, University of California, Berkeley.
67. Letter from Thomas Cunningham to Roger Heyns, 10/6/65, in Chancellor's Records, University Archives, University of California, Berkeley.
68. Chancellor's Records, University Archives, University of California, Berkeley.
69. Memo from Dick Hafner to Earl Cheit, 10/15/65, in Chancellor's Records, University Archives, University of California, Berkeley.
70. Berkeley City Manager report on "Activities of Vietnam Day Committee", 12/10/65.
71. Steve Weissman interview, 1/26/65.
72. Steve Smale interviews, 7/20/94 and 1/4/95; Jack Weinberg interview, 11/13/94.
73. W. J. Rorabaugh, *Berkeley At War*, Oxford University Press, Oxford, 1989, p. 97; and *New York Times*, 10/17/65; and *San Francisco Chronicle*, 10/17/65.
74. *New York Times*, 10/16/65 and 10/17/65; and Fred Halstead, *Out Now!*, Monad Press, New York, 1978.

75. Steve Smale, phone interview, 8/8/95.
76. Frank Bardacke interview, 7/23/94.
77. Jerry Rubin interview, 7/22/94.
78. Letter from John Morin to Steve Smale et al., 11/12/65, in Smale archives; and 11/17/65 Preliminary Injunction by W. T. Seigert in University Archives, University of California, Berkeley.
79. *San Francisco Examiner*, 11/21/65.
80. VDC News, 11/15/65, pro and con articles by Steve Weissman and Duncan Stewart in University Archives, University of California, Berkeley.
81. W. J. Rorabaugh, *Berkeley At War*, Oxford University Press, Oxford, 1989, p. 99.
82. Fred Halstead, *Out Now!*, Monad Press, New York, 1978.
83. Jerry Rubin interview, 7/22/94.
84. Fred Halstead, *Out Now!*, Monad Press, New York, 1978; and Tom Wells, *The War Within*, University of California Press, Berkeley, 1994.
85. Marilyn Milligan interview, 7/21/94.
86. Robert McNamara, *In Retrospect*, Times Books, New York, 1995, p. 266.
87. Tom Wells, *The War Within*, University of California Press, Berkeley, 1994.
88. Jerry Rubin interview, 7/22/94.
89. Tom Wells, *The War Within*, University of California Press, Berkeley, 1994.
90. Robert McNamara, *In Retrospect*, Times Books, New York, 1995.
91. Steve Smale interview, 5/26/93.
92. Tom Wells, *The War Within*, University of California Press, Berkeley, 1994.

93. See Tom Wells, *The War Within*, University of California Press, Berkeley, 1994; Fred Halstead, *Out Now!*, Monad Press, New York, 1978; and Nancy Zaroulis and Gerald Sullivan, *Who Spoke Up*, Doubleday, Garden City, New York, 1984.

94. Jerry Rubin interview, 7/22/94.

95. Jerry Rubin interview, 7/22/94.

Chapter 7. The Summer of 1966

1. Donald J. Albers, G. L. Alexanderson, and Constance Reid, *International Mathematical Congresses*, Springer-Verlag, New York, 1986.

2. Henry S. Tropp, *The origins and history of the Fields Medal*, Historia Mathematica, **3** (1976), 167-181.

3. Elizabeth Crawford, *The Beginnings of the Nobel Institution*, Cambridge University Press, Cambridge, 1984, Appendix A.

4. In "Appendix I" of Henry S. Tropp, *The Origins and History of the Fields Medal*, Historia Mathematica, **3** (1976), 167-181.

5. Donald J. Albers, G. L. Alexanderson, and Constance Reid, *International Mathematical Congresses*, Springer-Verlag, New York, 1986.

6. Lars Gårding and Lars Hörmander, *Why is there no Nobel Prize in mathematics?*, The Mathematical Intelligencer, **7** (1983), 73-74.

7. See Peter Ross, *Why isn't there a Nobel Prize in mathematics?*, Math. Horizons, Nov. 1995, p. 9.

8. See also Nicholas Wade, "Nobel Follies," *Science*, vol. 211 (1981), p. 1404; and Howard Eves, *In Mathematical Circles, Quadrants III and IV*, Prindle, Weber, and Schmidt, Boston, MA, 1969, p. 130.

9. Elizabeth Crawford, *The Beginnings of the Nobel Institution*, Cambridge University Press, Cambridge, 1984, p. 251.

10. From a 1/14/72 letter Tropp received from Synge, Henry S. Tropp, *The origins and history of the Fields Medal*, Historia Mathematica, **3** (1976), 167-181.

11. Elizabeth Crawford, *The Beginnings of the Nobel Institution*, Cambridge University Press, Cambridge, 1984, p. 36.

12. Lars Hörmander, "The First Woman Professor and Her Male Colleague," *Miscellanea Mathematica* (P. Hilton et al., eds.), Springer-Verlag, Berlin, 1991.

13. Elizabeth Crawford, *The Beginnings of the Nobel Institution*, Cambridge University Press, Cambridge, 1984, p. 251.

14. Elizabeth Crawford, *The Beginnings of the Nobel Institution*, Cambridge University Press, Cambridge, 1984, pp. 51-52.

15. Elizabeth Crawford, *The Beginnings of the Nobel Institution*, Cambridge University Press, Cambridge, 1984, p. 251.

16. Kenne Font, *Alfred Nobel*, Arcade Publishing, New York, 1993; and H. Schuck, R. Sohlman, et al., *Nobel, The Man and his Prizes*, University of Oklahoma Press, Norman, OK, 1951.

17. H. Schuck, R. Sohlman, et al., *Nobel, The Man and his Prizes*, University of Oklahoma Press, Norman OK, 1951.

18. Lars Gårding and Lars Hörmander, *Why is there no Nobel Prize in mathematics?*, The Mathematical Intelligencer, **7** (1983), 73-74.

19. Elizabeth Crawford, *The Beginnings of the Nobel Institution*, Cambridge University Press, Cambridge, 1984, pp. 52-53.

20. Elizabeth Crawford, *The Beginnings of the Nobel Institution*, Cambridge University Press, Cambridge, 1984, p. 113.

21. Elizabeth Crawford, *The Beginnings of the Nobel Institution*, Cambridge University Press, Cambridge, 1984.

22. Elizabeth Crawford, *The Beginnings of the Nobel Institution*, Cambridge University Press, Cambridge, 1984.

23. Elizabeth Crawford, *The Beginnings of the Nobel Institution*, Cambridge University Press, Cambridge, 1984.

24. For an interesting perspective see "The European Mathematicians' Migration to America" by Lipman Bers in *A Century of Mathematics in America, Part I* (Peter Duren, ed.), History of Mathematics, vol. 1, American Mathematical Society, Providence, RI, 1988, pp. 231-243.

25. Donald J. Albers, G. L. Alexanderson, and Constance Reid, *International Mathematical Congresses*, Springer-Verlag, New York, 1986.

26. J. Milnor, *On manifolds homeomorphic to the 7-sphere*, Annals of Mathematics, **64** (1956), 395-405.

27. Steve Smale interview, 1/4/95.

28. Steve Smale interview, 1/4/95.

29. See "Choosing up sides" by Joseph Barry in *Village Voice*, 6/2/66.

30. Steve Smale, *On the steps of Moscow University*, The Mathematical Intelligencer, **6** (1984), p. 21.

31. "Choosing up sides" by Joseph Barry in *Village Voice*, 6/2/66.

32. For Steve's personal account of the summer see *On the steps of Moscow University*, by Steve Smale, The Mathematical Intelligencer, **6** (1984), pp. 21-27.

33. W. J. Rorabaugh, *Berkeley At War*, Oxford University Press, Oxford, 1989, p. 103.

34. *San Francisco Examiner*, 8/5/66.

35. Steve Smale, *On the steps of Moscow University*, The Mathematical Intelligencer, **6** (1984), p. 22.

36. Steve Smale, *On the steps of Moscow University*, The Mathematical Intelligencer, **6** (1984), p. 23.

37. Donald J. Albers, G. L. Alexanderson, and Constance Reid, *International Mathematical Congresses*, Springer-Verlag, New York, 1986, p. 34.

38. *New York Times*, 2/15/66.

39. Steve Smale email, 1/29/96.

40. Steve Smale, *On the steps of Moscow University*, The Mathematical Intelligencer, **6** (1984).

41. Chandler Davis interview, 1/6/95.

42. Chandler Davis interview, 1/6/95.

43. Chandler Davis interview, 1/6/95.

44. Steve Smale, *On the steps of Moscow University*, The Mathematical Intelligencer, **6** (1984).

45. Steve Smale, *On the steps of Moscow University*, The Mathematical Intelligencer, **6** (1984).

46. Chandler Davis interview, 1/6/95.
47. Steve Smale, *On the steps of Moscow University*, The Mathematical Intelligencer, **6** (1984).
48. *New York Times*, 8/27/66.
49. Steve Smale, *On the steps of Moscow University*, The Mathematical Intelligencer, **6** (1984).
50. Moe Hirsch interview, 7/18/94.
51. *Washington Post*, 8/27/66.
52. Investigative Hearings before the Committee on Un-American Activities, Washington, D.C.: U.S. Government Printing Office, 1966.
53. *New York Times*, 8/16/66, 8/17/66.
54. *New York Times*, 8/18/66.
55. Investigative Hearings before the Committee on Un-American Activities, Washington, D.C.: U.S. Government Printing Office, 1966, pp. 1078-1079.
56. Investigative Hearings before the Committee on Un-American Activities, Washington, D.C.: U.S. Government Printing Office, 1966, pp. 1109.
57. *New York Times*, 8/16/66, 8/17/66, 8/18/66, 8/19/66.
58. *New York Times*, 8/20/66.
59. Steve Smale interview, 5/24/93.
60. *New York Times*, 8/21/66.
61. *New York Times*, 8/22/66.

Chapter 8. Smale Versus the National Science Foundation

1. More details on the history of the NSF can be found in Milton Lomask, *A Minor Miracle*, NSF, Washington, 1976; and Daniel Greenberg, *The Politics of Pure Science*, New American Library, New York, 1967.
2. Daniel Greenberg, *The Politics of Pure Science*, New American Library, New York, 1967, p. 79.

3. Milton Lomask, *A Minor Miracle*, NSF, Washington, 1976, p. 69.

4. Milton Lomask, *A Minor Miracle*, NSF, Washington, 1976.

5. First Annual Report of the National Science Foundation, 1950-51.

6. Milton Lomask, *A Minor Miracle*, NSF, Washington, 1976, p. 91.

7. Second Annual Report of the National Science Foundation, 1951-52.

8. Daniel Greenberg, *The Politics of Pure Science*, New American Library, New York, 1967.

9. Gordon Lill and Arthur Maxwell, "The Earth's Mantle," *Science*, vol. 129, May 22, 1959.

10. Robert Caro, *The Path to Power*, Knopf, New York, 1982; and Robert Caro, *Means of Ascent*, Knopf, New York, 1990.

11. Milton Lomask, *A Minor Miracle*, NSF, Washington, 1976, p. 149.

12. Herbert Solow, "How NSF Got Lost in Mohole," *Fortune*, vol. 67, May 1963.

13. Primary sources for the remainder of this chapter include the following collections of papers:

Office of the Director, Subject Files, National Science Foundation, Washington National Records Center, Suitland, Maryland

Leland Haworth Chronology File, NSF Historian's File, National Science Foundation, Arlington, VA

Smale Folders #1, 2, 3, NSF Historian's File, National Science Foundation, Arlington, VA

University Archives, University of California, Berkeley

Richard Roudebush papers, Indiana State Library

Stephen Smale archive.

14. *New York Times*, 8/22/66.

15. Transcript of 8/26/66 call in Office of the Director, Subject Files, NSF; and Cheit notes, University Archives, University of California, Berkeley.

16. *Washington Post*, 8/27/66.

17. Connick notes, University Archives, University of California, Berkeley.

18. Office of the Director, Subject Files, NSF.
19. 9/13/66 transcript, Office of the Director, Subject Files, NSF.
20. Office of the Director, Subject Files, NSF.
21. Smale letter to Connick, 9/16/66, Stephen Smale archive.
22. Helen Lasota letter to Matthew Gaffney, 5/11/66, and Gaffney message to Helson, 7/14/66, Stephen Smale archive.
23. Connick memo to Vice President Angus Taylor, 1/16/67, University Archives, University of California, Berkeley.
24. Haworth memo to Joseph Califano, 9/8/67, Leland J. Haworth Chronology File, NSF Historian's File; and Frank Sheppard letter to J. Edgar Hoover, 8/22/67 referring to previous reports, Stephen Smale archive.
25. Stephen Smale, *Differentiable dynamical systems*, Bull. Amer. Math. Soc. (N.S.), **73** (1967), 747-817.
26. Application and award letter, Office of the Director, Subject Files, NSF.
27. Ray Wilder letter to Smale, 10/7/66, Stephen Smale archive.
28. Connick letter to NSF Comptroller Aaron Rosenthal, 2/1/67, Office of the Director, Subject Files, NSF.
29. Wright memo to Haworth, 5/31/67, Office of the Director, Subject Files, NSF.
30. Elberg letter to Wright, 6/2/67, and Wright memo about discussion, 6/7/67, both in Office of the Director, Subject Files, NSF.
31. Robertson memo to Haworth, 6/28/67, in Office of the Director, Subject Files, NSF. While Robertson, Wright, and Haworth each opposed funding the Smale proposal, it possibly had the support of Deputy Director John Wilson. The Office of the Director, Subject Files, NSF, contain no record of Wilson's position, but his approval was mentioned in the tipoff letter to Roudebush of 6/16/67, Roudebush papers.
32. Stephen Smale archive.
33. Smale letter to Wright, 9/8/67, Stephen Smale archive.
34. Milton Lomask, *A Minor Miracle*, NSF, Washington, 1976, p. 163.

35. Stephen Smale archive.
36. Smale letter to Elberg, 9/27/67, Stephen Smale archive.
37. Haworth letter to Elberg, 10/12/67, Stephen Smale archive.
38. Unpublished letter by Smale to *Science*, 10/15/67, Stephen Smale archive.
39. Haworth letter to Roudebush, 11/3/67, Roudebush papers.
40. Congressional Records of 11/28/67 and 12/15/67.
41. Haworth statement, 1/10/68, Smale Folder 3, NSF Historian's File, NSF.
42. Milton Lomask, *A Minor Miracle*, NSF, Washington, 1976, p. 161.
43. Sanford Elberg interview, 10/22/93.

Chapter 9. The Aesthetic Side: Minerals and Photography

1. Wayne Thompson interview, 2/9/96.
2. Wayne Thompson interview, 2/9/96.
3. Steve Smale interview, 5/19/96.
4. Paul Desautels, *The Mineral Kingdom*, Madison Square Press, New York, 1968.
5. Steve Smale, email, 9/26/96.
6. Brian Lloyd, *Mineral Auctions at Sotheby's of London*, The Mineralogical Record Jan.-Feb. 1971.
7. *Dynamical Systems*, edited by M. M. Peixoto, Academic Press, New York, 1973.
8. Paul Desautels, *The Mineral Kingdom*, Madison Square Press, New York, 1968, p. 79.
9. Steve Smale email, 1/11/97.
10. "The Annual Meeting of the National Academy of Sciences", July 1, 1971.
11. Mike Shub interview, 10/27/95.
12. See Michael Shub's letter to the *New York Review of Books*, 12/30/71.
13. Statement by Elon Lima, Jacob Palis, and Maurcio Peixoto, 4/28/99.

14. Michael Shub's letter to the *New York Review of Books*, 12/30/71.
15. Mike Shub interview, 10/27/95.
16. Dave Wilber interview, 2/10/96.
17. Steve Smale email, 9/26/96 and 1/21/97.
18. Dave Wilber interview, 2/10/96.
19. Bill Larson interview, 2/10/96.
20. Steve Smale interview, 5/19/96, and email 1/21, 27/97.
21. Wendell E. Wilson, *What's New in Minerals? The Tucson Show*, The Mineralogical Record, May-June 1976, pp. 129-137.
22. Steve Smale interview, 5/19/96.
23. Wayne Thompson phone interview, 2/23/97.
24. Steve Smale interview, 5/19/96.
25. Wayne Thompson phone interview, 2/23/97.
26. Joel Bartsch interview, 2/10/96.
27. Bill Larson interview, 2/10/96.
28. Steve Smale interview, 5/19/96.
29. Steve Smale interview, 5/19/96.
30. Wayne Thompson phone interview, 2/23/97.
31. Steve Smale email, 12/29/97.
32. Steve Smale interview, 8/16/96.
33. For more information on mineral photography see "*Photographing Minerals, Fossils, & Lapidary Materials* " by Jeffrey Scovil, Geoscience Press, Inc., Tucson, Arizona, 1996.
34. Steve Smale email, 5/28/97.
35. Steve Smale email, 6/4/97.
36. Steve Smale email, 6/4/97.
37. Jane Jackson interview, 5/20/97.
38. Stephen Scheer interview, 5/23/97.

Chapter 10. Adventure and Physical Risks

1. Vince Giuliano interview, 3/19/94.
2. Steve Smale interview, 5/23/93.
3. Jon Krakauer, *Into Thin Air*, Anchor, 1998.
4. Steve Smale interview, 5/23/93.
5. Robert Rossiter, *Teton Classics*, Chockstone Press, Evergreen, CO, 1994, p. 46.
6. Steve Smale email, 6/25/97.
7. Steve Smale interview, 5/23/93.
8. Steve Smale interview, 5/24/93.
9. Steve Smale interview, 8/6/96.
10. Charles Pugh interview, 8/6/96.
11. "Steve Smale" in *More Mathematical People. Contemporary Conversations* by Albers et al., Harcourt, Brace, Jovanovich, 1990, Boston, MA, p. 318.
12. Steve Smale interview, 8/6/96.
13. Steve Smale interview, 8/6/96.
14. Steve Smale interview, 8/6/96.
15. Charles Pugh interview, 8/6/96.
16. Charles Pugh interview, 8/6/96.
17. Steve Smale interview, 8/6/96.
18. Steve Smale interview, 8/6/96.
19. Stardust Log, 5/3/87.
20. Charles Pugh interview, 8/6/96.
21. Steve Smale interview, 8/6/96.
22. Stardust Log, 5/14/87.
23. Stardust Log, 5/20/87.
24. Charles Pugh interview, 8/6/96.
25. Charles Pugh interview, 8/6/96.

26. Steve Smale email, 7/18/97.
27. Steve Smale email, 7/18/97.
28. Steve Smale email, 7/21/97.
29. Charles Pugh interview, 8/6/96.
30. Stardust Log, 7/28/87.
31. Charles Pugh interview, 8/6/96.
32. Steve Smale interviews, 5/19/96, 8/11/97.

Chapter 11. Other People

1. Steve Smale, "Some Autobiographical Notes," *From Topology to Computation: Proceedings of the Smalefest* (M. W. Hirsch et al., eds.), Springer-Verlag, New York, 1993, p. 4.
2. Laura Smale interview, 4/25/96, and Nat Smale interview, 8/7/96.
3. Laura Smale interview, 4/25/96.
4. Erik Molvar and Tamara Martin, *Hiking Zion & Bryce Canyon National Parks*, Falcon Press, Helena, Montana, p. 68.
5. Nat Smale interview, 8/7/96, and Steve Smale interview, 8/11/97.
6. From police report in the Stephen Smale archive.
7. Mario Vargas Llosa, "Inquest in the Andes," *The New York Times Magazine*, 7/31/83; and *Shining Path of Peru*, Edited by David Scott Palmer, St. Martin's Press, New York, NY, 1992, especially p. 177.
8. Steve Smale interview, 8/11/97.
9. Steve Smale interview, 8/11/97.
10. Mike Shub interview, 10/27/95.
11. Mike Shub, "Appendix: Personal Reminiscences," *From Topology to Computation: Proceedings of the Smalefest* (M. W. Hirsch et al., eds.), Springer-Verlag, New York, 1993, p. 298.
12. Mike Shub interview, 10/27/95.
13. "Discussion," *From Topology to Computation: Proceedings of the Smalefest* (M. W. Hirsch et al., eds.), Springer-Verlag, New York, 1993, p. 184.

14. Margy Rochlin, "The Mathematics of Discrimination," *Los Angeles Times Sunday Magazine*, 5/2/93.

15. Jenny Harrison, *Unsmoothable Diffeomorphisms*, Annals of Mathematics, **102** (1975), pp. 85-94.

16. See Allyn Jackson, *Fighting for Tenure*, Notices Amer. Math. Soc., **141** (1994), pp. 187-194; and Lenore Blum, *Breaking the Silence*, unpublished manuscript.

17. Rob Kirby, *A History and Critique of Virginia Harrison 'vs' U.C. Berkeley Math Dept.*, unpublished manuscript, September, 1993.

18. Steve Smale interview, 5/19/96.

19. Steve Smale interview, 5/19/96.

20. Brady Kahn, *The Gender Factor*, East Bay Express, 3/8/91.

21. Allyn Jackson, *Fighting for Tenure*, Notices Amer. Math. Soc., **141** (1994), p. 188.

22. P&T Committee to Chancellor Ira Heyman, report, 9/5/89.

23. Margy Rochlin, "The Mathematics of Discrimination," *Los Angeles Times Sunday Magazine*, 5/2/93.

24. Paul Selvin, "Does the Harrison case reveal sexism in math?" *Science*, vol. 252, 1991, pp. 1781-1783.

25. Steve Smale interview, 5/19/96.

26. Steve Smale interview, 5/19/96.

27. Paul Selvin, "Harrison case: no calm after storm," *Science*, vol. 262, 10/15/93, pp. 324-327; see also the Notices of American Mathematical Society, March 1994.

Chapter 12. Smale the Mathematician

1. *From Topology to Computation: Proceedings of the Smalefest* (M. W. Hirsch et al., eds.), Springer-Verlag, New York, 1993.

2. Stephen Smale, *A convergent process of price adjustment and global Newton methods*, Journal of Mathematical Economics, **3** (1976), pp. 107-120.

3. Stephen Smale, *The fundamental theorem of algebra and complexity theory*, Bull. Amer. Math. Soc. (N.S.), **4** (1981), pp. 1-36.

4. Stephen Smale, *Efficiency of algorithms of analysis*, Bull. Amer. Math. Soc. (N.S.), **13** (1985), pp. 87-121.
5. Stephen Smale, *On the topology of algorithms* I, Journal of Complexity, **3** (1987), pp 81-89.
6. Lenore Blum, Mike Shub, Stephen Smale, *On a theory of computation and complexity over the real numbers: NP-completeness, recursive functions and universal machines*, Bull. Amer. Math. Soc. (N.S.), **21** (1989), pp. 1-46.
7. Lenore Blum, Felipe Cucker, Mike Shub, Stephen Smale, *Complexity and Real Computation*, Springer-Verlag, New York, 1998.
8. Steve Smith interview, 2/9/96.
9. Steve Smale interview, 8/11/97.
10. Steve Smale interview, 8/11/97.

Chapter 13. Six Years at City University

1. Scmp.com/lifestyle/travel-leisure/article/2171675/kowloon-tong-history-hong-kongs-original-garden-city-and
2. Info.gov.hk/gia/general/199807/10/0710062.htm
3. Zolima CityMag, 11/7/18.
4. Steve Smale interview, 1/30/24.
5. Steve Smale interview, 1/8/24.
6. Leonore Blum, Felipe Cucker, Michael Shub, and Steve Smale, *Complexity and Real Computation*, Springer, New York, 1998.
7. Lenore Blum interview, 1/25/24.
8. Felipe Cucker interview, 4/5/24.
9. Stephen Smale, "Theory of computation," *Mathematical Research Today and Tomorrow: Viewpoints of Seven Fields Medalists. Lectures given at the Institut d'Estudis Catalans, Barcelona, Spain, June 1991* (Carles Casacuberta and Manuel Castellet, eds.) Springer, Berlin (1992), pp. 59-69.
10. Felipe Cucker interview, 3/5/24.
11. Felipe Cucker interview, 4/5/24.

12. Felipe Cucker interview, 3/5/24.
13. Lenore Blum interview, 1/25/24.
14. Felipe Cucker interview, 3/5/24.
15. Lenore Blum interview, 1/25/24.
16. Steve Smale interview, 12/14/23.
17. Scholar.google.com/citations?user=rKyTmpkAAAAJ&hl=en
18. Felipe Cucker interview, 3/5/24.
19. Lenore Blum interview, 1/25/24.
20. Felipe Cucker and Steve Smale, *Complexity Estimates Depending on Condition and Round-Off Error*, Journal of the ACM, **46** (1999), no. 1, pp. 113-184.
21. Stephen Smale, *Mathematical problems for the next century*, The Mathematical Intelligencer, **20** (1998), pp. 7-15.
22. Peter Bürgisser and Felipe Cucker, *On a problem posed by Steve Smale*, Annals of Mathematics, **174** (2011), no. 3, pp. 1785-1836.
23. Buffalo.edu/cubs/research/pioneers-in-ai-systems.html
24. News.usps.com/2024/04/23/the-postal-services-digital-detectives-are-marking-a-milestone
25. Felipe Cucker interview, 4/5/24.
26. The paper is in the top 10% of all research outputs ever tracked by Altmetric, cityu.altmetric.com/details/3226266#score.
27. Ding-Xuan Zhou interview, 1/24/24.
28. Macrotrends.net/global-metrics/countries/HKG/hong-kong/population
29. Yuhong Tian, C.Y. Jim, Yan Tao, and Tao Shi, *Landscape ecological assessment of green space fragmentation in Hong Kong*, Urban Forestry & Urban Greening, **10** (2011), no. 2, pp. 79-86.
30. Geopark.gov.hk/en/discover/attractions/tai-long-wan
31. Stephen Smale, *The Smale Collection: Beauty in Natural Crystals,* Lithographie LLC, Connecticut, 2006; Stephen Smale, *The Smale Collection II,* Lithographie LTD, Colorado, 2021.
32. Ding-Xuan Zhou interview, 2/25/24.

33. *South China Morning Post*, 12/30/05.
34. William Wang email, 5/27/24.
35. Ding-Xuan Zhou interview, 2/25/24.
36. Ding-Xuan Zhou email, 4/17/24.
37. Steve Smale interview, 12/14/23.
38. Laura Smale interview, 1/17/24.
39. Steve Smale interview, 1/7/24.
40. Steve Smale interview, 1/18/24.
41. Laura Smale interview, 1/17/24.
42. *The Washington Post,* 9/1/02.
43. Nat Smale interview, 1/31/24.
44. Steve Smale interview, 1/4/24.
45. Steve Smale interview, 4/8/24.
46. Felipe Cucker email, 3/30/24.
47. Ding-Xuan Zhou interview, 2/25/24.
48. Steve Smale interview, 4/8/24.

Chapter 14. New Challenges in Chicago

1. Ttic.edu/about
2. David McAllester email, 4/25/24.
3. Steve Smale interview, 1/11/24; 4/8/24.
4. David McAllester email, 4/25/24.
5. Steve Smale interview, 4/8/24.
6. Steve Smale interview, 1/8/24.
7. Steve Smale interview, 1/11/24.
8. Shmuel Weinberger email, 4/15/24.
9. Steve Smale interview, 1/11/24.
10. Ding-Xuan Zhou email, 4/17/24.
11. Ding-Xuan Zhou email, 4/18/24.

12. Britannica.com/topic/universal-grammar
13. Felipe Cucker interview, 4/5/24.
14. Felipe Cucker, Steve Smale, and Ding-Xuan Zhou, *Modelling Language Evolution,* Foundations of Computational Mathematics, **4** (2004), pp. 315-343.
15. Steve Smale interview, Mathmedia, Institute of Mathematics, Academia Sinica, Taipei, 12/19/04.
16. Felipe Cucker interview, 3/5/24.
17. Nat Smale interview, 1/31/24.
18. Steve Smale interview, 1/8/24.
19. Dylan Smale interview, 1/17/24.
20. Nat Smale interview, 1/31/24.
21. Steve Smale interview, 1/4/24.
22. Dylan Smale interview, 1/17/24.
23. Steve Smale interview, 1/12/24.
24. Nat Smale interview, 1/31/24.
25. Laura Smale interview, 1/17/24.
26. Dylan Smale interview, 1/17/24.
27. Laura Smale interview, 1/17/24.
28. Dylan Smale interview, 1/17/24.
29. Ding-Xuan Zhou email, 4/17/24.
30. Steve Smale and Ding-Xuan Zhou, *Shannon sampling and function reconstruction from point values*, Bulletin of the American Mathematical Society, **41** (2004), pp. 279-305.
31. Steve Smale and Ding-Xuan Zhou, *Shannon sampling II: Connections to learning theory*, Applied and Computational Harmonic Analysis, **19** (2005), no. 3, pp. 285-302.
32. Ding-Xuan Zhou email, 4/17/24.
33. Tomaso Poggio and Steve Smale, *The mathematics of learning: Dealing with data*, Notices of the American Mathematical Society, **50** (2003), no. 5, pp. 537-544.

34. Steve Smale interview, MathemediaFilms, 11/15/11.
35. Felipe Cucker interview, 3/5/24.
36. Felipe Cucker and Steve Smale, *Emergent behavior in flocks*, IEEE Transactions on Automatic Control, **52** (2007), no. 5, pp. 852-862.
37. Felipe Cucker interview, 3/5/24.
38. Steve Smale interview, MathemediaFilms, 11/15/11.
39. Stephen Smale, *The Smale Collection: Beauty in Natural Crystals*, Lithographie LLC, Connecticut, 2006.
40. Wolffund.org.il/stephen-smale
41. Steve Smale interview, 1/4/24.
42. Nat Smale interview, 1/31/24.
43. Laura Smale interview, 1/17/24.
44. Nat Smale interview, 1/31/24.
45. Laura Smale interview, 1/17/24.
46. Steve Smale and Ding-Xuan Zhou, *Geometry on probability spaces*, Constructive Approximation, **30** (2009), pp. 311-323.
47. Ding-Xuan Zhou interview, 2/25/24.
48. Partha Niyogi, Stephen Smale, and Shmuel Weinberger, *Finding the homology of submanifolds with high confidence from random samples*, Discrete & Computational Geometry, **39** (2008), pp. 419-441.
49. Shmuel Weinberger email, 4/14/24.
50. Partha Niyogi, Stephen Smale, and Shmuel Weinberger, *A topological view of unsupervised learning from noisy data*, SIAM Journal of Computing, **40** (2011), no. 3, pp. 646-663.
51. Shmuel Weinberger email, 4/14/24.

Chapter 15. Back to Hong Kong

1. Ding-Xuan Zhou interview, 1/24/24.
2. Felipe Cucker interview, 3/5/24.
3. Way Kuo interview, 1/29/24.
4. Steve Smale interview, 1/12/24.

5. Ding-Xuan Zhou interview, 1/24/24.
6. Laura Smale interview, 1/17/24.
7. Steve Smale interview, 4/8/24.
8. Way Kuo interview, 1/29/24.
9. Topuniversities.com/world-university-rankings; Qschina.cn/en/universities/city-university-hong-kong
10. Timeshighereducation.com/student/best-universities/most-international-universities-world
11. Cityu.edu.hk/media/news/2022/10/21/cityu-shines-again-stanfords-list-worlds-top-2-scientists-explores-government-plans-increase-funded-postgraduate-places
12. Clarivate.com/highly-cited-researchers
13. Steve Smale interview, *Plus* magazine, 2017.
14. Steve Smale, Lorenzo Rosasco, Jake Bouvrie, Andrea Caponnetto, and Tomaso Poggio, *Mathematics of the Neural Response*, Foundations of Computational Mathematics, **10** (2009), pp. 67-91.
15. Indika Rajapakse, *Conversation with Dr. Steve Smale and Dr. Lee Hartwell*, American Mathematical Society, Notices of the AMS, October 2021.
16. George Szpiro, *Interview with Stephen Smale*, American Mathematical Society, Notices of the AMS, September 2007.
17. Nat Smale interview, 1/31/24.
18. Laurent Bartholdi, Thomas Schick, Nat Smale, and Steve Smale, *Hodge Theory on Metric Spaces*, Foundations of Computational Mathematics, **12** (2012), no. 1, pp. 1-48.
19. Nat Smale interview, 1/31/24.
20. Steve Smale interview, 4/8/24.
21. Morris W. Hirsch, Stephen Smale, and Robert L. Devaney, *Differential Equations, Dynamical Systems, and an Introduction to Chaos*, Third Edition, Academic Press, 2012.
22. Robert Devaney interview, 5/20/24.
23. News.mit.edu/2008/obit-lorenz-0416

24. Robert Devaney email, 5/20/24.
25. Steve Smale interview, *Plus* magazine, 2017.
26. Ding-Xuan Zhou interview, 4/17/24.
27. Laureate interview: Stephen Smale, Heidelberg Laureate Forum, September 2017.
28. Felipe Cucker interview, 3/5/24.
29. Felipe Cucker email, 4/9/24.
30. Felipe Cucker interview, 3/5/24.
31. Lenore Blum interview, 1/25/24.
32. Felipe Cucker interview, 3/5/24.
33. Nat Smale interview, 1/31/24.
34. Claradavissmale.blogspot.com
35. Steve Smale interview, 1/12/24.
36. Laura Smale interview, 4/23/24.
37. Laura Smale email, 4/17/24.
38. Steve Smale interview, 2/15/24.
39. Steve Smale interview, 4/8/24.
40. Steve Smale interview, 1/18/24.
41. Claradavissmale.blogspot.com
42. Felipe Cucker interview, 5/31/24.
43. Laura Smale interview, 1/17/24.
44. Steve Smale interview, 3/14/24.
45. Steve Smale interview, 1/11/24.
46. Laura Smale interview, 4/23/24.
47. Claradavissmale.blogspot.com
48. Laura Smale email, 4/26/24.
49. Laura Smale interview, 4/23/24.
50. Laura Smale interview, 4/23/24.

Chapter 16. Biology, Berkeley, and a Lasting Legacy

1. Laura Smale email, 4/17/24.
2. Focm-society.org/christiansen.php
3. Laura Smale email, 4/17/24.
4. Dylan Smale interview, 1/17/24; Nat Smale interview, 1/31/24.
5. Laura Smale interview, 1/17/24.
6. Wen-Jun Shen, Hau-San Wong, Quan-Wu Xiao, Xin Guo, and Stephen Smale, *Introduction to the peptide binding problem of computational immunology: New results*, Foundations of Computational Mathematics, **14** (2014), no. 5, pp. 951-984.
7. W. J. Shen et al., *MHC binding prediction with KernelRLSpan and its variations*, Journal of Immunological Methods, **406** (2014), pp. 10-20.
8. Steve Smale interview, 1/4/24.
9. Felipe Cucker interview, 3/5/24.
10. Steve Smale interview, 1/4/24.
11. Indika Rajapakse interview, 1/24/24.
12. H. Chen et al., *Functional organization of the human 4D nucleome*, Proc. Natl. Acad. Sci. U.S.A., **112** (2015), no. 26, pp. 8002-8007.
13. Indika Rajapakse interview, 1/24/24.
14. Indika Rajapakse and Steve Smale, *Mathematics of the genome*, Foundations of Computational Mathematics, **17** (2017), pp. 1195-1217.
15. Way Kuo interview, 1/29/24.
16. Indika Rajapakse interview, 1/24/24.
17. Steve Smale interview, 4/8/24.
18. Steve Smale interview, 2/14/24.
19. Laura Smale interview, 4/23/24.
20. Nat Smale interview, 1/31/24.
21. Indika Rajapakse interview, 1/24/24.

22. Indika Rajapakse and Steve Smale, *The pitchfork bifurcation*, International Journal of Bifurcation and Chaos, **27** (2017), no. 9.

23. S. Liu et al., *Genome architecture mediates transcriptional control of human myogenic reprogramming*, iScience, **6** (2018), pp. 232-246.

24. Indika Rajapakse interview, 5/21/24.

25. Phys.org/news/2016-03-huygens-pendulum-synchronization.html

26. Newsroom.hlf-foundation.org/newsroom/lectures/video/lecture-the-mathematics-of-the-heart-beat.html

27. Way Kuo interview, 1/29/24.

28. Charles Pugh email, 6/8/24.

29. Indika Rajapakse interview, 5/21/24.

30. Indika Rajapakse interview, 3/7/24.

31. Indika Rajapakse, *Conversation with Dr. Steve Smale and Dr. Lee Hartwell*, American Mathematical Society, Notices of the AMS, October 2021.

32. Stephen Smale, *The Smale Collection II*, Lithographie LTD, Colorado, 2021.

33. Judy Smale interview, 2/2/24.

34. Indika Rajapakse interview, 3/7/24.

35. Laura Smale interview, 1/17/24.

36. Nat Smale email, 2/24/24.

37. Felipe Cucker interview, 4/5/24.

38. George Szpiro, *Interview with Stephen Smale*, American Mathematical Society, Notices of the AMS, September 2007.

39. Indika Rajapakse interview, 5/21/24.

Mathematical Appendix A: Smale's Thesis

1. Hassler Whitney, *On regular closed curves in the plane*, Composito Mathematica, **4** (1937), pp. 276-284.

2. Hassler Whitney, *On regular closed curves in the plane*, Composito Mathematica, **4** (1937), pp. 276-284.

3. Stephen Smale, *Regular curves on Riemannian manifolds*, Trans. Amer. Math. Soc. **87** (1958), pp. 492-512.
4. For a more technical discussion see Morris Hirsch, "The work of Stephen Smale in differential topology," *From Topology to Computation: Proceedings of the Smalefest* (M. W. Hirsch et al., eds.) Springer-Verlag, New York (1993), pp. 83-106.
5. Raoul Bott interview, 3/22/94.

Mathematical Appendix B: Everting the Sphere

1. More details are available in "The work of Stephen Smale in differential topology," by Morris Hirsch in *From Topology to Computation: Proceedings of the Smalefest* (M. W. Hirsch et al., eds.), Springer-Verlag, New York (1993), pp. 83-106.
2. "Turning a surface inside out" by Anthony Phillips in *Scientific American.*
3. The first film of sphere eversion was made by Nelson Max.
4. Constructed by Charles Pugh and later stolen from the Berkeley mathematics department.
5. Stephen Smale, *The classification of immersions of spheres in Euclidean spaces*, Annals of Mathematics, **69** (1959), 327-344.

Mathematical Appendix C: Chaos and the Horseshoe

1. Robert Devaney, *An Introduction to Chaotic Dynamical Systems*, Addison-Wesley, Redwood City, CA, 1989.
2. Steve Smale, "On How I Got Started in Dynamical Systems," *The Mathematics of Time*, Springer-Verlag, New York, 1980, 147-151.
3. Steve Smale, "Diffeomorphisms with Many Periodic Points," *Differential and Combinatorial Topology*, (Stewart Cairns, ed.), Princeton University Press, Princeton 1965.

Mathematical Appendix D: The Higher Dimensional Poincaré Conjecture

1. Stephen Smale, *On gradient dynamical systems*, Annals of Mathematics, **74** (1961), pp. 199-206.
2. Stephen Smale, *Generalized Poincaré's conjecture in dimensions greater than four*, Annals of Mathematics, **74** (1961), pp. 391-406.

3. M. W. Hirsch, "The work of Smale in differential topology," *From Topology to Computation: Proceedings of the Smalefest* (M. W. Hirsch et al., eds.), Springer-Verlag, New York (1993) pp. 83-106.